International Conference on Geotechnical Engineering-Iraq (ICGE-Iraq 2020)

IOP Conference Series: Materials Science and Engineering
Volume 901

Baghdad, Iraq
19 - 20 February 2020

ISBN: 978-1-7138-3185-3
ISSN: 1757-8981

Printed from e-media with permission by:

Curran Associates, Inc.
57 Morehouse Lane
Red Hook, NY 12571

Some format issues inherent in the e-media version may also appear in this print version.

Institute of Physics
Dirac House, Temple Back
Bristol BS1 6BE UK

Phone: 44 1 17 929 7481
Fax: 44 1 17 920 0979

techtracking@iop.org

Additional copies of this publication are available from:

Curran Associates, Inc.
57 Morehouse Lane
Red Hook, NY 12571 USA
Phone: 845-758-0400
Fax: 845-758-2633
Email: curran@proceedings.com
Web: www.proceedings.com

TABLE OF CONTENTS

IOP Conf. Series: Materials Science and Engineering **901** (2020) 011001 doi:10.1088/1757-899X/901/1/011001

Preface

The international conference on geotechnical engineering-Iraq (ICGE-Iraq 2020) had finished its work successfully in Baghdad, Iraq to discuss the challenges, opportunities, and problems of geotechnical engineering in Civil engineering problems. The ICGE-Iraq was organized by the Iraqi Scientific Society of Soil Mechanics and Foundation Engineering (ISSSMFE) with the cooperation of Civil Engineering Department/University of Baghdad and the ICGE-Iraq hold at the University of Baghdad. ICGE-Iraq 2020 received about 138 research manuscript in 11 themes: soil improvement; shallow and deep foundation; soil mechanics and dynamics; unsaturated soils; soil-structure interaction; engineering geology; geoenvironmental engineering; applications of sustainability in geotechnical engineering; application of remote sensing in geotechnical engineering; transportation geotechnique; and construction management.

ICGE-Iraq 2020 is a forum to bring together scientists, researchers, business people, and industry professionals to discuss innovative ideas and diverse topics on the next generation of sustainable geotechnical and environmental engineering. The papers submitted to the conference received from Iraq, Iran, Kazakhstan, Russia, Germany, USA, India, Turkey, Oman, Jordan, and Algeria. More than 800 participants attended the conference in the opening ceremony and plenary session.

Hundreds of attendees, paper presenters, keynote speakers, exhibitors, and tutorial participants, students have benefited in many ways from this conference. More than 118 scientific papers were presented by authors in oral presentations and posters from several countries. Five invited keynote speakers from different countries addressed to the audience and shared knowledge and rich experience of applications of the Iraqi scientific society of soil mechanics and foundation engineering in their universities and companies.

Prof. Mahdi Karkush
President of ISSSMFE
Conference Chair

IOP Conf. Series: Materials Science and Engineering **901** (2020) 011002 doi:10.1088/1757-899X/901/1/011002

Peer review declaration

All papers published in this volume of IOP Conference Series: Materials Science and Engineering have been peer reviewed through processes administered by the Editors. Reviews were conducted by expert referees to the professional and scientific standards expected of a proceedings journal published by IOP Publishing.

- **Type of peer review: Single-blind**
  The review process is single-blind, the authors did not know reviewers, but reviewers know the authors.

- **Conference submission management system:** The papers submitted online via website devoted for conference.

- **Number of submissions received:** 139

- **Number of submissions sent for review:** 120

- **Number of submissions accepted:** 118

- **Acceptance Rate (Number of Submissions Accepted / Number of Submissions Received X 100):** 84%

- **Average number of reviews per paper:** 2-3

- **Total number of reviewers involved:** 142

- **Any additional info on review process:** English language editing and formatting are included in the review process.

- **Contact person for queries:** Prof. Mahdi Karkush (e-mail: mahdi_karkush@coeng.uobaghdad.edu.iq), Mobile: 009647801058893

Improving Shallow-Deep Foundation with Reinforcement by Geogrids for Tall Building

H. E. Al-Laban[1,a *] **and M. K. Fakhraldin**[1,b]

[1]Civil Engineering Department, University of Kufa, Kufa, Iraq

[a]haiderehsan22@yahoo.com, **b**mohammed.fakhruddin@uokufa.edu.iq

Abstract. This research has studied the effect of the vertical load, it has acted at the center of the raft foundation and piled raft foundation. Stresses resistance and displacements were significant parts to examine the behavior of improving foundation. These are important parts of life the structure. The study focuses on different cases of the embedded layer in tension zone with concrete grade B20 of (MD/XMD) Tensor SS2 geogrids in raft and piled raft foundation for a tall building in Incheon Tower in South Korea, it was treated to solve those problems. Stresses resistance of raft and piled raft foundation were greater than vertical stress applied of dead and live loads, also the stress value of XMD was greater than the value of MD.

Keywords: Improvement, shallow foundation, deep foundation, reinforcement, geogrid.

1. Introduction

Shallow foundations, where feasible, are generally more economical than deep foundations if they do not have to be installed deep into the ground and extensive ground improvement works are not required. They are often used to support structures at sites where subsurface materials are sufficiently strong. Unless a shallow foundation can be founded on a strong rock, some noticeable settlement will occur. The design of shallow foundations should ensure that there is an adequate factor of safety against bearing failure of the ground and that the settlements, including total and differential settlement, are limited to allowable values. The structural design of piles should be carried out under the requirements in local structural codes and regulations. The piles should be capable of withstanding both the stresses induced during handling and installation as well as during their service life [1]. In the calculation of vertical stress, it generally assumes that the foundation of a structure is flexible [2]. In practice, this is not the case; no foundation is perfectly flexible, nor is it infinitely rigid. The actual nature of the distribution of contact stress will depend on the elastic properties of the foundation and the soil on which the foundation is resting [2].

The most common materials from which most structures are built are wood, steel, reinforced (including prestressed) concrete, and masonry. Lightweight materials such as aluminum and advanced composite materials, such as fiber-reinforced plastics (FRB) are also becoming more common in use. Plain concrete is made by mixing cement, fine aggregate, coarse aggregate, water, and frequent admixtures. The strength of concrete depends on many factors: notably the proportion of the ingredients and the conditions of temperature and moisture under which it is placed and cured. Reference to concrete strength means uniaxial compressive strength measured by a compression test of a standard test cylinder. The most important variable in determining concrete strength is the water to cement (w/c) ratio. The lower the water/cement ratio, the higher the compressive strength [3]. Geosynthetics include exclusively manmade polymeric products such as geotextile, geogrids, geonets, geomembranes, geosynthetic clay

IOP Conf. Series: Materials Science and Engineering **901** (2020) 012001 doi:10.1088/1757-899X/901/1/012001

liners, and geocomposites. Geogrids are mainly used as tensile reinforcement. In manufacturing uniaxial geogrids, circular holes are punched on the polymer sheet, which is subsequently drawn to improve the mechanical properties [4].

2. Materials Properties

2.1. Study Model

In the current study, a full three-dimensional (3D) FE model is created on Abaqus CAE v.6.14 software [5]. Two sets of the raft and piled raft foundation with geogrid embedded in the tension zone of the foundation by concrete grade B20, as shown in Figs. 1 and 2. The foundation comprises a 5.5 m thick concrete grade B20 both sets. Piled distribution of piled raft foundation was 172 piles of 2.5 m diameter with lengths below the base of the raft different from (36 m, 50 m, and 66 m) row by row gradually, without basement walls and soil properties in analysis for both sets of models.

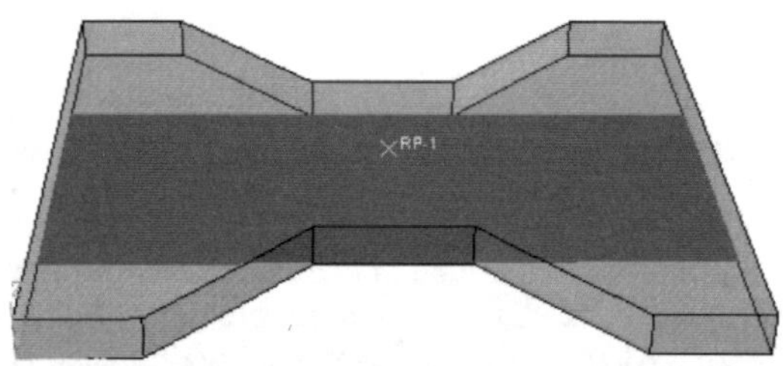

Figure 1. Description of raft foundation.

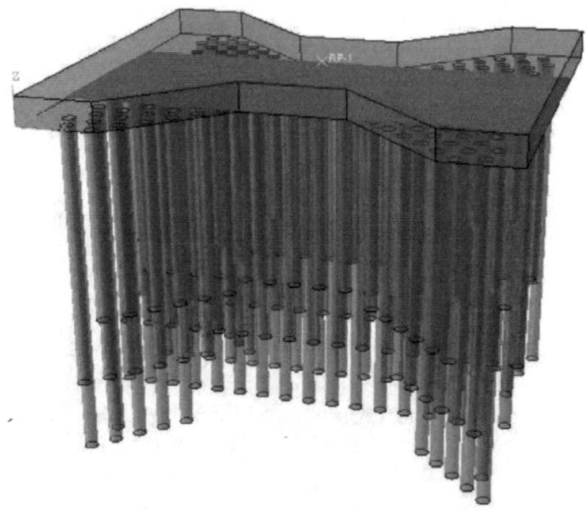

Figure 2. Description of piled raft foundation.

2.2. Finite Element Model

The physical and mechanical properties of Tensor geogrid (SS2) are summarized in Table 1. Material properties for concrete with simplified concrete damage plasticity model in class B20 are listed in Table 2. The 10-node quadratic tetrahedron elements were used to mesh the raft and piled raft foundation. The 2-node linear 3-D truss was used to mesh Tensor SS2 (MD/XMD) geogrids. Boundary conditions at the base of the raft and piles at the piled raft were fixed.

Table 1. Properties of SS2 [6].

Dimensional Properties		
Property	Unit	Data
Aperture size(MD/XMD)	mm	28/40
Mass per unit area	kg/m^2	0.3
Rib thickness MD/XMD	mm	1.2/1.1
Junction thickness	mm	3.9
Longitudinal rib width lw	mm	3
Transverse rib width tw	mm	3
Roll width	m	4
The Mechanical Properties		
Peak Tensile Strength MD/XMD	kN/m	14.4/28.2
Elastic modules MD/XMD	GPa	0.57/0.99
Upper yield strength MD/XMD	MPa	1 /3
Lower yield strength MD/XMD	MPa	1/3
Tensile strength MD/XMD	MPa	24/30.7

The physical properties	
Property	Data
Mesh type	square
Standard color	Black
Polymer type	PP
Packaging	Rolls

IOP Conf. Series: Materials Science and Engineering **901** (2020) 012001 doi:10.1088/1757-899X/901/1/012001

Table 2. Properties of concrete [7].

Material's parameters	B20	Plasticity parameters	
Concrete elasticity		Dilation angle	31
Modulus of elasticity, E (GPa)	21.2	Eccentricity	0.1
		Fb0/fc0	1.16
Poisson's ratio (υ)	0.2	K	0.67
		Viscosity parameter	0
Concrete compressive behavior		Concrete compression damage	
Yield stress (MPa)	Inelastic strain	Damage parameter C	Inelastic strain
10.2	0	0	0
12.8	7.73585E-05	0	7.73585E-05
15	0.000173585	0	0.000173585
16.8	0.000288679	0	0.000288679
18.2	0.000422642	0	0.000422642
19.2	0.000575472	0	0.000575472
19.8	0.00074717	0	0.00074717
20	0.000937736	0	0.000937736
19.8	0.00114717	0.01	0.00114717
19.2	0.001375472	0.04	0.001375472
18.2	0.001622642	0.09	0.001622642
16.8	0.001888679	0.16	0.001888679
15	0.002173585	0.25	0.002173585
12.8	0.002477358	0.36	0.002477358
10.2	0.0028	0.49	0.0028
7.2	0.003141509	0.64	0.003141509
3.8	0.003501887	0.81	0.003501887
Concrete tensile behavior		Concrete tension damage	
Yield stress (MPa)	Cracking strain	Damage parameter T	Cracking strain
2	0	0	0
0.02	0.000943396	0.99	0.000943396

2.3.Vertical Loading

The vertical load (6687 MN) was applied incrementally at the center of the raft, it computed of dead and live load from Table 3.

Table 3. Load components for Incheon Tower [8].

Load component	Value	Load component	Value
Dead Load	6036 MN	Horizontal load x-direction	149 MN
Live Load	651 MN	Horizontal load z-direction	115 MN
Moment in x-direction	21,600MN-m	Earthquake load x-direction	110 MN
Moment in z-direction	12,710MN-m	Earthquake load z-direction	110 MN

3. Verification Results with Program

The numerical results obtained from the FE analysis of piled raft foundation were compared with the results by [8] as shown in Fig. 3, where the vertical load applied at the center point of the raft foundation versus the vertical displacement is plotted. The percentage of the average vertical displacement of Fig. 3 was 60%. Figure 4 was plotted the vertical load versus the vertical displacement of the raft foundation,

IOP Conf. Series: Materials Science and Engineering **901** (2020) 012001 doi:10.1088/1757-899X/901/1/012001

the reduction of displacements was very small. Figs. 5 and 6 were described the comparison of vertical loads of raft foundation and piled raft foundation with [8], where vertical loads were given equal values in the middle of analysis of raft and piled raft but piled raft had taken more steps to complete his analysis to reach finally at the same value of [8]. The percentage of the average vertical load in Fig. 5 and Fig. 6 were 6.5% and 64% respectively.

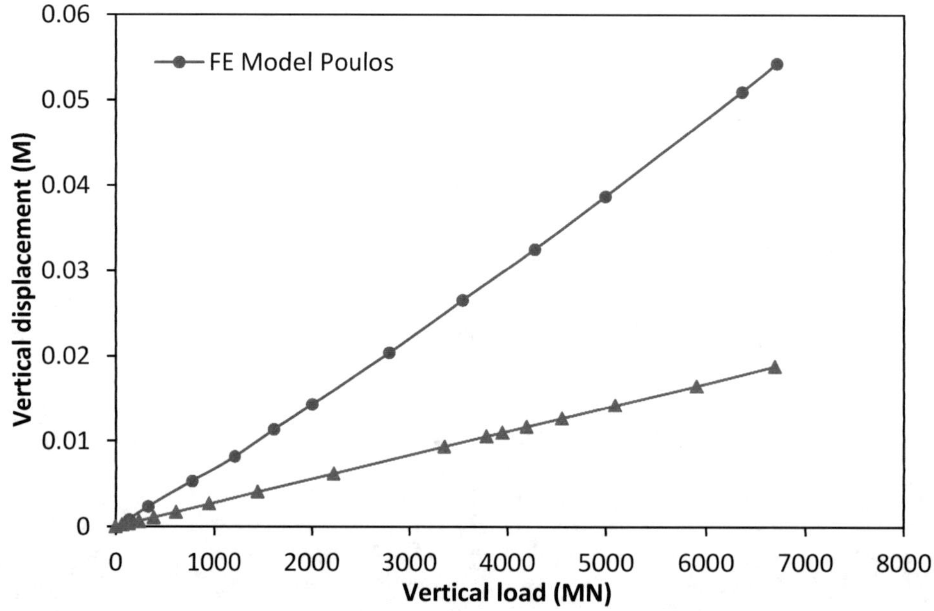

Figure 3. Vertical load versus vertical displacement of the piled raft foundation and FE model Poulos.

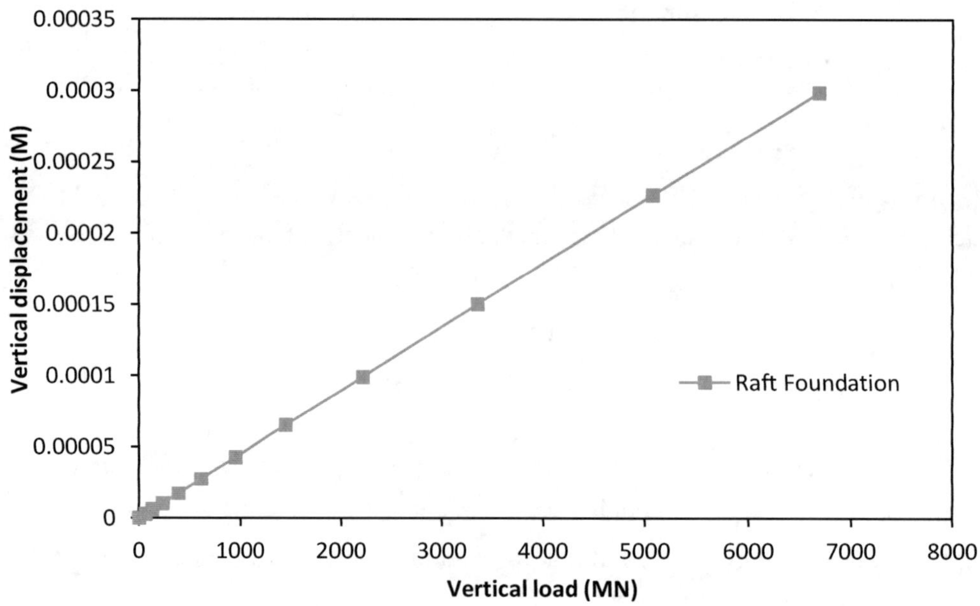

Figure 4. Vertical load versus vertical displacement of raft foundation.

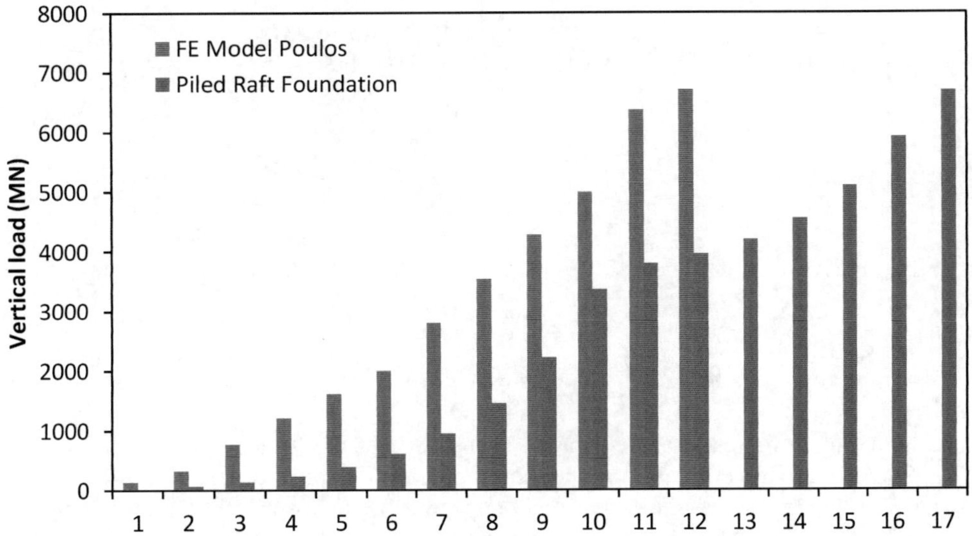

Figure 5. Comparison of Vertical loads of piled raft foundation and FE model Poulos.

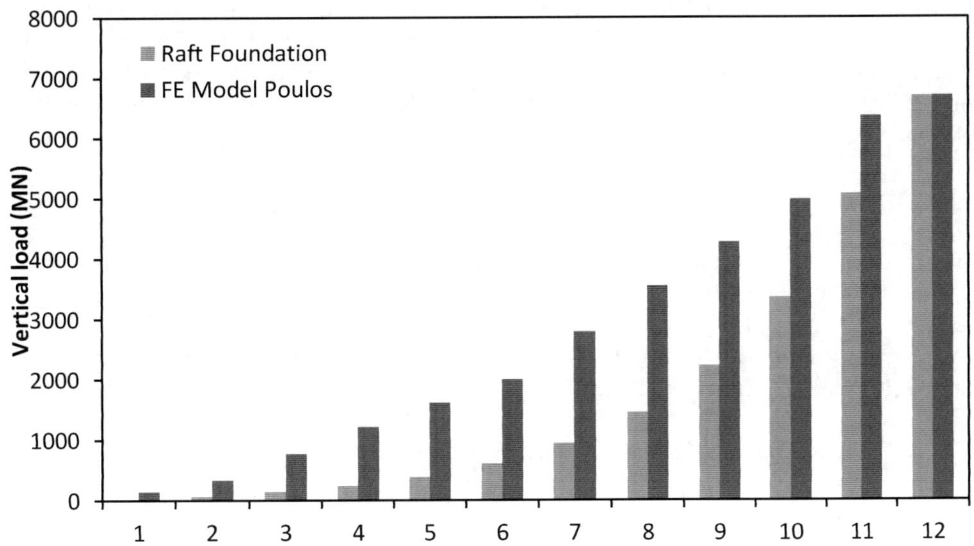

Figure 6. Comparison of Vertical loads of raft foundation and FE model Poulos.

4. Results: Set 1 (Raft Foundation) and Set 2 (Piled Raft Foundation)

4.1 Stress Distribution

Von-Mises described that the yield surface has an open cylinder around the hydrostatic axis in the space of the principal stresses as shown in Fig. 7. The Von-Mises is the most common yield criterion for structure. It is a combination of stresses and represents very well the yielding and ultimate stresses [9]. Thus, the value was got from Abaqus CAE v.6.14 software [5] for set 1 and set 2 (1.49 MPa and 19.98 MPa) respectively. Therefore, it can easily say that magnitude of stresses obtained higher than the value of stress applied at the center of the foundation (1.2661 MPa).

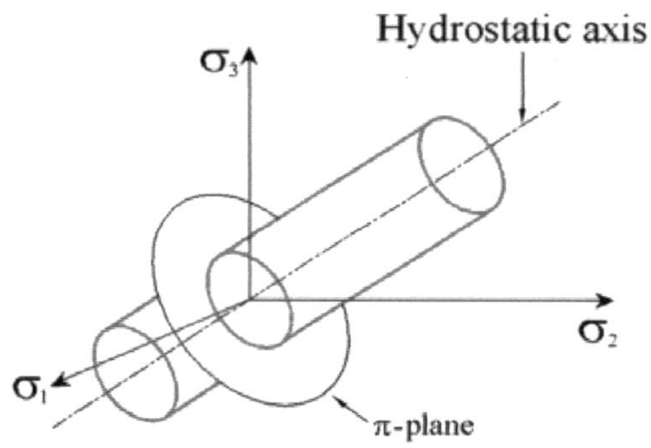

Figure 7. Yield surface of von-Mises.

The maximum value of stress was distributed in the edge of the foundation for set 1 and different places for set 2 referred to the critical region at the foundation Figs. 8 and 12, as summarized in Tables 2 and 4. The high value of stress (19.98 MPa) appeared at the connection between pile and raft foundation as shown in Fig. 12 but the result of the stress in Fig. 8 (1.49 MPa) was observed at all edges of the raft foundation. Stress resistance of piled raft is too much higher than the raft foundation. The Von-Mises stress of (MD and XMD) Tensor SS2 for raft foundation in set 1 and piled raft in set 2 were (0.001364 MPa and 0.002367 MPa) and (0.007011 MPa and 0.02862 MPa), respectively summarized in (Tables 2 and 4). The regions of the plastic deformation effect were symmetrical especially in the edge of the layer as shown in Fig.10 and Fig.11 but Figs. 14 and 15 were observed very close to the connection between pile and raft foundation. In general, XMD Tensor SS2 is given a high value of stress, it depends on mechanical properties of elastic modulus, yield strength, and tensile strength.

4.2 Displacement Effect

Displacements value for raft foundation in the set were summarized in Tables 4 and 5 approximately (0.3 mm) while they indicated in Tables 6 and 7 approximately (20 mm) of piled raft foundation in set 2, as noticed in (Figs. 9 and 13). The value of displacement of raft foundation (0.299 mm) was noticed along the foundation as shown in Fig. 9. It compares with the value of displacement of piled raft foundation that it was (19.99 mm) as shown in Fig. 13. It depends on distributed lengths of piled at the foundation.

Table 4. Summarizing table of finite element for set 1.

Specimen	Case	Aperture size (MD/XMD)	Material Properties
Raft Foundation with MD Tensor SS2	Reinforced	28 by 28 mm C/C	Concrete Grade B20 and One layer in Tension zone of the MD Tensor SS2
Raft Foundation with XMD Tensor SS2	Reinforced	40 by 40 mm C/C	Concrete Grade B20 and One layer in Tension zone of the XMD Tensor SS2

Table 5. Results summary table for set 1.

Specimen	Ultimate Stress (MPa)	Ultimate Displacement (mm)	Ultimate Stress of MD/XMD (MPa)
Raft Foundation with MD Tensor SS2	1.49	0.299	0.001364
Raft Foundation with XMD Tensor SS2	1.49	0.299	0.002367

IOP Conf. Series: Materials Science and Engineering **901** (2020) 012001 doi:10.1088/1757-899X/901/1/012001

Table 6. Summarizing table of finite element for set 2.

Specimen	Case	Aperture size (MD/XMD)	Material Properties
Piled Raft Foundation with MD tensor SS2	Reinforced	28 by 28 mm C/C	Concrete Grade B20 and One layer in Tension zone of the MD tensor SS2
Piled Raft Foundation with XMD tensor SS2	Reinforced	40 by 40 mm C/C	Concrete Grade B20 and One layer in Tension zone of the XMD tensor SS2

Table 7. Results summary table for set 2.

Specimen	Ultimate Stress (MPa)	Ultimate Displacement (mm)	Ultimate Stress of MD/XMD (MPa)
Piled Raft Foundation with MD tensor SS2	19.98	19.99	0.007011
Piled Raft Foundation with XMD tensor SS2	19.98	19.99	0.02862

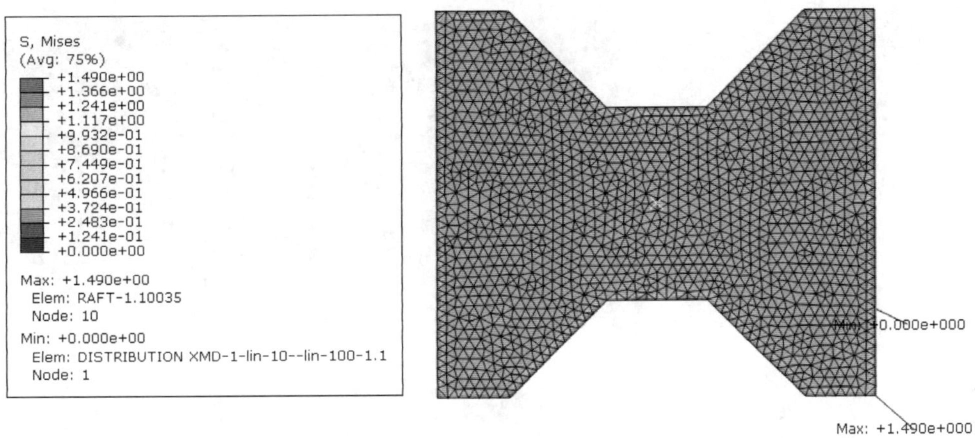

Figure 8. Von-Mises stress contour plots of raft foundation with XMD tensor SS2.

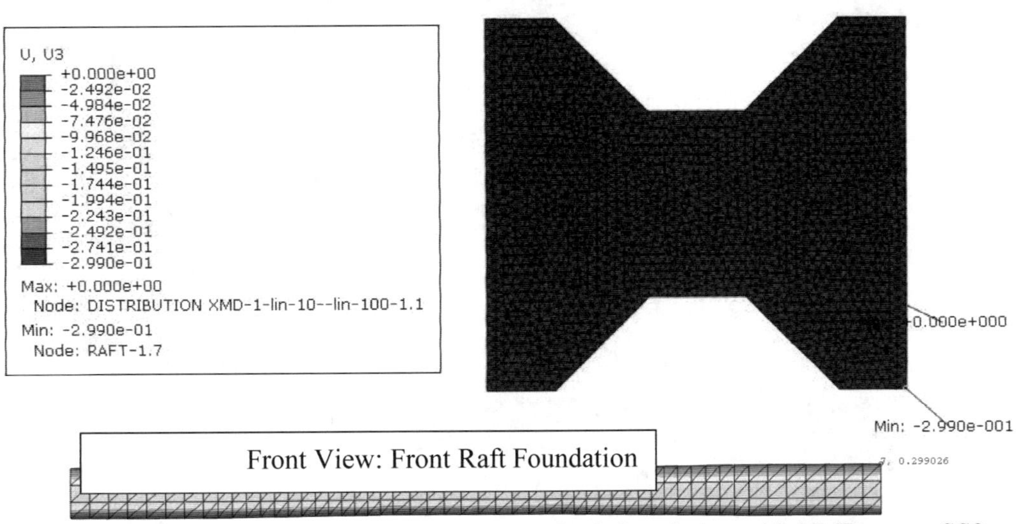

Figure 9. Vertical displacement contour plots of raft foundation with XMD tensor SS2.

IOP Conf. Series: Materials Science and Engineering **901** (2020) 012001 doi:10.1088/1757-899X/901/1/012001

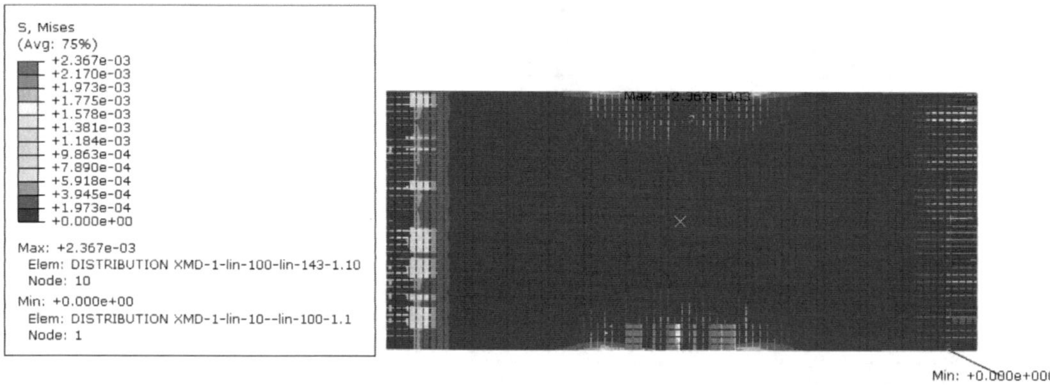

Figure 10. Von-Mises stress contour plots of the XMD tensor SS2 in the raft foundation.

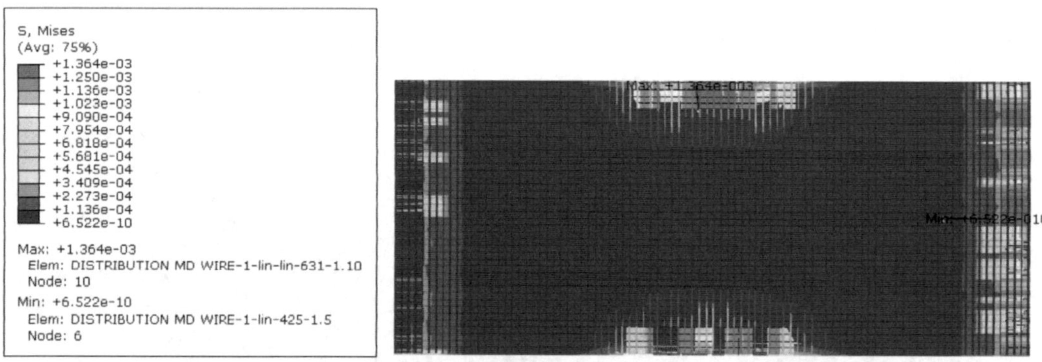

Figure 11. Von-Mises stress contour plots of the MD tensor SS2 in Raft Foundation.

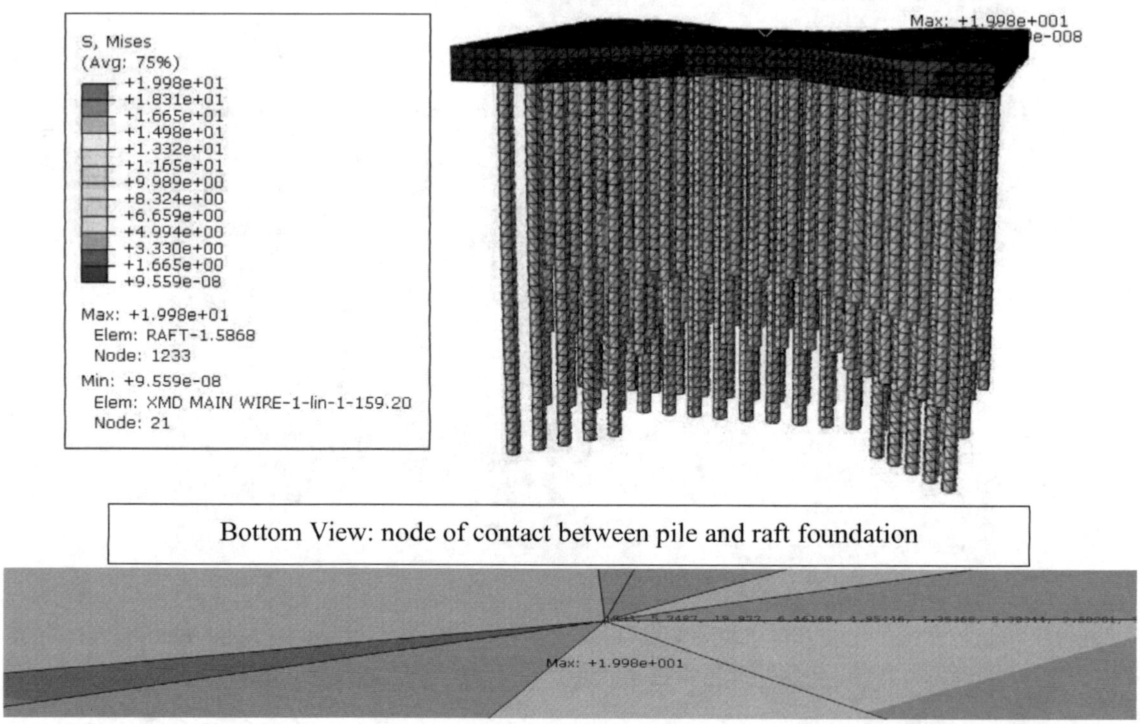

Figure 12. Von-Mises stress contour plots of piled raft foundation with XMD Tensor SS2.

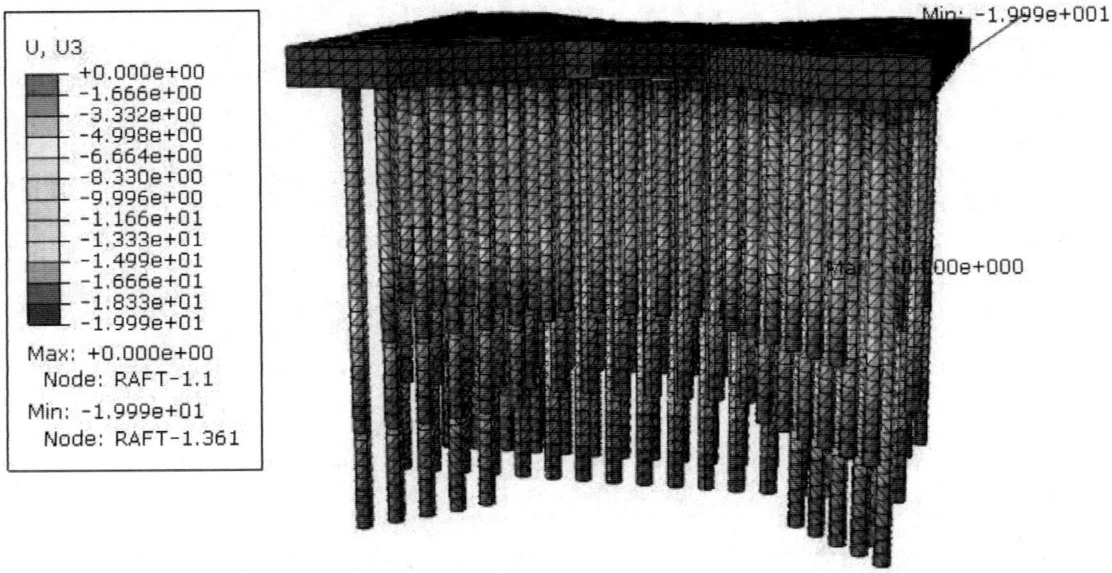

Figure 13. Vertical displacement contour plots of piled raft foundation with XMD tensor SS2.

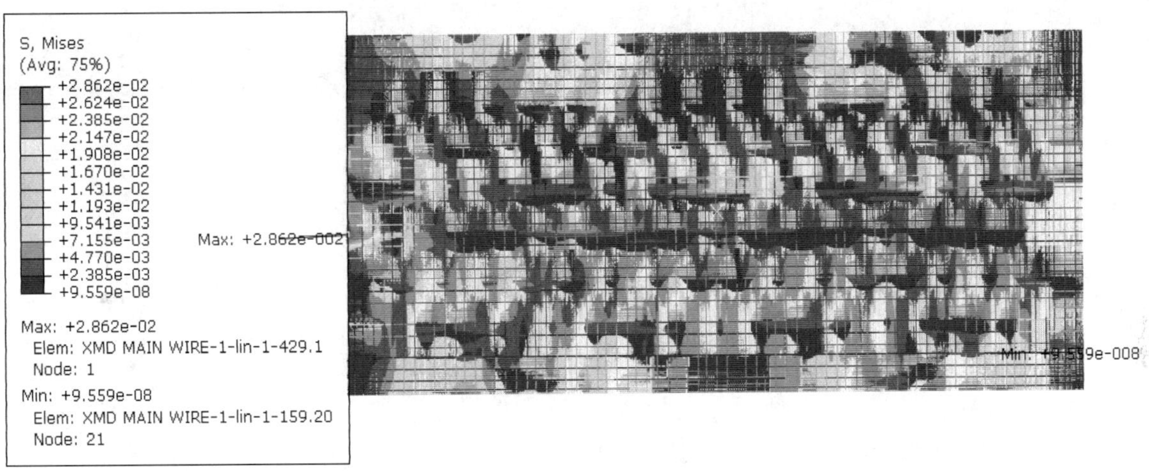

Figure 14. Von-Mises stress contour plots of the XMD tensor SS2 in piled raft foundation.

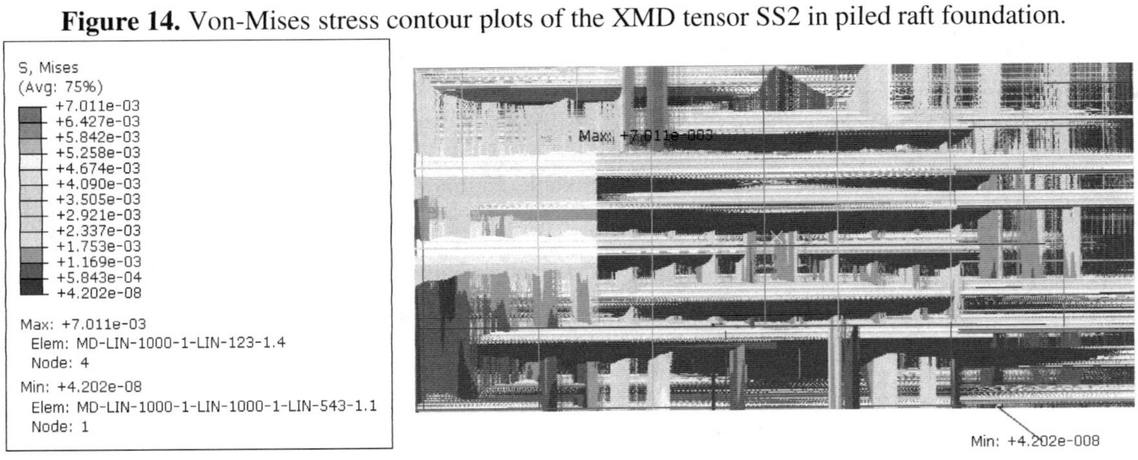

Figure 15. Von-Mises stress contour plots of the MD Tensor SS2 in piled raft foundation.

5. Conclusion

- Different values of the Von-Mises stress of (MD and XMD) Tensor SS2 referred to the mechanical properties of elastic modulus, yield strength and tensile strength of XMD tensor had bigger values than MD tensor.
- Displacements value for raft foundation in set 1 were smaller than values of piled raft foundation in set 2 because the boundary conditions of the raft foundation base were fixed and material properties of soil neglected too.
- The magnitude of stresses obtained from the piled raft foundation bigger than the raft foundation in the same value of the applied load.
- The applied load was (6687 MN) that it leads of the applied stress (1.2661 MPa) the results of resistance stresses come of Abaqus software were of raft foundation (1.49 MPa), piled raft foundation (19.98 MPa). The magnitude of resistance stresses was greater than the applied stress.

Recommendations

The strength of the foundation should be improved by embedded more layers of XMD Tensor SS2 geogrid.

References

[1] Geotechnical Engineering Office, Civil Engineering and Development Department, The Government of the Hong Kong Special Administrative Region, Foundation Design and Construction, 1st Published, GEO Publication No. 1/2006, Hong Kong, 2006.

[2] B.M. Das, Stresses and Displacements in A Soil Mass, Three-Dimensional Problems, in: Advanced Soil Mechanics, 5th edition, Library of Congress Cataloging, Taylor & Francis Group: Publisher's, New York, 2019 151-208.

[3] C.K. Wang, C.G. Salmon, and J.A. Pincheira, Reinforced Concrete Design, 7th edition, John Wiley & Sons, Inc., United States of America, 2007.

[4] H.I. Ling, D. Leshchinsky, and F. Tatsuoka, (Eds.), Reinforced soil engineering: advances in research and practice, CRC Press (2003).

[5] Simulia Dassault Systems. ABAQUS 6.14 analysis user's guide. ABAQUS 6.14 Documentation (2014), Information on: (http://130.149.89.49:2080/v6.14/pdf_books/ANALYSIS_4.pdf).

[6] M.K. Fakharaldin, Properties Measurement and Applications of Some Geogrid in Sand, Ph.D. Thesis, AlNahrain University, Iraq (2013).

[7] M. Hafezolghorani, F. Hejazi, R. Vaghei, M.S.B. Jaafar, and K. Karimzade, Simplified damage plasticity model for concrete, Structural Engineering International, 27(1) 2017 68-78.

[8] H.G. Poulos, J.C. Small, and H. Chow, Piled raft foundations for tall buildings. Geotechnical Engineering Journal of the SEAGS & AGSSEA, 42(2) 2011 78-84.

[9] L.S. Srinath, Theories of Failure or Yield Criteria and Introduction to Ideally Plastic Solid, in: Advanced Mechanics of solids, third edition, Tata McGraw-Hill: Publisher's, New Delhi, 2007, pp. 109-140.

IOP Conf. Series: Materials Science and Engineering **901** (2020) 012002 doi:10.1088/1757-899X/901/1/012002

Comparison of Single and Group Jet Grout Columns Settlement Based on Field Test and Theoretical Methods

Ali M. Al-Kinani [1,a*] and Mahmood D. Ahmed [2,b]

[1]Department of Civil Engineering, University of Thi-Qar, Thi-Qar, Iraq.
[b]Department of Civil Engineering, University of Baghdad, Baghdad, Iraq.
[a]al1984kinani@gmail.com, [b]mahmoud_baghdad @yahoo.com

Abstract. The settlement evaluation for the jet grouted columns (JGC) in soft soils is a problematic matter, because it is influenced by the number of aspects such as soil type, effect mixture between soil and grouting materials, nozzle energy, jet grouting, water flow rate, rotation and lifting speed. Most methods of design the jet-grouting column based on experience. In this study, a prototype single and group jet grouting models (single, 1*2, and 2*2) with the total length and diameter were (2000 and 150 mm) respectively and clear spacing (3D) has been constructed in soft clay and subjected to vertical axial loads. Furthermore, different theoretical methods have been used for the estimation of (JGC) settlement. Pile load settlement analysis of the jet grout columns showed that the average settlement values were (0.41, 0.663, and 1.5 mm) for the single, group (1*2) and group (2*2) jet grouted columns respectively. While, in the theoretical methods give a higher value of the settlement (2.0, 3.48, and 5.24 mm) for the single, group (1*2) and group (2*2) jet grouted columns compared with the settlement results acquired from field pile load test data. Therefore, it is not recommended to be used for soft clay. On the other hand, Fuller and Hoy's, Hansen's 90%, and Butler and Hoy's results may be considered faithful interpretation methods for the single and group (1*2 and 2*2) (JGC).

Keywords: Settlement, jet grouted columns, theoretical analysis. Pile load test.

1. Introduction

The construction activities in the south of Iraq often request deep foundations for the reason of the insignificant engineering properties of the subsoil and the associated difficulties arising from weak soil at surface depths. The structures erected on such soils may suffer from over settlement. Consequently, the soil improvement by Jet Grouting is one of the suitable techniques in improving the weak soil. The major purpose of grouting is to fulfill the space of the formatting material by replacing the current fluids with the grout and in that way improving the engineering properties of the soil, in this technique, the cement slurry is injected into the ground and column of hardened soil is obtained with required diameter and depth. Therefore, the strength of soft soil is increased when using this method, and the load is transferred from the foundation to both the jet grout columns and the original soils [1]. Jet grouting is a soil treatment technique adoption of a high hydraulic energy to destroy the surrounding soil. Ground improvement by Jet grouting technology can be used in varied and difficult geological and geotechnical situations, showing its practical and economic advantages, in addition to the importance of the performance control on-site, before and during the construction of the final foundations [2].

The Jet grouting has increased uses in the last years in the ground improvement that effort of low strength, permeability and seepage problems. The jet grouting has a high potential application to decrease the settlement in embankments [3]. Jet grouting technology has become popular because of installation it's very easy of, low cost, low noise, and no vibration accrues. So, it's best suitable, especially for the city constructions, noise, vibration, and site limitation area is limited [4,5]. Bustamante [7] conducted the back analysis of a large number of results from axial load tests performed on jet grouting soil-cement columns prepared with removable extensometers [6].

According to the interaction between the jet grout column installations of soil-cement reinforcing elements significantly increase the serviceability of the foundation utilizing increasing its strain characteristics and active protection of soils from the influence of the seismic vibrations, providing for the safe operation of the sports structure [7]. Juzwa and Bzówka [8] showed that the highest settlements of jet grouting columns from the center of the group and if the group is larger (more columns in the group) than the values of settlements are increasing [8].

The jet grouting method is one of the greatest common techniques for strengthening soft soils. Jet grouting columns allow for transmitting substantial loads through the reinforced soft soils and reduce the construction's settlement [9,10]. Different techniques have been used to prepare the (JGC) and measure their properties, but those methods are site- reliant, and involve numerous limitations in terms of jet grouting techniques, soil types, and jet grouting effective parameters. So that, the greatest way to estimate jet grouting execution and soilcrete properties is by conducting a prototype model to simulate the jet grouting or pile foundation in a field by the real soil state and with real jet grouting parameters and apparatus. The real jet grouting equipment is very large to be used. A choice was made for this mission to design and construct machines of jet grouting apparatus that would have almost a similar capability as a real field apparatus but with less size and cost.

2. Field Investigation and Soil Sampling

The site of the conducted study was located at the distance (100) m from the Euphrates River, in Al-Nasiriyah city (375) km southeast of Baghdad as shown in Fig. 1. The site investigation included drilling borehole 10 m in length and the basic physical and mechanical properties of the soil are represents in Tables 1 to 5.

Figure 1. Location of the study area.

Table 1. Results for corrected SPT at the specified depth.

Depth (m)	SPT	q_u (kPa) based on Terzaghi and Peck, (1967) [19]
1.0-1.5	5	60
5.5-6.0	7	87
7.5-8.0	8	100

Table 2. Results of undrained shear strength and consistency.

Depth (m)	$q_{unconfined}$ (kPa)	c_u (kPa)	Consistency based on Terzaghi and Peck, (1967)[19]
1.5-2.0	73	36	Soft
2.5-3.0	42	21	Very soft
3.5-4.0	76	38	soft
4.5-5.0	86	43	Medium

IOP Conf. Series: Materials Science and Engineering **901** (2020) 012002 doi:10.1088/1757-899X/901/1/012002

Table 3. Soil Properties.

Depth (m)	Sample type	LL %	PL %	PI %	Gs	Water content %
0.0-1.0	DS	32	15	17	2.61	34.73
1.5-2.0	US	34	19	15	2.62	35.26
2.5-3.0	US	60	29	31	2.78	37.95
3.5-4.0	US	59	30	29	2.76	38.22
4.5-5.0	US	35	21	14	2.58	33.71
8.0-10.0	DS	36	21	15	2.61	34.78

Table 4. Summary of consolidation test results.

Depth (m)	P_c (kPa)	Cc	Cr	Cv (mm^2/sec)	e_o	e_f
0.0-2.0	70	0.529	0.0598	0.02	0.63	0.475

Table 5. Soil description and classification.

Depth (m)	Groundwater table (m)	γ_{wet} (kN/m^3)	Clay (%)	Silt (%)	USCS
0.0-1.0		----	----	----	Fill
2.0-2.5		20.48	61	39	CL
2.5-3.0	0.75	21.02	70	30	CH
3.5-4.0		21.98	72	28	CH
4.5-7.5		21.5	60	40	CL
7.5-8.0		21.4	62	38	CL

3. Jet Grouting Field Design

To implement the jet grouting technique, it is necessary to design and manufacture an actual jet grouting system to simulate the actual process to great soilcrete column. The main equipment of the system includes, as shown in Fig.2 and Table 6.

High-pressure water jetting pump.

Mixer unit.

Jet grout machine.

High-pressure jet grouting pump.

Nozzles geometry.

Figure 2. Jet grout machine and parts.

Table 6. The main equipment and parts of the jet grouting system.

Jet Grouting Technique	Important performance
Jetting Grouting Machine	-
High-pressure water pump	The water pump rate is (11 L/min) with maximum pressure (200 bar) at a maximum speed of 1600 rpm
High-Pressure Grouting Pump	a progressive cavity pump with rated capacity (5.0 m3/hr.), a rated differential pressure (3.0 bar), rated speed (1450 rpm) and maximum allowable casing pressure (10 bar)
Mixer Unit	The cylindrical mixer is made from an iron with 800 (mm)-in height and 500 (mm) in diameter.
Nozzle	Electroformed Nickel and Brass nozzles with the diameter 2(mm), and 3 (mm).

4. Material and Methods

The construction theory of jet grouting was double fluid system (water, grout, and zero air pressure to reduce the diameter of the column) can be represented as:

- The water jet was at first used to create borehole to the required depth under pressure 75 bar. The drilling tools generally rotated at a continuous rate of 30 rpm to erode the soil and create column geometry.
- After the erosion process has done, the high pressure of grout slurry (Portland cement and water) was 6 (bar) with water-cement ratio (1:1) injected into the hole. In general, the setting time of the hardening slurry and soil mixture was 24 hr. On the other hand, the total time of the process to complete the drill the borehole by water jetting and slurry injection to build up the soilcrete was 2 hr. As a result, the shape of the soilcrete column was regular and looked as the cylinder and the cross-sectional diameter of the column was measured at the top (0.5-0.6 m) from the column head. All construction works of the group are represented in Fig. 3.

Figure 3. Construction steps of single, group (1*2), and (2*2) jet grout soil-cement column (D=150 mm and L=2000 mm).

4.1. *Construction of Concrete Cap for Jet Grouted Column*

A reinforced concrete cap was required for transferring the loading forces from the assembly (Structure of kentledge blocks) to the soil-cement column during the pile load test. Figure 4 depicts the schematic of the reinforcement concrete cap head connection. The dimensions of cap and steel reinforcement for concrete cap was calculated according to meet standard code requirements (ACI 318-14) [11]. The formwork fabricated for casting the concrete cap is displayed in Fig. 5.

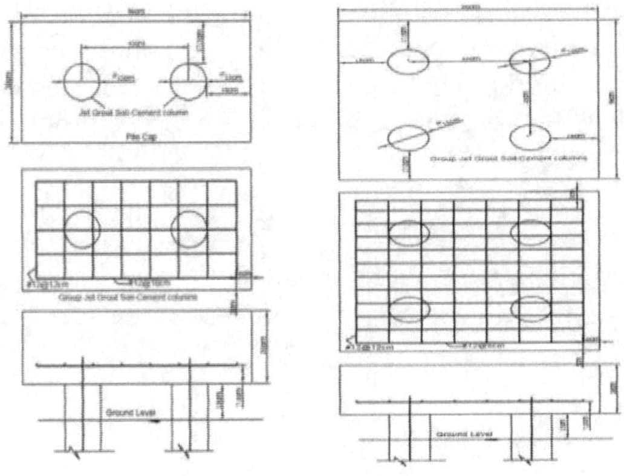

a. Group (1×2) b. Group (2×2)

Figure 4. Schematic representation of pile cap of jet grout soil-cement column.

a. Single b. Group (1×2) c. Group (2×2)

Figure 5. Cast in place pile cap of soil-cement column.

4.2. Pile Load Test

The two important features considered in the design of jet grout soil-cement columns are the bearing capacity and expected settlements due to affected axial loads. The jet grout capacity can be estimated using a theoretical method based on Poulos and Davis (1980) [12]. Usually, these methods tend to calculate approximately the jet grout column capacities with variable degrees of accuracy. The best accurate method to evaluation of the jet capacity and settlements are to make full-scale load tests representing the real column behavior normally expressed in terms of a load-settlement relationship. Although, the shortcomings of this method contain the relatively long time required and high budgets undertake in performing such tests. "Furthermore, it would not be feasible to perform pile load tests in the stage of the project planning stage. They are frequently controlled during the construction phase on construction piles which cannot be loaded to failure [13].

4.2.1. Quick Pile Load Test. The Quick pile load test procedure presented in (ASTM-D1143, 2007) [13]. Use the test load for an increment of 5 % of the expected failure load. Add load increments until accomplishing a failure load but do not exceed the safe structural capability of the pile, pile group, or loading apparatus. Through each load interval, save the load continual for a time interval of not more than 4 min and not more than 15 min, by the same periods for all loading increments during the test (ASTM-D1143, 2007).

4.2.2. Evaluation of the Jet Grout Load- Settlement Results-Based to Interpretation Methods. To compare the total settlement results of the pile load test for jet grouting soil-cement column four designated interpretation methods were chosen to inspect their accuracy for the calculated total settlement. In general, no exact method or criteria that can clearly describe the exact ultimate load or total settlement. However, these methods including settlement requirements mathematical and graphical methods [14,15].

- Brinch Hansen's 90% Method (1963).
- Butler and Hoy's Method (1977).
- Fuller and Hoy's Method (1970).

4.2.3. Single and Group Jet Grout Column Settlement. Most deep foundations requirements for the designing method will have the total accepted settlements of no more than about 12 mm. The total settlement estimations of a single and group jet grout column settlement are it's treated as the pile and consisting of the elastic settlement of a single pile and consolidation settlement below the pile group. Furthermore, the elastic settlement of single pile always related to many factors, such as; the relative stiffness of the pile material and the surrounding soil (Kps= EP/ES), the ratio between length to diameter ratio (L/D). On the other words, the relative stiffness of the base (Esb) and over the pile length and the modulus of elasticity of the soil and distribution ranges along the depth of the pile. Therefore, various methods have proposed to analyze and calculate the total settlement for the single pile and pile group as described by:

Method 1: Described by Budhu [16]:

Poulos (1989) as cited by [16] explained an excellent discussion to calculate settlement based on various numerical procedures, the total settlement of a single pile consists of settlement based on skin friction as explained in Eq. (1):

$$\rho_{ES} = \frac{Q_{all,friction}}{E_S.L} I \tag{1}$$

$$I = 0.5 + \log(\frac{L}{D})$$

Poulos, (1989) developed estimation for elastic settlement for floating pile as explained in Eq. (2):

$$\rho_{ES} = \frac{Q_{all\ friction}}{E_S.D}.I_P \tag{2}$$

I_p: influence factor depends on the L/D and K_{ps}, [16].

Settlement based on end bearing assuming the pile base is a rigid and punches on the surface of the soil transferred at a depth. The base settlement based on Timoshenko, and Groodier [16] is explained in Eq. (3):

$$\rho_b = \frac{Q_{base}}{r_b.G_b}.\frac{1-v}{4} \tag{3}$$

Where; **v**, Poisson's ratio of soil, r_b, and G_b are the radius and shear modulus at base. The elastic shortening of a single pile is presented in Eq. (4):

$$\rho_p = C.\frac{Q_{all\ friction}}{E_p.A_p} \tag{4}$$

Where C is the reduction factor (C≈0.5 for most soil, and C≈0.7 for soft soil). The shortening settlement calculates only when ($E_P/E_S < 500$). The total elastic settlement can be shown in Eq. (5);

$$\rho_{ET} = \rho_{ES} + \rho_b + \rho_p \tag{5}$$

Pile group settlement calculations based on settlement of single pile through a group the settlement factor Rs, as;

$$Rs = \frac{\text{settlement of group}}{\text{settlement of single pile at same average load}}$$

Fleming, et al. [16] gives an empirical solution of Rs: Rs= n^{φ} where φ=0.4-0.6 and n= number of the pile in the group. The consolidation settlement under the pile group based on the pile group may be embedded in soft soil and transfer the load causes consolidation settlement, the full load design act at depth 2/3L and the distribution according to the 2:1.

Method 2: Described by Das [17] explained that the total settlement of single pile caused by a vertical working load is:

$$S_{ESt} = S_{E1} + S_{E2} + S_{E3}$$

Where S_{E1} is the elastic settlement of the pile, S_{E2} is settlement caused by the vertical load at the pile tip, and S_{E3} is settlement caused by the transmitted load along the pile shaft. The total settlement can be represented in Eqs. (6, 7, and 8):

$$\mathbf{S_{E1}} = \frac{(Q_{WP}+\xi Q_{WS})L}{A_P.E_P} \tag{6}$$

$$S_{E2} = \frac{(Q_{WP}D)(1-\mu_S{}^2)I_{WP}}{E_S} \quad (7)$$

$$S_{E3} = \frac{(Q_{WS}D)(1-\mu_S{}^2)I_{WS}}{P.L.E_S} \quad (8)$$

Vesic (1977) explained another semi-empirical solution to obtaining the magnitude of the settlement as in Eq. (9 and 10) [17].

$$S_{E2} = \frac{Q_{WP}.C_P}{D.q_p} \quad (9)$$

$$S_{E3} = \frac{Q_{WS}.C_S}{L.q_p} \quad (10$$

Where Q_{wp} is the load carried at the pile point, Q_{ws} is the load carried by frictional (skin) resistance, A_p: area of a cross section of pile, L is the length of pile, Ep: modulus of elasticity of the pile material, ξ is the amount of varies between 0.5 and 0.67 and will depend on the nature of the distribution of the unit friction resistance along the pile shaft, D is the diameter of pile, Q_{wp} is the point load per unit area at the pile = Q_{wp}/A_p, Es is the modulus of elasticity of soil at or below the pile point, μ_s is Poisson's ratio of soil, I_{wp} is the influence factor = 0.85, q_p is the point resistance of the pile, C_p is the empirical coefficient. Representative values for various soils are given below, P: perimeter of the pile, L is the length of pile, I_{ws} is the influence factor = $2+0.35\,(L/D)^{0.5}$, Cs is an empirical constant = $(0.93+0.16(L/D)^{0.5}$. C_p: values (0.03-0.06) for bored pile. In general, a group pile elastic settlement at a similar working load on each pile increases with the width of the group (Bg) and the center-to-center spacing of the piles (*d*). The simplest relation for the settlement of group piles was given by Vesic, Eq. (11) [16]:

$$Se\,(G) = (Bg/d)^{0.5}.\ Se_{\,(single\ pile)} \quad (11)$$

Bg: width of pile group.

The consolidation settlement of a group jet grout column in clay can be estimated by using the 2:1 stress distribution method. The elastic jet grout column settlement, E_p= E_j, is determined according to the uniaxial compression strength (UCS). After completing the pile load test process, a series of core barrel samples were taken from different depths of jet grouting columns using the continued Coring method. The uniaxial compressive strength (Fc = 4.57MPa), E_J:=175*$q_{u(jet)}$= 175*4.57=800 MPa, the modulus of soil is equals (Es=9000), Poisson ratio for soft clay soil, $\boldsymbol{v}$ =0.5 [18].

5. Determination of Jet Column Load – Settlement Capacity

In performing a pile load test, two settlement dial gauges were used to record the average vertical settlement of the jet grouting columns. The dial gauges were fixed to the reference beam. The arrangement of the load reaction was consist of a platform consist of iron H-beam section, the end of platform rested over timber grabbing and then overreaction support, the applied load on the head of pile consist of the load of the platform and dead load of heavy material (a kentledge) were supplied by using a concrete blocks Fig.6.

(A) single (B) Group (1×2) (C) (2×2) jet grout soil-cement column

Figure 6. Pile load test in the field.

The subsequent results supported the observation of full-scale models of in situ check results of the jet grouting soil-cement column. Several limits influenced the depth and diameter of jet grout soil-cement columns like jetting pressure, radial rotation, and vertical motion of rod, the operation values and results will be listed in Table 7 and illustrated in Fig 7. Many interpretation theories have been proposed for estimation ultimate axial bearing capacity of jet grouted soil columns. A comparison was made between the measured settlement of single and group jet grouting soil-cement column were listed in Table 8. According to the load settlement test results of the jet grouted column, the total head settlement and for single and group were increases with increasing numbers of jet grout columns in a group as shown in Fig. 8.

Table 7. Operational parameters and dimensions were used in constructing the jet grouting soil-cement columns (D=150 mm).

Category of JGC	Pile Dimensions		Pile Cap Dimension (mm)	Water Jetting Pressure (bar)	Grouting Jetting Pressure (bar)	Rode Rotation Motion (rpm)	Rod Vertical Motion cm/min
	D(mm)	L (mm)					
Single	150	2000	D=150	75	6	40	30
Group (2*1)	150	2000	(500*960)	75	6	40	30
Group (2*2)	150	2000	(960*960)	75	6	40	30

Table 8. Summary of theoretical and measured settlement of jet grouting column (D=150 mm).

Jet grout references	Category of Measured Settlement (mm)				Category of Theoretical Settlement (mm)		
	Hansen's 90% Method	Butler and Hoy's	Fuller and Hoy's	Average	Poulos	Vesic's	Average
Single	0.45	0.47	0.32	0.41	1.625	2.41	2.0178
Group (2*1)	0.62	0.73	0.64	0.663	3.423	3.533	3.48
Group (2*2)	1.5	1.65	1.35	1.5	4.08	6.4	5.24

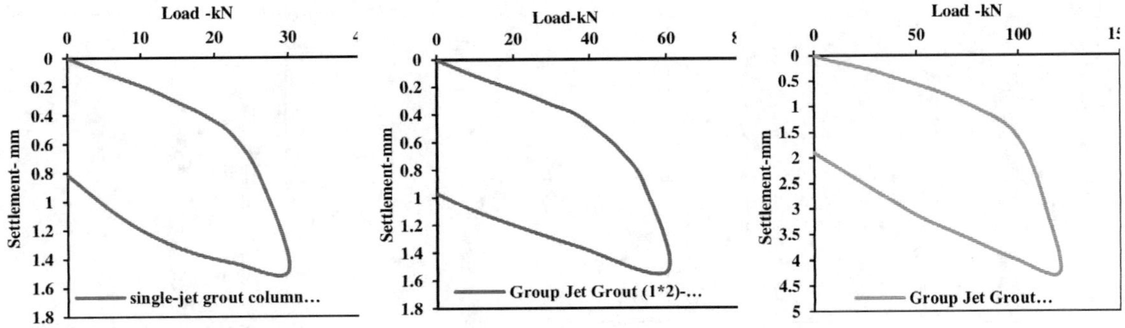

Figure 7. Load settlement curves of single and groups of jet grout soil-cement columns (D=150 mm and L=2000 mm and S/D=3).

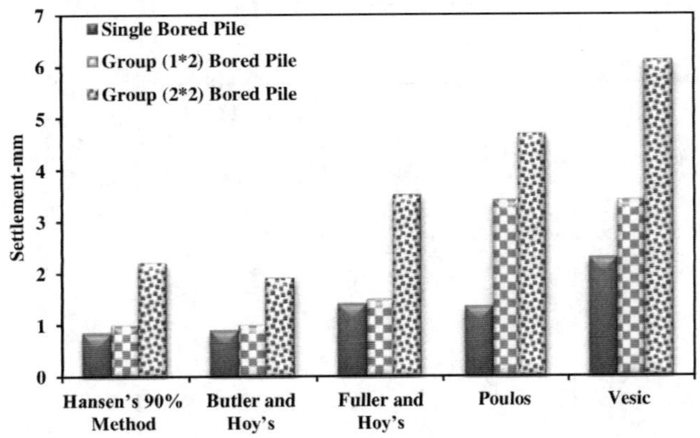

Figure 8. Comparison of Field and Theoretical Settlement.

Comparison of Interpreted and Measured Settlement. To comparison of interpreted total settlement and measured settlement based on Poulos and Das, the ratio of interpreted total to measured settlement (ST/Sm), expressed in percent were illustrated in histograms from Fig. 9 and Fig. 10 for single and group jet grout soil-cement column. The line 100% of Poulos and Vesic's represents a basis for comparison, the value of (ST/Sm) higher than 100% indicate to overestimation, whereas, lower than 100% indicate an underestimation of the total settlement, from the figures, it may be noted that:

- The settlement of a single and group jet grout soil-cement column estimation by Hansen's 90%, Fuller, and Hoy's and Butler and Hoy's Method given underestimation value to measure value by Poulos about (18.11 to 40.44%).
- The settlement of a single and group jet grout soil-cement column estimation by Hansen's 90%, Fuller, and Hoy's and Butler and Hoy's Method given underestimation value to measure value by Poulos about (13.28 to 25.78%).

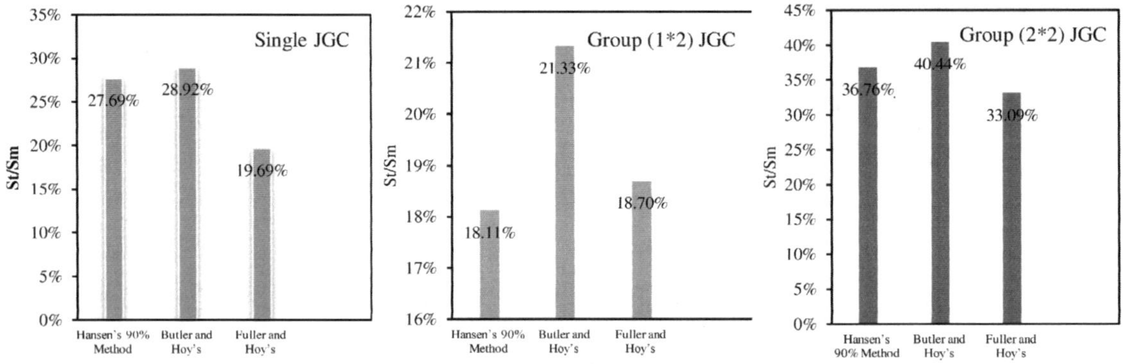

Figure 9. Percentage of S_t/S_m for comparison interpreted total and measured settlement based on Poulos (1989) [16].

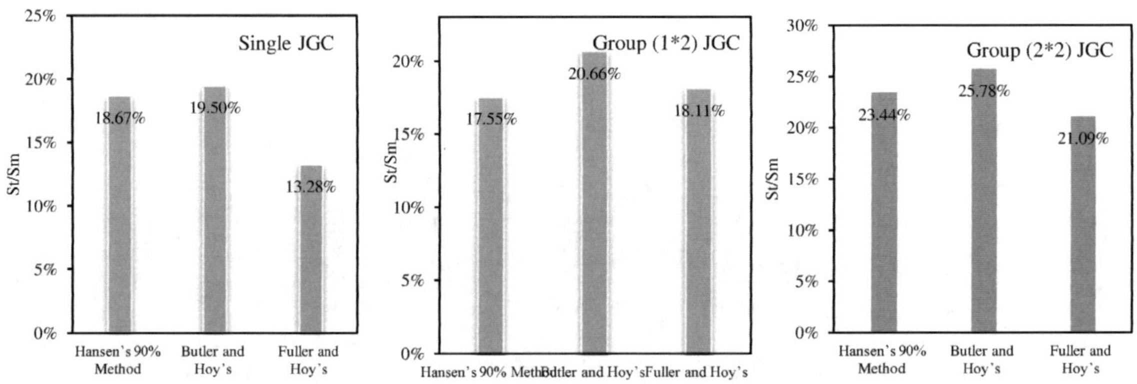

Figure 10. Percentage of S_t/S_m for comparison interpreted total and measured settlement based on Das (2011) [17].

6. Conclusion

The analysis of deep foundation settlement was performed employing the most widely used standards and approaches. To compare the results of the pile load test for jet grout soil-cement column settlement, three selected interpretation methods such as Hansen's 90%, Butler and Hoy's, and Fuller and Hoy's were chosen to examine their accuracy for calculating settlement. According to the pile load test analysis, it has been exposed that the jet grout soil-cement column settlement increase with an increase in the number of jet grout column. The comparative analysis results of methods indicate that the settlement values are similar for interpreted methods for the single and the group jet grout column while, the largest value of jet grout column settlement was obtained based on the Das methods. The reliable results of pile settlements were obtained from Fuller and Hoy's, and Hansen's 90% method. At last, the best way for computation and prediction of jet grout column settlement in soft clayey soils is a pile load test and use the average value of the settlement from the adequate interpretation methods.

References

[1] S.L. Shen, Z.F. Wang, W.J. Sun, L.B. Wang, and S. Horpibulsuk, A field trial of horizontal jet grouting using the composite-pipe method in the soft deposits of Shanghai, Tunnelling and Underground Space Technology, 35 2013142-151.

[2] A. Pinto, R. Tomásio, and G. Marques, Ground improvement with jet grouting solutions at the new cruise terminal in Lisbon, Portugal, Procedia engineering 143 2016 1495-1502.

IOP Conf. Series: Materials Science and Engineering **901** (2020) 012002 doi:10.1088/1757-899X/901/1/012002

[3] J.J. Chen, L Zhang, J.F. Zhang, Y.F. Zhu, and J.H. Wang, Field tests, modification, and application of deep soil mixing method in soft clay, Journal of Geotechnical and Geoenvironmental Engineering, 139(1) 2013 24-34.

[4] S. Horpibulsuk, W. Phojan, A. Suddeepong, A. Chinkulkijniwat, and M.D. Liu, Strength development in blended cement admixed saline clay. Applied clay science, 55 2012 44-52.

[5] S.L. Shen, Z.F. Wang, S. Horpibulsuk, and Y.H. Kim, Jet grouting with a newly developed technology: the twin-jet method. Engineering Geology, 152(1) 2013 87-95.

[6] M. Bustamante, Jet grouting columns. Report of the Seminar: Pathology of soils and foundations, Hammamet (2002).

[7] O.A. Makovetskiy, Application of "Jet Grouting" for installation of substructures of estates, Procedia Engineering 150 2016 2228-2231.

[8] J. Anna, and J. Bzówka, Numerical simulations of settlement of jet grouting columns, Transactions of the VŠB–Technical University of Ostrava, Civil Engineering Series. 16(1) 2016 1-6.

[9] J. Fan, D. Wang, and D, Qian, Soil-cement mixture properties and design considerations for reinforced excavation, Journal of Rock Mechanics and Geotechnical Engineering 10(4) 2018 791-797.

[10] J. Bzówka, A. Juzwa, K. Wanik, L. Wanik, and T. Żyrek, T., Discussion on the influence of various technological parameters on jet grouting columns geometry, Transactions of the VŠB–Technical University of Ostrava, Civil Engineering Series, 15(1) 2015 11-16.

[11] ACI Committee 318, Building Code Requirements for Structural Concrete (ACI 318-14): an ACI Standard: Commentary on Building Code Requirements for Structural Concrete (ACI 318R-14): an ACI Report. American Concrete Institute (2014).

[12] H.G. Poulos and E.H. Davis, Pile Foundation Analysis and Design, (1980).

[13] ASTM, D 1143/D 1143M – 07, "Standard Test Methods for Deep Foundations under Static Axial Compressive Load".

[14] A.K.M. Zein and E.M. Ayoub. "Evaluation of measured and interpreted failure loads of bored piles in alluvial soil deposits, International journal of GEOMATE: geotechnique, construction materials and environment 10(1) 2016 1636-1643.

[15] [S. Prakash and H.D. Sharma, Pile Foundations in Engineering Practice, John Wiley & Sons, 1990.

[16] M. Budhu, Soil Mechanics and Foundations, John Wiley & Sons (2008).

[17] B.M. Das, Principles of Foundation Engineering, Global Engineering–Christopher M (2011).

[18] G.A. Lorenzo and D.T. Bergado, Fundamental characteristics of cement-admixed clay in deep mixing, Journal of materials in civil engineering 18(2) 2006 161-174.

[19] K. Terzaghi, R.B. Peck, and G. Mesri, Soil Mechanics in Engineering Practice, John Wiley & Sons, 1996.

IOP Conf. Series: Materials Science and Engineering **901** (2020) 012003 doi:10.1088/1757-899X/901/1/012003

Experimental Investigation of Under Reamed Pile Subjected to Dynamic Loading in Sandy Soil

Maher M. Jebur[1,a*] and Mahmood D. Ahmed[1,b]

[1]Department of Civil Engineering, University of Baghdad, Baghdad, Iraq.
[a]mmg.maher@yahoo.com, [b]mahmoud_baghdad@yahoo.com

Abstract. This paper presents an experimental study between uniform pile and different types of under-reamed pile, single bulb. The under-reamed piles are piles with enlarged bases that are suitable to resist considerable movement of the ground, filed up ground, soft clay, and loose sand which have advantages to increase the soil strength, uplift capacity, and decrease the displacement. In the present study, there are experimental analyze to performance the suitable under-reamed type under sinusoidal load from vertical vibration (motor-oscillator was mounted directly on the pile cap. The main finding of this work is that the pile capacity increases with the ream and that all stress values of soil-structure interaction beneath the pile tip and under-reamed pile bulb due to external load can be founded. The under-reamed pile is made of reinforced concrete and the bulb is located at different positions of the embedded length of the pile. The type of soil used in the present work was loose sand and obtained from Karbala city. The Dynamic time-settlement curve has been obtained from the analysis and it was showed that the reduction in settlement is about 223% in case of using a single bulb from L/2 from pile tip which is the best results in comparison with the uniform pile.

Keywords. Under-reamed pile, dynamic load, machines foundation, sandy soil.

1. Introduction

The present-day trend of modern civil engineering structures tends to rise high into the sky i.e., skyscrapers, transmission tower of electricity, water tanks and tall structures subjected to high axial loads combined with large bending moments due to lateral forces such as wind, have needed superior types of foundation such as deep foundation. Under-reamed piles are cast in-situ and made of concrete piles with one or more bulbs in their stems. The bulbs help to distribute the loads in soil and then lead to an increase in the carrying capacity of the piles in both pull-out and compressive loadings. Also, under reamed piles provide better anchoring along the embedded depth of the pile. The first methodical studies on under-reamed piles were conducted in South Africa in 1949 by Jenkins and Henkel. Qian et al. [1] presented that the behavior of under reamed piles inserted in sandy soils. The authors determined the pull-out and axial static loading carrying capacity of the piles, which description and calculation the stresses of soil mass and provide a theoretical basis for a future study of this type of pile. Bharathi and Dubey [2] implemented 2D linear finite element analyses of vertical piles (with and without under-ream) subjected to vertical sinusoidal dynamic load using ABAQUS software.

The soil was silty sand (SM) has an angle of internal friction ($\phi = 32^o$) and dry unit weight ($\gamma = 15.38$ kN/m^3). Based on the results of the numerical study, it was observed that; as the eccentricity of the oscillator (i.e. force level) increases, the vertical displacement amplitude increases. It was also found that the maximum vertical displacement amplitude of the under-ream pile is less than that of a vertical pile. Hence, it is favorable to use under-ream pile in situations where the system is subjected to heavy vertical dynamic sinusoidal loads. Jebur et al. [3] studied the influence of the geometrical shape of bulbs and subjected to machine load on the carrying capacity of under reamed piles by PLAXIS 3D software-2013 version. The result show that shapes of bulbs have significant effects on the failure of the under-reamed pile in the soil mass. The geometrical shape of the reamed pile can be changing the soil pile interaction and change the type of soil mass failure from punching occurred in a pile of uniform cross-section to local failure or general shear failure which occurred in under-reamed piles.

Christopher and Gopinath [4] carried out two sets of mild steel under reamed piles models in sand soil to study the behavior of under-reamed piles subjected to vertical loads by an experimental work, the pile, and under-reamed diameter equal to 25 and 75 mm respectively. From the tests, the results show that, the load capacity of piles with ream is higher in comparison with the straight shaft. In the case of a prismatic pile of diameter 25 mm the ultimate capacity increased about 305% in compression, when the pile is under-reamed. Al-Suhaily et al. [5] used an open pile, which consists of a square steel section for a pile of square with dimensions of (24×24) mm with 480 mm in length and 1.5 mm in thickness. A steel frame was planned to be used in motivating the piles, the steel structure was located above the pile. The results showed that gates are opening at an angles ranged from 20–80° the changed types of substrates, thus increasing the loading area from (B2 to 9.8 B2) by an increase of 880% and increasing the increment ratio in pull out resistance from (53 N to 430 N) by an increase of 711%.

The present study devoted to studying the behavior of under-reamed piles inserted in sandy soil and subjected to vibrating loading through obtaining the settlement-dynamic time variation for conventional pile and under-reamed pile; understanding the behavior of reamed position under sinusoidal load, and to investigate the relationship between net pile friction force and bulb position and bearing pile tip.

2. Materials Used

2.1. Sand

The soil used in the model tests is natural sand but passing through sieve No. 10 and retaining on sieve No. 100 obtained from Karbala city. The site is located in station (0+0) km for Al-Rufae road which leads to Razaza lake. The useful physical and mechanical properties of used sandy soil are listed in Table 1

Table 1. Geotechnical properties of soil used in the test.

Property	Unit	Value	Specification
Specific gravity, Gs	-	2.66	ASTM (D854)
Coefficient of uniformity, Cu	-	2.0	
Coefficient of curvature, Cc	-	0.96	ASTM (D422)
Classification (USCS)	-	SP	
Maximum, $\gamma_{d,max.}$	kN/m^3	16.80	ASTM (D4253)
Minimum, $\gamma_{d,min.}$	kN/m^3	14.90	ASTM (D4254)
Dry unit weight (Used)	kN/m^3	15.21	-
Young's modulus, E	kPa	1200	ASTM (D4253)
Poisson's ratio, ν	-	0.3	Assumed
Angle of internal friction, φ	degree	32	ASTM D 3080)
Dilation angle, ψ	degree	2.0	ASTM (D3080)

2.2.Model Pile

A pile length 570 mm with a diameter 23 mm, the pile is kept embedded in the sand so that the pile top is 20 mm above the soil surface. The model piles used in this study are made of an admixture of reactive powder concrete. This reactive powder allows producing a high performance and workability mortar, increase mortar toughness, ductility, and impact resistance Mohammed et al. [6]. To construct the under reamed bulb at specific locations of the model pile, the wall of uniform circular cross-sectional area pile is coated with a structural epoxy bonding agent (Sikadur-23) which is a moisture tolerant to bond the surface of the pile at the desired level of reamed with new mortar of under reamed bulb. Figure 1 shows the plastic molds of cone shape have diameter equals 23 mm at top and diameters equal to 46 mm at the bottom are used to construct the form of under reamed and filed by mortar slurry by special injection needle.

Figure 1. Under reamed construction.

2.3.Piled Cap Models

The piled cap used in this study is made from steel plate of dimensions (150×150×10) mm

3. Experimental Setup and Methodology

The test was arranged in a steel container (1000×1000×1000 mm), these dimensions were chosen to meet the boundary effects of physical models subjected to harmonic vertical vibration. Sand was filled in layers using the raining system which was manufactured to obtain a relatively homogeneous sand bed depending on the falling rate and height. Several trials with different heights of fall were performed to achieve the desired relative density as shown in Fig 2.

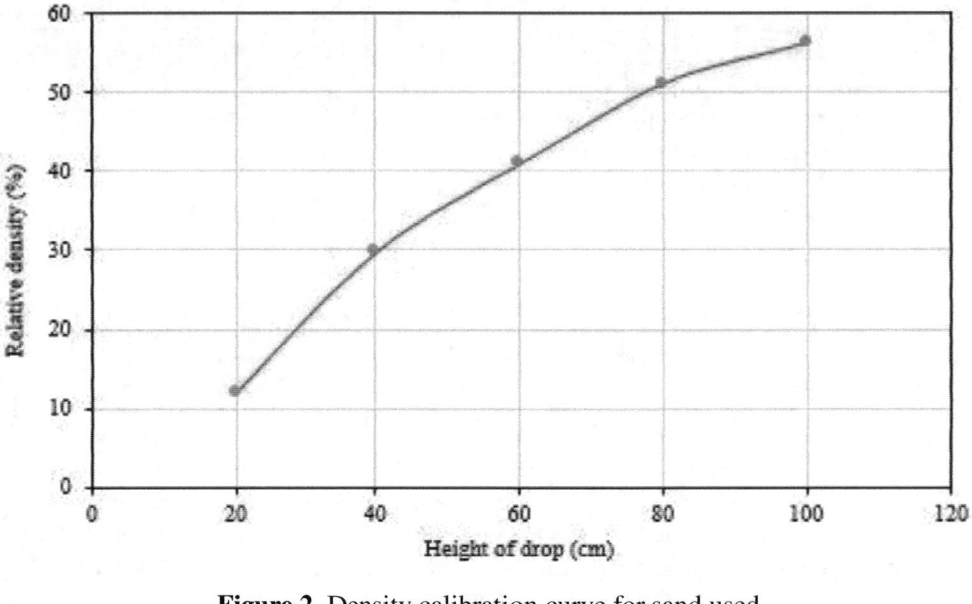

Figure 2. Density calibration curve for sand used.

The sinusoidal excitation load reciprocating machines and unwell balanced rotating equipment cause periodic dynamic forces. The maximum amplitude of the vertical dynamic force produced (Fmax) is:

$$F_{max} = m_e \, e \, \omega_r^2 \quad (N) \tag{1}$$

Where ω_r is the circular operating frequency of the machine, r/s; e is the eccentric distance from the center of shaft to the unbalanced mass m_e; and m_e is the unbalanced mass, g. Small motor was used to simulate a rotary machine with a total mass of 2.11 kg while the eccentric mass was 0.0145 kg and the eccentric distance was 22 mm, from the above data and for (30 Hz) operating frequency, the induced dynamic load for the machine from the above formula is 11.3 N. The corresponding settlement of pile was recorded by Computerize linear variable differential transducer (LVDT). The force sensor is used to measure the forces directly beneath under reamed pile (pile tip and bulbs), and the number of load senores has been used equal to the number of pile bulbs plus 1 Fig. 3.

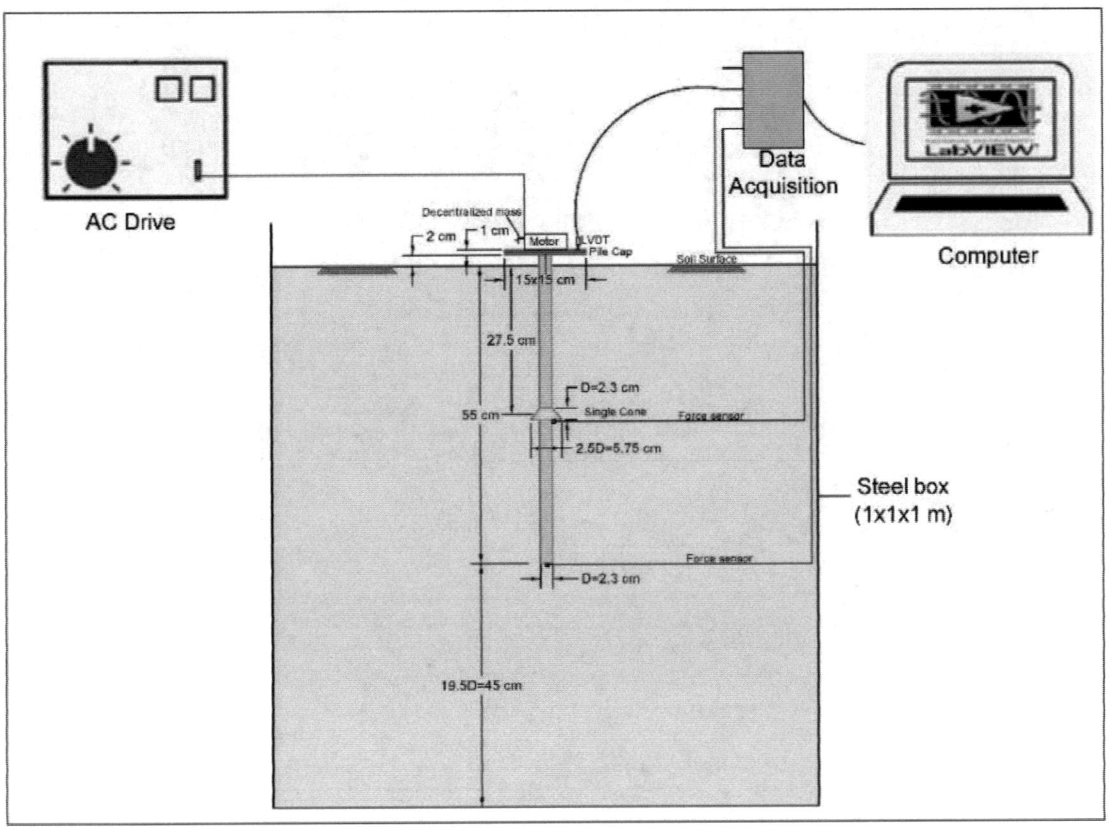

Figure 3. Schematic diagram of the physical model.

4. Results and Discussion

In this study, the position of reamed from the pile tip used was 0, L/4, L/2, and 3L/4 as shown in Fig. 4. Figure 5 shows the relation between bulbs position and maximum settlement whit the time, therefore, it can be noted that when bulb position equal to L/2 from pile tip, there is minimum settlement 0.55 mm compare with uniform pile 1.8 mm see Fig. 6, the motive of reducing the settlement return to there is a good interaction between bulb position and confined soil pressure against external dynamic force.

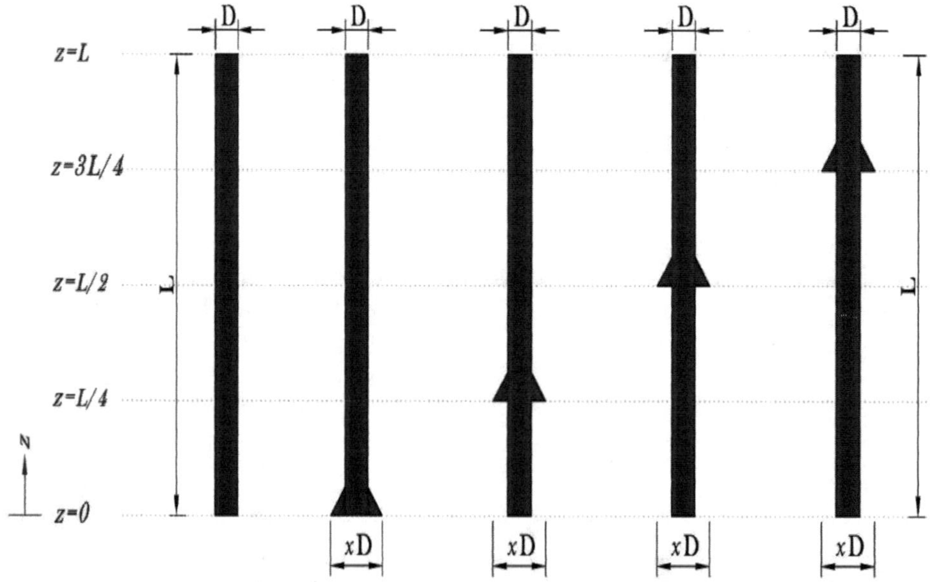

Figure 4. the different positions of the bulb.

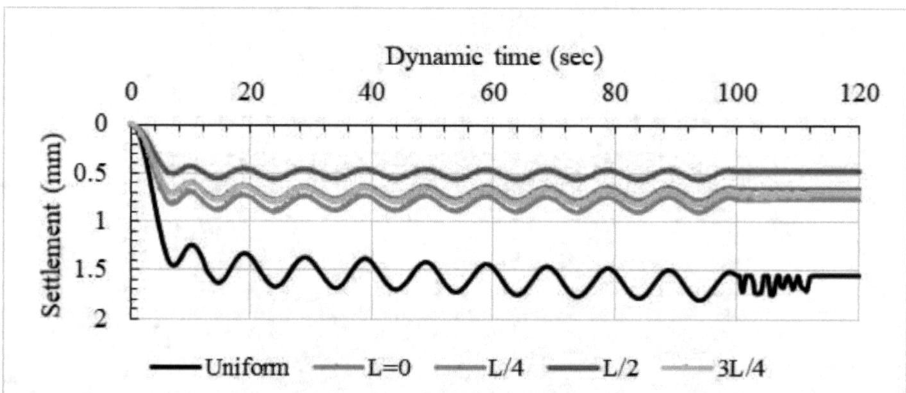

a

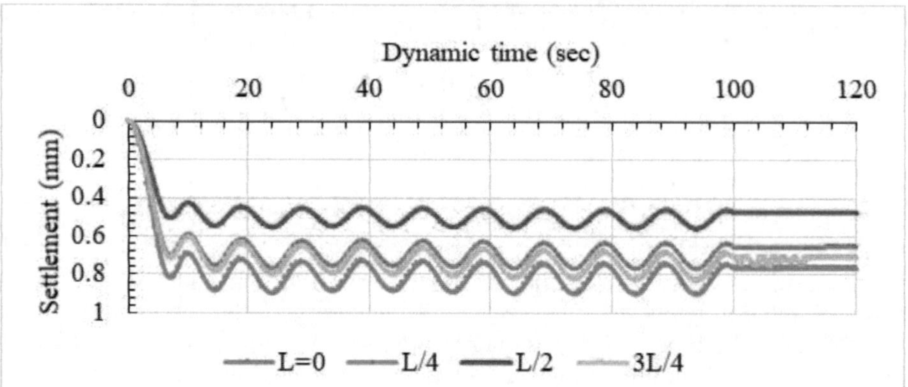

b

Figure 5. Time-Settlement for (30 HZ-RD 35%).

IOP Conf. Series: Materials Science and Engineering **901** (2020) 012003 doi:10.1088/1757-899X/901/1/012003

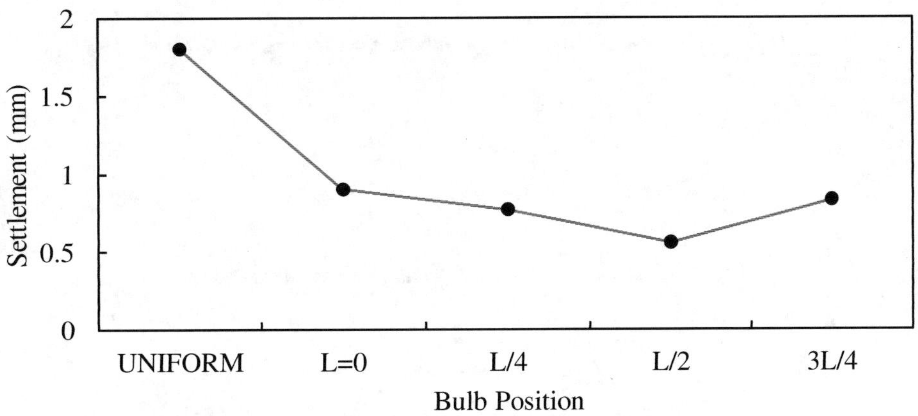

Figure 6. Settlement-Bulb position for (30 HZ-RD 35%).

Figures 7 to 11 display the relationship of dynamic bearing force in (pile tip and bulb)-time curve with RD 35% and 30 Hz that means dynamic load put on pile cap was 11.13 N. From these Figures, it is intended to study the average friction force along under-reamed pile (F), the average force under bulbs (BU) and the pile tip force in time (120) sec motor stop (PT).

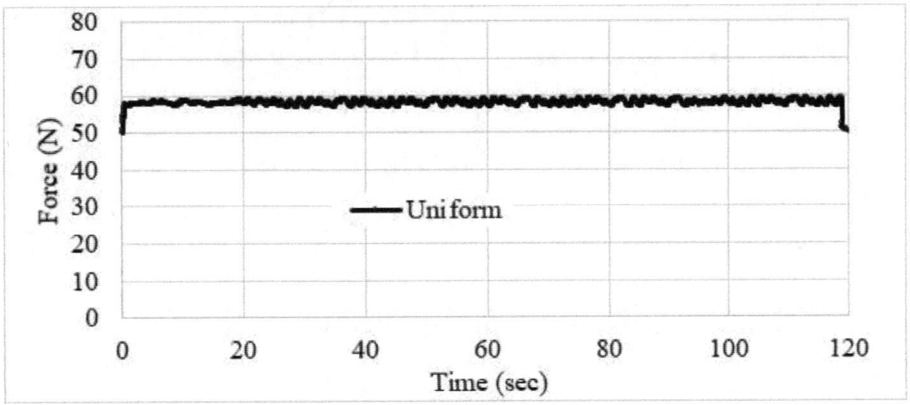

Figure 7. Force-Time Curve for (30 HZ-RD 35%- Uniform).

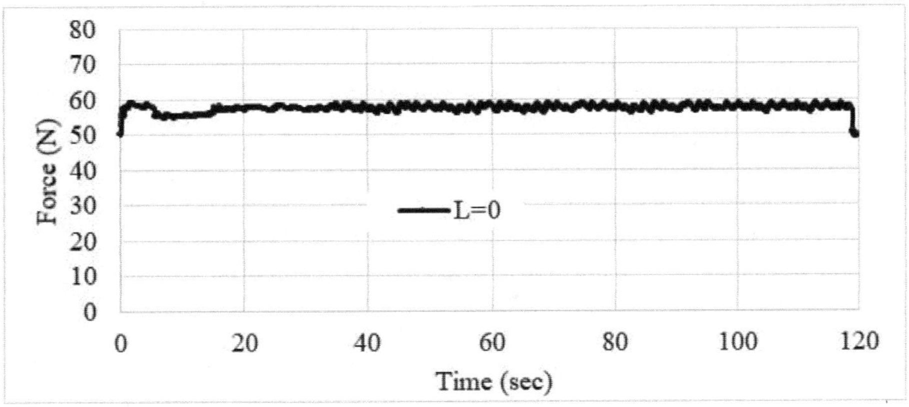

Figure 8. Force-Time Curve for (30 HZ-RD 35%- L=0).

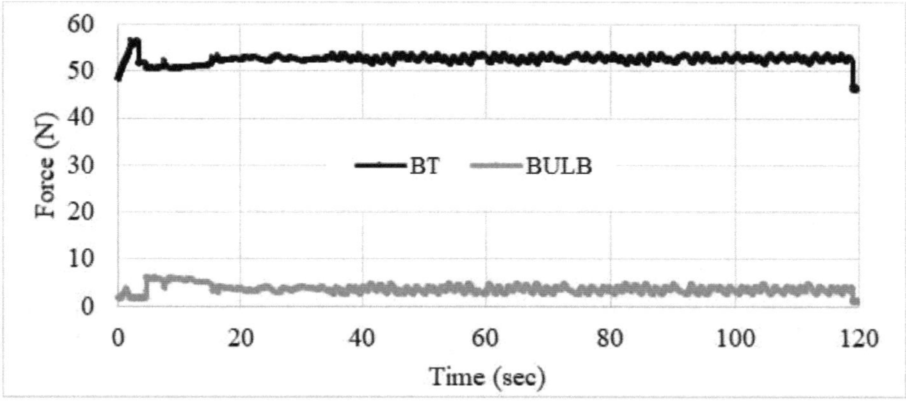

Figure 9. Force-Time Curve for (30 HZ-RD 35%- L/4).

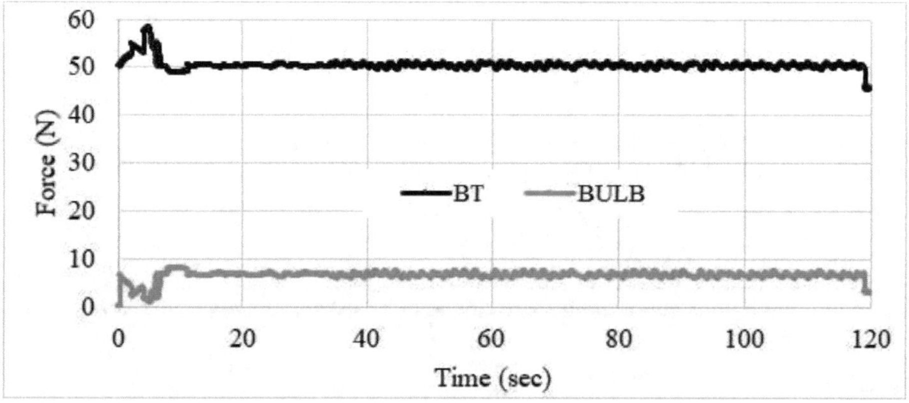

Figure 10. Force-Time Curve for (30 HZ-RD 35%- L/2).

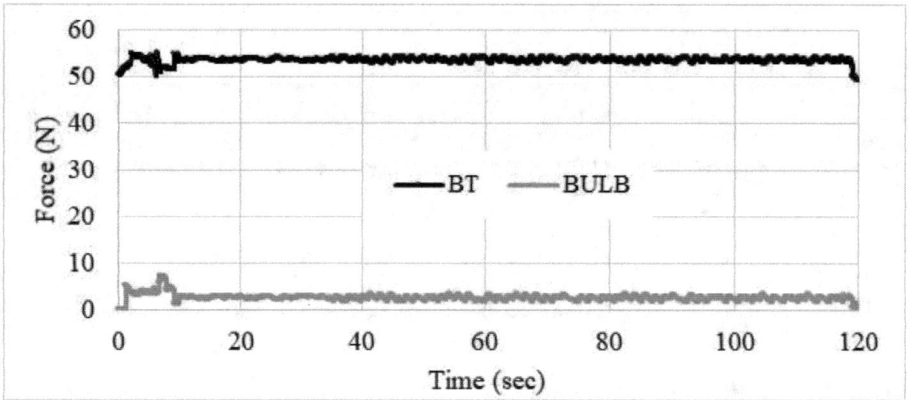

Figure 11. Force-Time Curve for (30 HZ-RD 35%- 3L/4).

Figure 12 shows that (F) value in (L/4, L/2 and 3L/4) was more than (Uniform and L=0) because bulb position makes its work together with the pile to encourage soil particular around bulb for involvement in total friction force, which is opposite to what happened in bearing force under bulb (BU) Fig. 13.

IOP Conf. Series: Materials Science and Engineering **901** (2020) 012003 doi:10.1088/1757-899X/901/1/012003

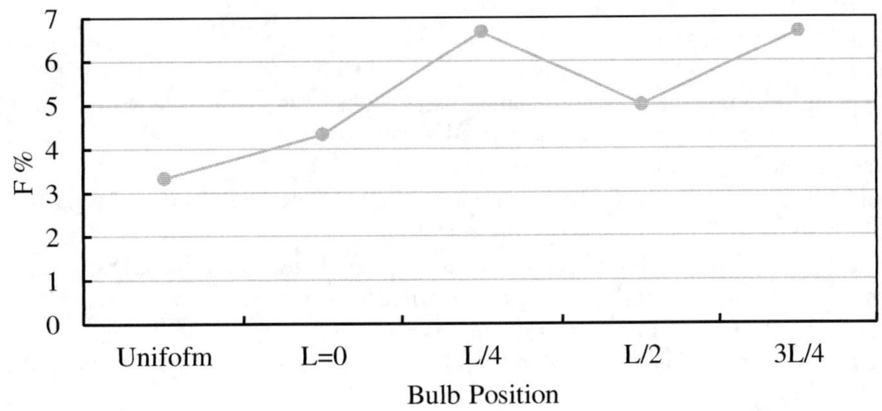

Figure 12. Friction force-blub position for (30 HZ-RD 35%).

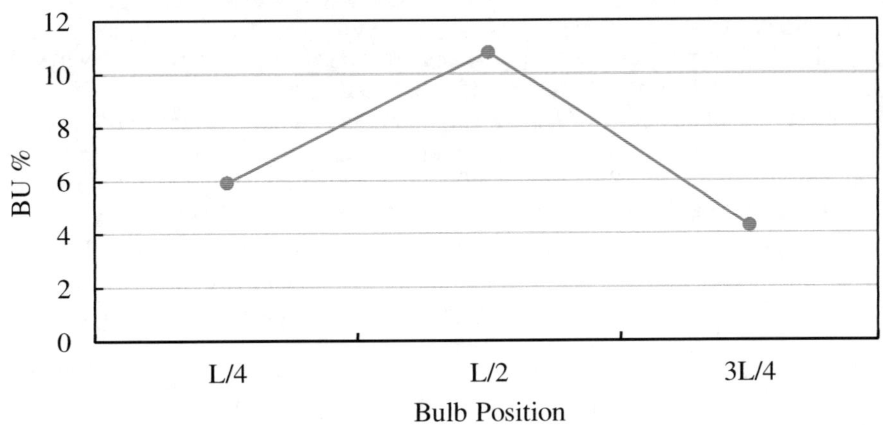

Figure 13. BU%-blub position for (30 HZ-RD 35%).

Finally, Fig. 14 shows that under-reamed pile can be reduced pile tip force under machine vibration compare with uniform pile at the final time which the demising vibration was stopped due to stick reamed with surround sand.

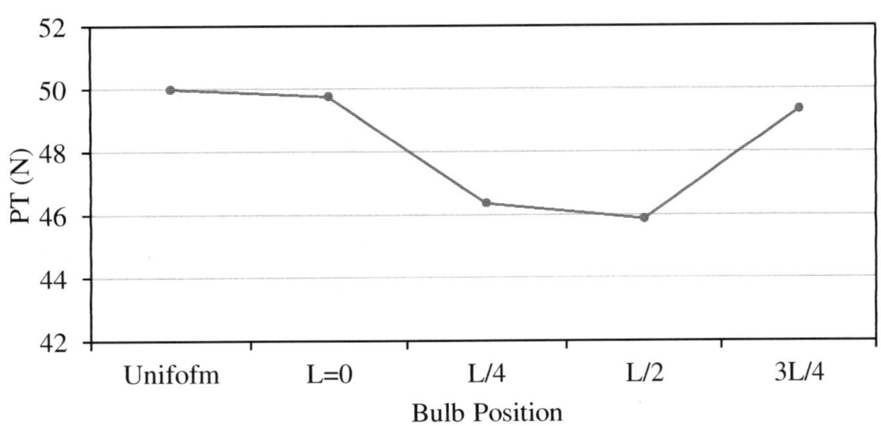

Figure 14. PT-Blub position for (30 HZ-RD 35%).

5. Conclusions

- In dynamic loading, the factor of safety against bearing failure increased during machine operation, during operation, the skin friction resistance mobilized along pile length led to decrease the bearing load.
- At time 120 sec for all cases of under-reamer pile end bearing less than initial time 0 sec and the minimum values appear in the position L/2.
- Using a single under-reamed pile can be increasing friction force instead of uniform pile, the maximum friction force was in position of reamed L/4 in loose sand.
- Finally, the optimum position of the reamed pile under dynamic load was at the position L/2 where it can be caring about 10.9% of total loading (external and dead load) and the settlement equal to 0.55 mm.

References

[1] Y.M. Qian, D.P. Zhao, and X.W. Xie, The research on the ultimate bearing capacity of soil around the push-extend multi-under-reamed pile at sliding failure state, Applied Mechanics and Materials 578-579 2014 232–235

[2] M. Bharathi and R.N. Dubey, Dynamic lateral response of under-reamed vertical and batter piles Construction and Building Materials 158 2018 910-920

[3] M.M. Jebur, M.D. Ahmed, and M.O. Karkush, numerical analysis of under-reamed pile subjected to dynamic loading in sandy soil, In IOP Conference Series: Materials Science and Engineering, 671(1) 2020 p. 012084.

[4] T. Christopher and M.B. Gopinath, Parametric study of under-reamed piles in sand, International Journal of Engineering Research & Technology (IJERT), 5(7) 2016 577-581.

[5] A.S. Al-Suhaily, A.S. Abood, and M.Y. Fattah, M. Y., (2018), Bearing capacity of uplift piles with end gates, Proceedings of China-Europe Conference on Geotechnical Engineering, 2018 893–897.

[6] S.D. Mohammed, W.Z. Majeed, N.B. Naji, and N.M. Fawzi, Investigating the influence of gamma ray energies and steel fiber on attenuation properties of reactive powder concrete. Nuclear Science and Techniques, 28(10) 2017 153.

Numerical Simulation of Predicting the Shaft Capacity of Single Pile in Unsaturated Soil

Bushra S. Albusoda[1,a], Mohammad F. Abbas[2,b *], and Durgham A. Al-Hamdani[3,c]

[1]Department of Civil Engineering, University of Baghdad, Baghdad, Iraq
[2]Department of Civil Engineering, University of Al-Muthanna, Muthanna, Iraq
[3]Al-Furat Al- Awsat Technical University, AL-Najaf, Iraq
[a]albusoda@yahoo.com, [b]engmoh863@gmail.com, [c]D.Alhamdani1901P@coeng.uobaghdad.edu.iq

Abstract. Pile foundations that placed above the groundwater table zone where the soil is typically in a state of the partially saturated case have been considered in this study. To appreciate the effect of the matric suction on the skin resistance of single pile foundations that placed in unsaturated clayey soil. A series of small scale and full-scale single model pile are simulated using the software Plaxis 3 D. Mohr-Coulomb model (Elastic-perfectly plastic model) that takes into account the effect of the matric suction has adopted in the analysis. Also, in this study comparison among numerical models and laboratory, small scale single and full-scale pile models have done. In general, the value of shaft capacity obtained by the numerical investigations are in good agreement with laboratory model pile test results.

Keywords: Unsaturated soil; matric suction; single pile; shaft capacity; Plaxis 3D.

1. Introduction

Generally, soil mechanics principles have two approaches dry or saturated soils and other approaches between them called unsaturated soils or partially saturated. However, most piles are placed in unsaturated soils. Unsaturated soils have three phases (solid, water, and gas) and contractile skin phase that made matric suction among particles of soil that could increase shear strength and other parameters such as stiffness of soil, so increase adhesion between soil and pile [1]. Variation of properties to many soils when it giving out or absorbing water. A variation of air pressure and pore water pressure effect on the stress state of the soil model [2,3]. Nearly all of the classical soil mechanics basic are appropriate for fully saturated soils, in a deal with unsaturated soils should be concerning by main point (changes of volume concerning changes degree of saturation [4].

The difference between the pore air pressure (u_a) and pore water pressure (u_w)that refers to matric suction. Many investigations have connected among shaft capacity of pile foundation, matric suction, and behavior of partially saturated soils. A considerable deal of interest has been toward to pile foundation design for engineering basics wherein the mechanics of partially saturated soils have been applied extensively [5-7]. Carrying load capacity and settlement are required information for pile foundation design in saturated and unsaturated soils. Shear strength and stiffness of the vicinity soil also called soil-pile interface supply considerable assistance into the load-settlement behavior of a pile foundation [8].

The theory of plasticity was used for simulating partially saturated soil behavior by using the proposed elastoplastic new model, which was an oval-shaped model similar to the modified Cam-Clay Model (MCC) [2]. The arrangement of this model differs based on the value of suction. This model is appropriate for a little to say the swelling partially saturated soils such as soils including silt, sand, and clay with low-plasticity. Many loaded piles are founded above the groundwater table zone and these piles are affected with matric suction. Vanapalli and Taylan [1] founded a new small model pile test in unsaturated soil and proposed using three a modified α, β and λ methods to look forward to the influence of matric suction on maximum skin resistance. Also, a single pile skin resistance influenced by the contribution of matric suction on along of pile. In this study numerical examinations of the small-scale single pile embedded in partially saturated fine-grain soil were carried out by Mohr-Coulomb model. The Mohr-Coulomb model significant adhesion strengths between the partially saturated soil and steel pile was employed to analyze skin resistance from small scale single pile load tests in the laboratory.

2. Testing Approach of laboratory Model

A group of small single pile model load tests under different conditions (fully saturated and partially saturated compacted clayey soil in short-term state) as shown in Fig. 1. A glacial till soil that tested by Vanapalli and Taylan,2012[1] was dependent on this study. The main goal of the test approach is to calculate the effect of matric suction on the skin resistance of small model pile and comparing between laboratories and numerical results.

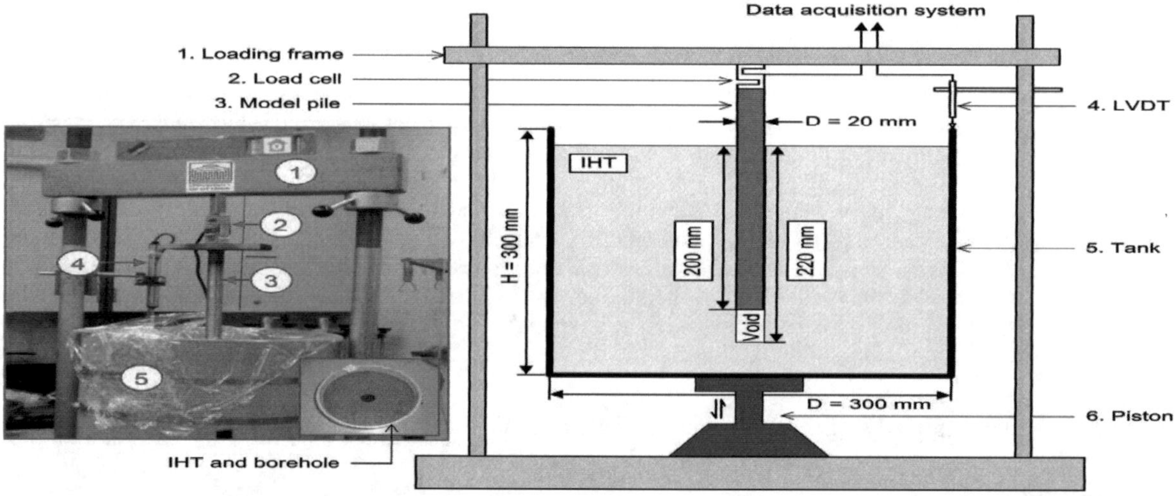

Figure 1. Laboratory model [1].

3. Physical and Mechanical Properties of Soil

The used soil properties (physical and mechanical) are tabulated in Table 1. The skin resistance of the small-scale model pile was suggested to be obtained at two moisture contents; 13% and16% in the dry side of the optimum. These moisture contents were selected from the density -water content curve data. The dry unit weight of the tested soil at these moisture contents was equal to 14.5 kN/m^3 and 16.1 kN/m^3 respectively. The values of matric suction of the laboratory model were evaluated by the axis-translation method [9]. The value of moisture content was (13% and 16%) for matric suction of (205 and 110) kPa respectively. The tests were taken for the moisture contents on the dry side of optimum at less than 90% of the degree of saturation values which generated in significantly high matric suction.

Table 1. Physical Properties of the tested soil [1].

Soil property	Value	Soil property	Value
Moisture content at optimum, M_{opt} (%)	18.6	Clay percentage	30
Maximum dry unit weight, γ_{dmax} (kN/m^3)	16.7	Liquid limit, LL (%)	33
Total unit weight, $\gamma_{sat.}$ (kN/m^3)	18.5	Plastic limit, PL (%)	17
Sand percentage	28	Plasticity index, PI (%)	16
Silt percentage	42	Undrained shear strength, cu (kPa)	11.5

The Soil Water Characteristic Curve *SWCC* of the samples designed with different moisture contents of 13% and 16% are shown in Fig.1. By using the pressure plate apparatus to prepared these *SWCC* following the drying path. More information on the matric suction tests employing both the pressure plate apparatus the modified null pressure plate and are explained in [10,3]. The laboratory samples were prepared by mixing with the quantity of water-limited and placed it in a tank with dimensions (300 mm in diameter and 300 mm in height). Compaction of samples statically by a specially designed compaction base plate with 350 kPa. The compaction and model pile load tests were carried out employing a standard triaxial test loading frame [1].

4. Estimation of the Ultimate Skin Resistance of Piles in Partially saturated Soils

4.1. Using Modified α Method

Many studies related to the skin resistance of a single pile to the undrained shear strength, c_u of the clayey soils [11,12]. In this study, for analyzing the results of the steel model pile tested and placed in partially saturated and saturated soil the modified α method was adapted. Also, back estimating the variation of adhesion factor between model steel piles and unsaturated soil concerning matric suction. Eq.1, it connected between undrained shear strength in saturated and partially saturated soils according to matric suction employing soil-water characteristic curve for partially saturated condition, $c_{u(unsat)}$.

$$Qf_{(unsat)} = \alpha\, c_{u(sat)} \left[1 + \frac{u_a - u_w}{P_a/101.3} S^{v}/\mu\right] \pi\, dl \qquad (1)$$

where, $c_{u(sat)}$, = saturated undrained shear strength, Pa = air pressure (i.e. 101.3 kPa), and ν, and μ = fitting parameters. The fitting parameter ν is based on the types of soils and it is 2 for the clayey soils. The fitting parameter μ is 9, however, it is a function for the plasticity index (PI).

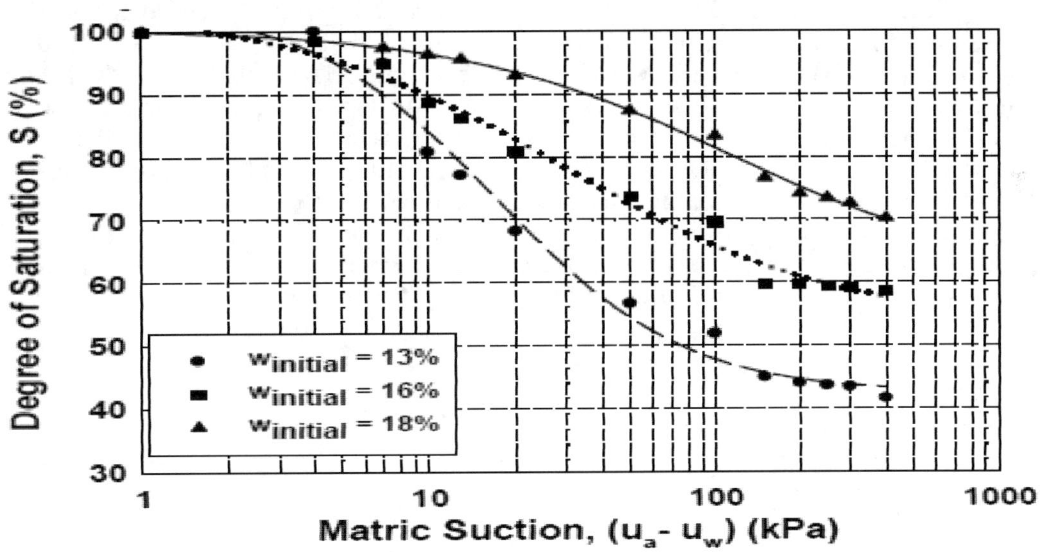

Figure 2. Soil water characteristic curve (SWCC) [1].

4.2. Numerical Modeling

In this section the numerical model will be defined. The Mohr-Coulomb model (MCM) is adopted in this study. Table 2 showed the essential parameters used in this model. The linear elastic-perfectly plastic constitutive model finite element extending MCM yield criterion was used in the numerical analysis include the conventional shear strength parameters (c_u) estimated from unconfined compression, the elastic modulus under saturated (E_{sat}) and unsaturated conditions (E_{unsat}) are required. Oh et al. [13] proposed Eq. 2 to evaluate the influence of respect to elastic modulus (E) at partially saturated condition (E_{unsat}) with soil suction employing modulus of elasticity at fully saturated condition

(E_{sat}) and two fitting parameters α and β values 1.5 and 2 respectively. The Plaxis 3D software is used in this research for modeling soil mechanical behavior. The non-soil material such as stainless steel should be taken into account during modeling processes. Concerning the model dimensions as cited by [1] width and height equal 300mm and the height of pile equal 200 mm. The full-scale model with height and width are 6 and 10 m respectively and the pile height is 4m and its diameter 0.4 m which present in this study. However, the carrying capacity of pile foundations can be obtained accurately by field tests. The field test of pile carrying capacity is time-consuming, needs trained professional and heavy equipment of expensive cost. All these reasons contribute to estimating the pile carrying capacity and settlement behavior using numerical methods.

$$E_{Unsat.} = E_{Sat.}\left[1 + \alpha\,\frac{u_a - u_w}{P_a/101.3}\,S^{\beta}\right] \qquad (2)$$

Table 2. Essential properties of the input parameter for Mohr-Coulomb (MCM).

Materials	Matric suction (u_a-u_w) kPa	Degree of saturation	Young modulus (E), kPa	Poisson's ratio (μ)	Undrained Strength, kPa
S1	0	90	2500	0.49	11.5
US2	205	43	15000	0.49	68
US3	110	65	8000	0.49	80
Steel pile*	-	-	2000000	0.2	-

* The model adopted for pile behavior is the linear elastic model.

5. Results and Discussion

In this part a comparison among the estimated, calculated and Plaxis 3D software results of carrying capacity of the pile models are shown in Figs. 3-8 and Tables 3-5. The figures show the ultimate skin resistance of single steel pile, the soil suctions value (measured and predicted) were comparable results at low of degree of saturation values (S1) with matric suction = 0 kPa, and US2 with matric suction = 205 kPa; US3.-16% with matric suction = 110 kPa). However, the predicted skin resistance values were low at the saturated sample with matric suction values equal 0 kPa at low shear strength 11.5 kPa with fully saturation that effect on factor of adhesion between pile surface and soil materials (shear strength decreased with the degree of saturation increased), also the ratio $(u_a-u_w)/\alpha$ and $(u_a-u_w)/c_u$ were opposites. The results of the MCM were lower than the experimental results may be because MCM neglected dilation value, and neglected variation of modulus of elasticity with depth, on the other hand, MCM results approximately in good agreement with laboratory model results which based on linearity theory the variation among these results in Tables 3-4 and Fig.7 based on different conditions.

Table 3. Ultimate shaft capacities (Qs) using different methods.

Degree of saturation, S%	Matric suction (u_a-u_w) kPa	c_u (Meas.) kPa	α	Q_s Eq. (1), kN	Q_s Lab. Model, kN	Q_S, Plaxis, kN
90	0	11.5	0.9	0.13	0.11	0.141
43	205	68	0.75	0.564	0.68	0.575
65	110	80	0.76	0.676	0.55	0.491

Table 4. Ultimate shaft capacities (Qs) using estimated and predicted methods

Degree of saturation, S%	Matric suction (u_a-u_w) kPa	c_u (Meas.), kPa	α	Q_s Eq. (1), kN	Q_S Plaxis 3D, kN
90	0	11.5	0.9	51.99	67
43	205	68	0.75	225.83	276
65	110	80	0.76	270.65	254

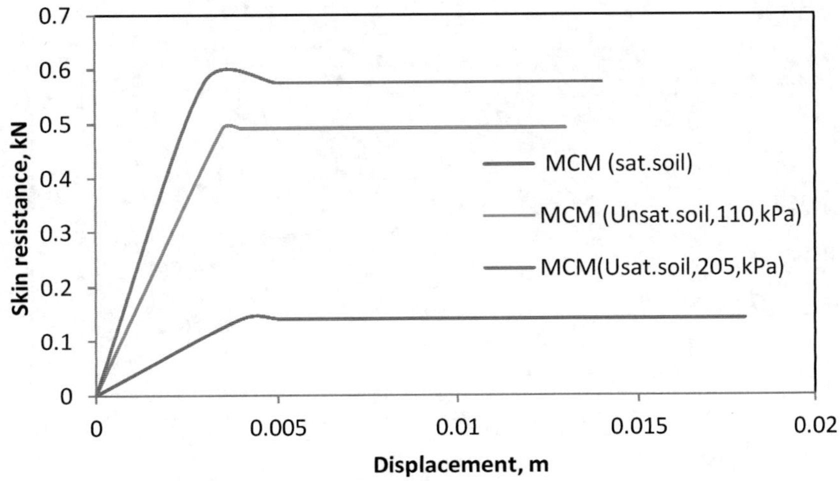

Figure 3. The relationship between skin resistance and vertical displacement at different matric suction (0, 205, and 110) kPa respectively by (MCM) small scale pile.

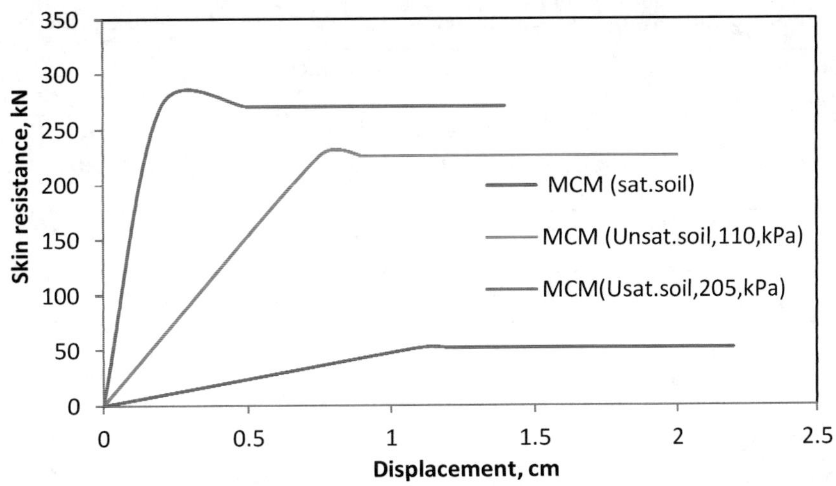

Figure 4. The relationship between skin resistance and vertical displacement at different matric suction (0, 205 and 110) kPa respectively by (MCM) full-scale pile

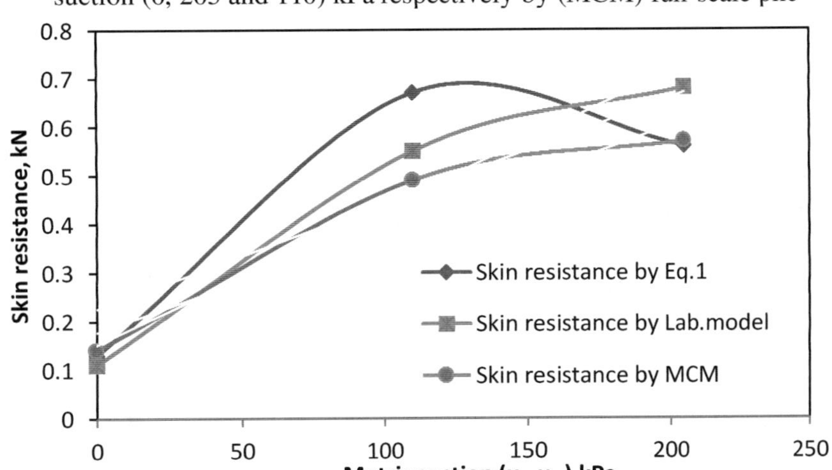

Figure 5. Comparison among the modeling results and matric suction.

IOP Conf. Series: Materials Science and Engineering **901** (2020) 012004 doi:10.1088/1757-899X/901/1/012004

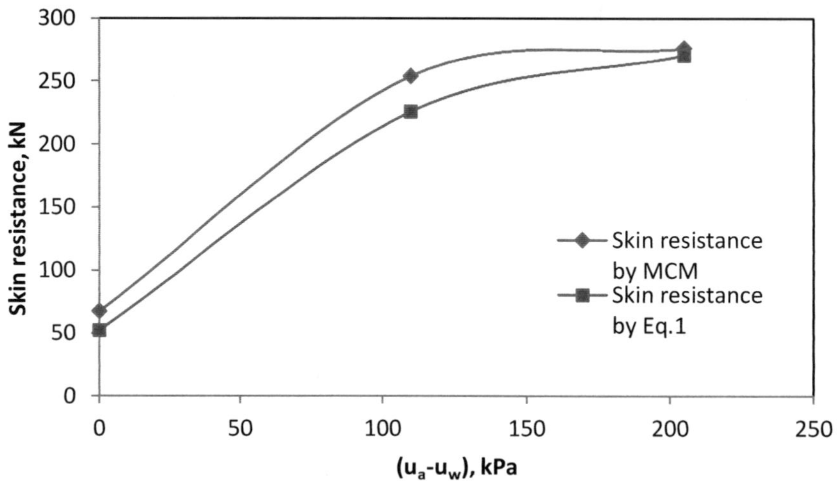

Figure 6. Comparison between the skin resistance of full-scale models and matric suction.

Table 5. Results of the back-calculated of adhesion factor by different methods.

Degree of saturation, %	Matric suction (u_a-u_w), kPa	Back calculated (α) Small scale Model	Back calculated (α) Full-scale Model
90	0	0.97	1.1
43	205	0.49	0.63
65	110	0.67	0.8

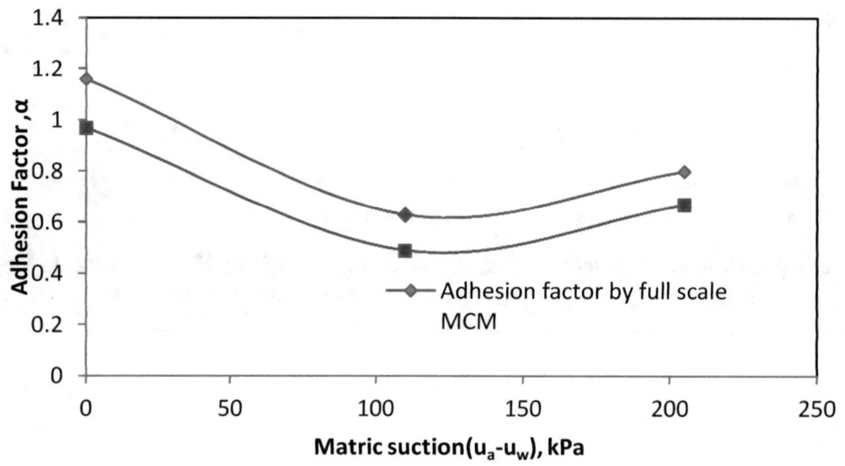

Figure 7. Relationship between the adhesion factors and matric suction.

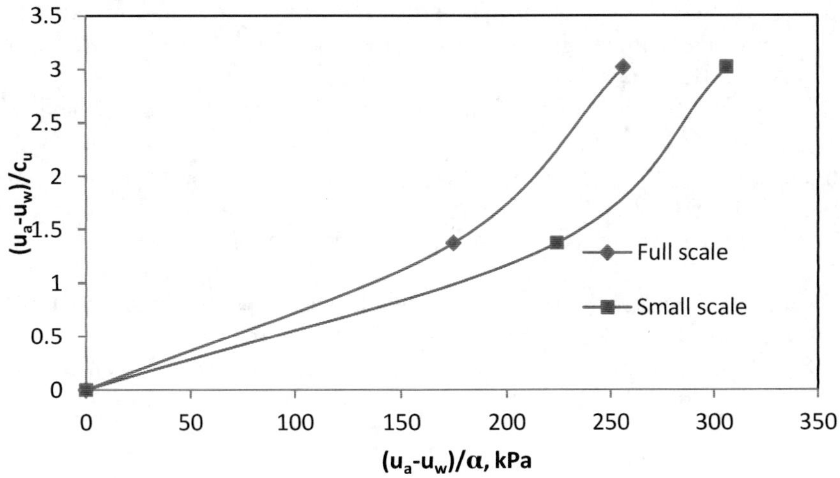

Figure 8. Relationship between the $(u_a-u_w)/cu$ and $(u_a-u_w)/\alpha$ of full scale and small-scale models.

6. Conclusions

In this study, laboratory, theoretical and numerical modeling studies were used for examining the behavior of skin resistance pile under undrained conditions in a saturated and unsaturated homogeneous layer that is prepared in the laboratory using two moisture contents. Numerical modeling (small and full scales) are made by different parameters founded or evaluated from the laboratory modeling process to simulate the behavior of the skin resistance pile. The main points may be expressed in the following:

- The laboratory modeling results explain the significant effect of matric suction on shear strength and stiffness of soil and interface between steel pile-soil.
- The carrying capacity of the friction pile model increase with increased matric suction under undrained loading conditions.
- Experimental and numerical models are compatible according to output results, then the unsaturated soil stiffness and soil suction changes with depth based on degree of saturation.
- The values of α were back-calculated to determine the difference between the calculated values from Eq. 1. These values obtained from the back calculations were very close to the first estimated values.
- The numerical studies based on the input parameters taking into account the effect of matric suction in the modeling of the analysis steel pile behavior.

References

[1] S.K. Vanapalli and Z.N. Taylan, Design of single piles using the mechanics of unsaturated soils. International Journal of GEOMATE, 2(1) 2012 197-204.
[2] E.E. Alonso, A. Gens, and A. Josa, A constitutive model for partially saturated soils, Géotechnique, 40(3) 1990 405-430.
[3] J.C. Vaca, The dry fraction of unsaturated soils. Acta Geotechnica Slovenia, 12(2) 2015 67-75.
[4] A. Gens, M. Sanchez, and D. Sheng, On constitutive modeling of unsaturated soils, Acta Geotech,1(3) 2006 137.
[5] K. Georgiadis, D.M. Potts, and L. Zdravković, The influence of partial soil saturation pile behavior, Géotechnique,53(1) 2003 11-25.
[6] N. Ravichandran, S. H. Krishnapillai, and B. Machmer, A novel procedure for physical modeling of unsaturated soil-pile system using geotechnical centrifuge, J. Earth Sci. Geotech. Eng., 3(1) 2013 119-134.
[7] M. Sheikhtaheri, Experimental and Numerical Modeling Studies for Interpreting and Estimating the p - δ Behavior of Single Piles in Unsaturated Sands. M.Sc. thesis, University of Ottawa, Ottawa, Canada, 2014.

[8] M. Al-Khazaali, and S. K. Vanapalli, Numerical modeling technique to predict the load versus settlement behavior of single piles in unsaturated coarse-grained soils, Proc. of GeoQuébec, Québec, Canada, 2015.

[9] J.W. Hilf, An investigation of pore water pressure in compacted cohesive soils, Ph.D. dissertation. 1956.

[10] T.B. Hamid and G.A. Miller, Shear strength of unsaturated soil interfaces, Canadian Geotechnical Journal, 46(5) 2009 595-606.

[11] R.B. Peck, A study of the comparative behavior of friction piles, Highway Research Board, no. 36, 1958.

[12] R. Woodward and J. Boitano, Pile loading tests in stiff clays, in Proc. 5th Int. Conf. on Soil Mechanics and Foundation Engineering, 1961 177-184.

[13] W.T. Oh, S.K. Vanapalli, and A.J. Puppala, Semi-empirical model for the prediction of modulus of elasticity for unsaturated soils, Canadian Geotechnical Journal, 46(8) 2009 903–914.

Numerical Analysis of Geogrid and Deep Mixing Column Supported Embankment

Bushra S. Albusoda[1,a], **Samir H. Hussein**[1,b], **Dhurgham A. Al-Hamdani**[2,c]*

[1]Department of Civil Engineering, University of Baghdad, Baghdad, Iraq
[2]Al-Furat Al-Awsat Technical University, 31001, Al-Najaf, Iraq
[a]Albusoda@yahoo.com, [b]Samir_40404@yahoo.com, [c]dergamalhamdany@yahoo.com

Abstract. Embankment is an essential component of civil engineering constructions. It is required for the implementation of many projects such as roads, railways, and dams. However, large deformations and lateral movements may occur due to the poor mechanical properties of the weak soils. To avoid that, the subgrade must be improved. One of the efficient and economic techniques that solve these problems is the column-supported embankment system. In this paper, a numerical model of the embankment was established using (ABAQUS) software program. Three different modes were studied to explain the behavior of these embankments. A parametric study was conducted to examine the various influencing factors, such as elastic modulus of columns, the effect of the geogrid layer on the performance of embankment, the settlements, the lateral movement, the shear strength of the soil, the stress concentration ratio, the soil arching ratio, and the tension force in geogrid layer were determined and analyzed for various embankment modes. The results show that the effect of the elastic modulus of columns plays a vital role in the performance of the embankment. Similarly, the geogrid layer also has significant effects on the lateral displacement, and the shear strength of the embankment. In general, for all modes, the decrease in the elastic modulus of the columns leads to an increase in the embankment loads that carried out by the soil between the columns. While, the absence of the geogrid layer leads to an increase in the embankment loads carried by columns.

Keywords: Embankment, Column-Supported Embankment, Geogrid, Settlement, and Soil Arching.

1. Introduction

The column supported embankments (CSE) with the geogrid layers have been confirmed to be one of the most effective and economic techniques for soft soils that may undergo large deformations and lateral movements [1]. In this system, the geogrid layer is acted as reinforcement with horizontal direction while the columns are acted as a vertical reinforcement. Many publications during the last years focused on using supporting embankments by geogrid column-supported technique. The designing of CSE with the geogrid reinforcement must include two determination steps. Step one evaluates the behavior of the arching of the column in the fill. Step two the behavior of the load-deflection of the geogrid reinforcement. Many researchers studying the analyses of the CSE and proposed a model for step one (the arching step). They proposed a limited state equilibrium model.

The arching is concentric in this model. This model is an extension of many models, such as these developed by Hewlett and Randolph, 1988 and Zaeske, 2001. The developed model represents the behavior obtains the arching for the relatively thin fill. The geogrid reinforcement in the column embankment obtains a more efficient method to transfer the load to the columns in a mechanism of arching. Many loading cases may be exerted on the model such as that affects the columns and geogrid reinforcement strips located between the columns. In the second case, the load is distributed as an inverse triangle. The behavior and mechanism of load distribution was explained to describe the situation with geogrid reinforcement. The fill height and the friction angle observations are presented with behavior concurs in these types of models. The quantity of arching increased with an increase in soil consolidation and geogrid reinforcement deflection [2]. The advantages of CSE were more rapid as a single construction phase for adjacent facilities protection and embankments. These advantages were limited because of the lack of overall design procedures of the CSE. CSE performance in the literature survey involves centrifuge testings, bench testing, instrumented test sections of case histories, numerical modeling, and some of field-scale testing. CSE with a field-scale testing facility that represented a unique approach where, it was used expanded polystyrene geofoam and its subsequence dissolution was used to model the soft soil below the CSE.

The geofoam properties were similar to the properties of soft clay. The geofoam was dissolved beyond the construction of the embankment by using a friendly environment solvent to modeling the worst case with zero support by the soft soil. Many researchers assumed that the geofoam used for temporary support of the CSE test, where later, the geofoam was dissolved and removed the support between columns focused. The facilities detail of the CSE test, material used, the test procedures, and instruments were focuses to study also of studying the CSE [3]. The behavior of conducted the CSE over the soft soil was studied. CSE encased and reinforced with geogrid. The tests conducting on multi cases with different parameters for the stone columns such as center to center spacing, two ratios of lengths to diameter (L/d), and the embankment heights. In this study 42 testing models were performed on soft soil with undrained shear strength about 10 kPa.

The tests involved the following parameters (s/d = 2.5, 3 and 4), (L/d= 5 and 8), (height of embankment H=200, 250, and 300 mm), (testing were conducted on both ordinary stone column (OSC), and encased stone columns (ESC). The vertical effective stress at the top, middle under the centerline, and the lower edge between two columns were measured by using three load cells. The results had shown that the key parameter ratio (h/s-d) of the height of embankment to the clear spacing between two columns less than 1.2 and about 1.4 for the two cases of OSC and ESC, respectively. (where d is the stone column diameter, s is the spacing between columns). Also, the researchers had founded that there is no effect of arching observed, very large settlement at the embankment surface, and virtually unmodified stress acting on the soil when the ratio of (h/s-d) grater than or equal to 2.2 for two cases of OSC and ESC, respectively [4].

2. Plane Strain Supposition

It is always used in geotechnical analysis when soil structures are very long with the same uniform cross-section area and with finite dimensions such as soil embankment, which have a uniform cross-section with limited dimensions in the x-y plane whilst having a very large dimension in the z-direction. In this analysis, the strains along z-direction are nil (i.e., $\varepsilon_{33} = \varepsilon_{13} = \varepsilon_{23} = 0$). Therefore, the 3-D embankment problem can be converted to a 2-D plane strain problem, by which, the x-y plane is assumed as the cross-section of the embankment, which can represent the whole embankment [5].

2.1.Composite Foundations

When the natural ground is reinforced or replaced partially by other materials, the new system which composed of ground soil and the reinforcement materials is called composite foundation; the types and stiffness of composite foundations depend on the reinforced materials [6]. In the present work, a composite foundation system with deep mixing columns is studied. The system was analyzed before and after loading by using two dimensions (ABAQUS) software program [7].

2.2.Load Transfer and Soil Arching Phenomena

The vertical stresses on columns and the soil between columns are increased with increase embankment height with time, the stress increment on columns is more than on the soil, and as a result of the stress concentration engendered on columns and the reduction engendered on the soil. The soil arching practically restrained by shear stresses of embankment fill, the shear conveys the embankment loads taken by soil to columns, the mechanism of load transfer called soil arching [8, 9].

Two global indexes descript the mechanism of soil arching phenomena; the first index is the degree of soil arching (ρ), and it can be expressed as the following in Eq. (1):

$$\rho = \frac{P_b}{\gamma h + q} \tag{1}$$

where (ρ) is soil arching ratio, (P_b) is applied vertical stress on soil, (γ) density of filling, (h) heigh of filling, and (q) is surcharge loading. Another factor is the stress concentration (n); which can explicate by the difference of elastic modulus of soil and columns which lead to transfer the loads from soil to columns. It is defined as the ratio of the vertical stress on the columns to the vertical stress on the soil between the columns; which can be expressed as the following in Eq. (2):

$$n = \frac{\sigma_p}{\sigma_s} \tag{2}$$

where: (n) is stress concentration ratio, (σ_p) is stress on column and (σ_s) is stress on soil between columns. Stresses concentration (n) and soil arching ratio (ρ) are used to evaluate the effect of the reinforce-column supported embankment technique in embankment performance [9].

3. Material Properties

Three modes were used to simulate the behavior of geogrid column-supported embankment by using the finite element software (ABAQUS). For comparison purposes, three modes are used with different elastic modulus of the column. The first mode is an embankment with reinforcement, it is modulus (Mode1), and the second mode is an embankment with reinforcement with different elastic modulus of the column that in mode 1, it is termed (Mode 2). These modes are compared with the third mode that includes the embankment without reinforcement, it is termed (Mode3). The details and the dimensions of the embankment are shown in Fig. 1. The material properties of the embankment, soil layers are presented in Table 1, while the column. and the geogrid layer are presented in Table 2. The geogrid layer is used as a horizontal reinforcement, which is embedded in the platform layer at 0.25m above the top of deep mixing columns.

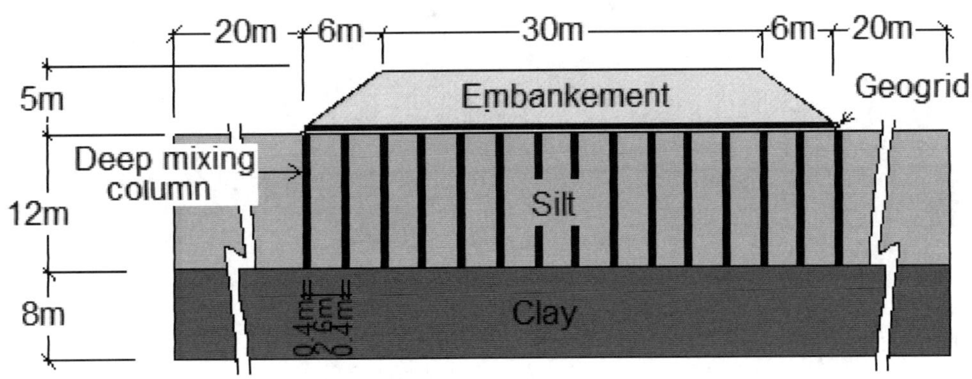

Figure 1. Embankment cross-section.

Table 1. Material properties.

Embankment layers	Depth (m)	Density ρ (kg/m³)	Elasticity E (MPa)	Internal friction φ′	Cohesion C′ (kPa)	Dilatancy angle, Ψ	Poisson ratio, μ
Fill soil	5	1939	20	24.6	16.8	0.1	0.3
Soft soil (silt)	12	1816	7.6	19.7	7.6	0.1	0.33
Clay	8	1867	22.1	21	15.7	0	0.3

IOP Conf. Series: Materials Science and Engineering **901** (2020) 012005 doi:10.1088/1757-899X/901/1/012005

Table 2. Deep mixing column properties.

Mode	Type	Length (m)	Diameter (m)	Density $\rho(kg/m^3)$	Elasticity E (MPa)	Poisson ratio μ
1	Deep mixing	12	0.4	2040	5000	0.2
2	Deep mixing	12	0.4	1837	30	0.3
Geogrid	-	42	0.02	2040	135	0.3

All embankment fills and the soil layers are modeled as a linear-elastic, and perfectly plastic model (Mohr-Coulomb failure criterion). The columns are modeled as linear elastic isotropic material embedded in the region under the embankment. The geogrid layer is modeled as a truss element embedded in the platform layer. After establishing the model, the soil properties, the surface interactions, and defining boundary conditions in the numerical model are assigned. For embankment filling and soil layers, the model is meshed by CPE4R elements type as shown in (Figure 2). While, the CPE3 is used to mish the columns. The CPE4R and CPE3 are plane strain elements. The geogrid layer is meshed by T2D2. T2D2 is defined as a 2D stress-displacement truss elements. The T2D2 element have 2 linear nodes [7]. Then, the analysis results are carried out for all models during loading. The verification problem is applied to the software, where the value of the program is close to the field data.

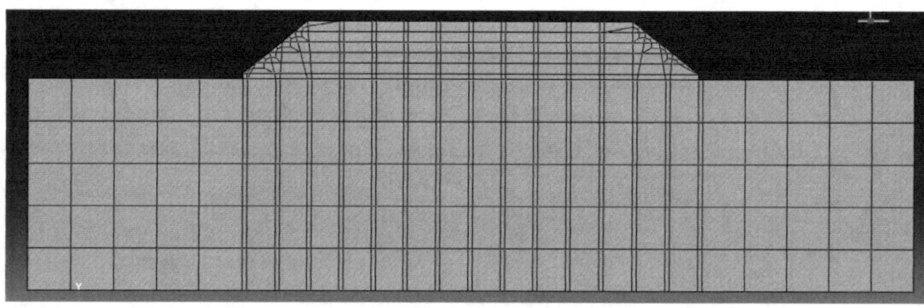

Figure 2. mesh of embankment filling and soil layers.

4. Analysis Results

4.1. Maximum settlements

The settlement results are presented in Table 3 shown below. The settlement for mode 2 is larger than mode 1. Also, the settlement of mode 2, takes much longer time to stabilize than the corresponding settlement of mode 1, while the settlement magnitude for mode 3 is much closed with settlement magnitude for mode 1. Also, lateral displacement in the toe of the embankment for the three modes are presented in Table 3. The results show that the lateral displacement for mode 2 and mode 3 is larger than that in mode 1.

Table 3. Maximum settlements of surface soil for all modes.

	Loading stages	Mode 1	Mode 2	Mode 3
Maximum settlements (mm)	Layer 1	9	12.3	9
	Layer 2	15.4	21	15.6
	Layer 3	21.6	29.4	22.2
	Layer 4	27.7	37.8	28.7
	Layer 5 (all layer)	33.6	47	35
	Layers with live load	39.82	58.54	42.10
Lateral displacements (mm)	Toe of embankment	6.73	10.99	11.80

IOP Conf. Series: Materials Science and Engineering **901** (2020) 012005 doi:10.1088/1757-899X/901/1/012005

4.2. Soil Shear Strength and Load Transfer

The shear strength of embankment for the three modes are presented in Table 4. As shown, mode 1 is much closed to mode 2, while the magnitude for mode 3 is larger than that in mode 1. The stress concentration ratio (n) and soil arching ratio (ρ) are used to study the effect of elastic modulus of columns and geogrid layer on embankment behavior. The magnitude of the stress by the column and arching that are depending on the percentage of embankment loads which columns carried to the loads carried by the soil between columns. The stresses concentration ratio and the arching ratio for the three modes are presented in Table 4. The stress concentration ratio and the arching ratio decreases significantly in mode 2 compared with mode 1. This means that the decrease of elastic modulus of deep mixing column is leading to an increase in the load that carried out by the soil between the deep mixing columns. Absence the geogrid layer in mode 3 comparing with mode 1 leads to increasing the stress concentration ratio and arching ratio. This means that the increase in deep mixing column loading and decreasing the load carried by soil between the deep mixing columns. The deflected shape is presented in Fig. 3. As shown in the figure the displacement in the left figures and the settlement in the right figures.

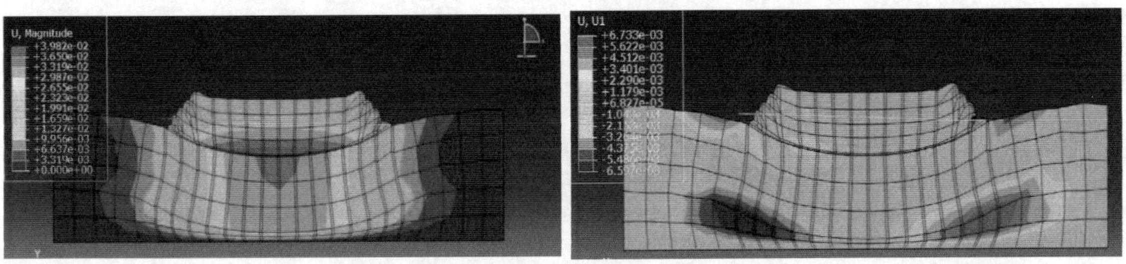

(a) Deflected shape of the mode 1.

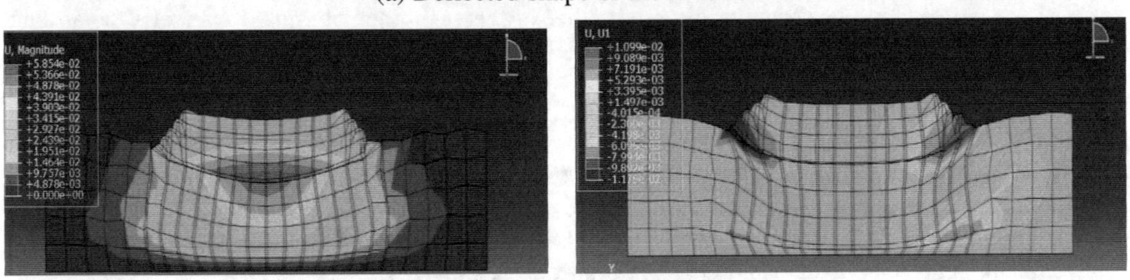

(b) Deflected shape of the mode 2.

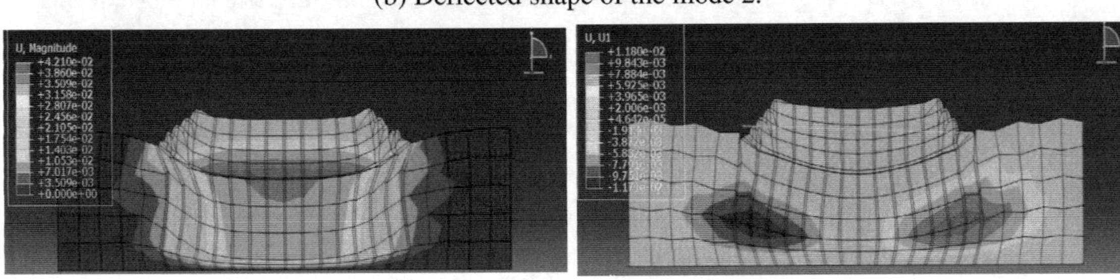

(c) Deflected shape of the mode 3.

Figure 3. Settlement and displacement of the three modes.

Table 4. Stresses concentration ratio and soil arching ratio.

Mode	Max. column stress (kPa)	Max. soil stress (kPa)	Stress concentration ratio	Arching ratio	Shear strength (kN)
1	860	41.3	20.8	28	30

2	120	146.7	0.82	3.9	30.9
3	1160	4.97	233.4	37.8	45

4.3. Tension in geogrid layer

The tension force in the geogrid layer is presented in (Table 5). The max tension for mode 2 is larger than that in mode 1. It means that the tension force in the geogrid layer is strongly influenced by the elastic modulus of deep mixing column Also, the tension force in the geogrid layer is decreased with an increase of the elastic modulus of columns. The decrease in the elastic modulus of the columns leads to an increase in the loads carried out by the soil between the columns. A single layer of the geogrid reinforcement behaves as a tensioned member due to its flexibility.

Table 5. Maximum tension force in the geogrid layer.

Mode	1	2
Shear strength (kN/m^2)	4.645	7.15

5. Conclusions

According to the results gained from the present work, several points may be concluded:

- The settlement values are influenced by the elastic modulus of the deep mixing column, but slightly influenced by the geogrid layer.
- The lateral displacement is mostly influenced by the elastic modulus of the deep mixing column and geogrid layer.
- The soil shear resistance is slightly influenced by the elastic modulus of columns, but strongly influenced by the geogrid layer.
- There are decreasing in the elastic modulus of the columns for all modes, that will increase the loads of the embankment that carried out by the soil between the columns. Reversely, the absence of the geogrid layer leads to an increase in the loads of the embankment that carried out by the columns

References

[1] S. W. Abusharar, Theoretical and Numerical Study of Column-Supported Embankments, Ph.D. dissertation, Huazhong University of Science & Technology Wuhan, China. 2008.

[2] V. Eekelen, S.J.M., Bezuijen, A., Van Tol, A.F., An analytical model for arching in piled embankments. Geotextiles and Geomembranes 39, 2013 78-102.

[3] J. A. Sloan, G. M. Filz, and J. G. Collin. Field-Scale Column-Supported Embankment Test Facility, Geotechnical Testing Journal, Available online at www.astm.org 36 2013 2-12. Fattah MY, Zabar BS, Hassan HA. Soil arching analysis in embankments on soft clays reinforced by stone columns. Structural Engineering and Mechanics. 2015 2015 507-534.

[4] H. Sam, Applied soil mechanics with ABAQUS Applications. Published simultaneously in Canada, John Wiley & Sons, Inc. 2007.

[5] J. J. Zheng, J. H. Ou, and N. Z. Yuan. Theory and practice of multi-column composite ground. Ch. J. of Geo. Eng. 24, 2002 208-212.

[6] Simulia Corp. ABAQUS 6.12 Analysis User's Manual. Volume IV: Elements. Product of Dassault Systems Simulia Corp. Providence RI, USA. 2012.

[7] J. W. McNulty. An experimental study of arching in sand, U.S. Army Engineer Waterways Experiment Station, Corps of Engineers, Report. No-I-674. Vicksburg, Miss., 170 (1965).

[8] J. Han, M.A. Gabr. Numerical analysis of geogrid-reinforced and column-supported earth platforms over soft soil. J. of Geo. and Geoen. Eng. 128 2002 44-53.

IOP Conf. Series: Materials Science and Engineering **901** (2020) 012006 doi:10.1088/1757-899X/901/1/012006

Strength and Compressibility of Screw Piles Constructed in Gypseous Soil

Omar J. Mukhlef[1,a], Mahdi O. Karkush[1,b*] and Askar Zhussupbekov [2,c]

[1]Department of Civil Engineering, University of Baghdad, Baghdad, Iraq
[2]Civil Engineering Department, Eurasian National University, Astana, Kazakhstan
[a]omar.fallujii@gmail.com, [b]mahdi_karkush@coeng.uobaghdad.edu.iq, [c]astana-geostroi@mail.ru

Abstract. The present work deals with the performance of screw piles constructed in gypseous soil of medium relative density; such piles are extensively used in piles foundations supported structures subjected to axial forces. The carrying capacity and settlement of a single screw pile model of several diameters (20, 30, and 40) mm inserted in gypseous soil is investigated in the present study. The gypsum content of soil used in tests was 40%. The bedding soil used in tests was prepared by raining technique with a relative density of 40%. A physical model was manufactured to demonstrate the tests in the laboratory. The model of screw pile has been manufactured of steel with a total length of 500 mm and helix diameters of 20, 30, and 40 mm continuous over the embedded depth of pile (400 mm). The results of tests showed decreasing the ratio of length to diameter (L/D) resulted in a higher axial carrying capacity of screw piles and low corresponding settlement.

Keywords: Screw piles; helix; gypseous soil; axial carrying capacity; settlement.

1. Introduction

The gypseous soils are spread in many areas of the world. The gypseous soils in Iraq covers about 30% of the area of the country [1]. Sometimes the gypsum content is high, causing serious problems for buildings and projects due to the dissolution of gypsum by running water through gypsum soil [2,3]. In general, gypseous soil is usually stiff when it is dry, but losses its strength upon wetting resulting from changing the water level or due to water leakage that may dissolve gypsum which causes large pores formation. The formation of cavities leads to increase the permeability of gypseous soils and these large cavities lead to collapses of buildings and failure of structures.

The screw piles differ from the usual piles, they are regularly made of high-quality steel. The screw piles consist of shaft and helix as the helix is fixed to the shaft at different spacing and has a pointed shape for easy installation in the ground [4]. Several elements can control the behavior of screw piles such as the diameters of shaft and helix and the spacing between the helices. Initially, the screw piles were mostly used as anchors and later used to support tensile loads such as power transmission towers. However, their use has been extended to include prone structures and parallel loading [5,6]. Screw piles provide essential structural resistance from the tensile, compression, and parallel forces along with the moments of inversion [7,8]. The basic components of the screw pile are illustrated in Fig. 1.

In the present study, the ultimate carrying capacity and settlement of screw piles made of steel of diameters (20, 30, and 40) mm inserted in gypseous soil of gypsum content and relative density of 40% has been investigated by conducting tests on laboratory scale physical model of piles. The ultimate carrying capacity of the screw pile is calculated from the load-settlement curve according to the failure criterion of 20% settlement that adopted in this study.

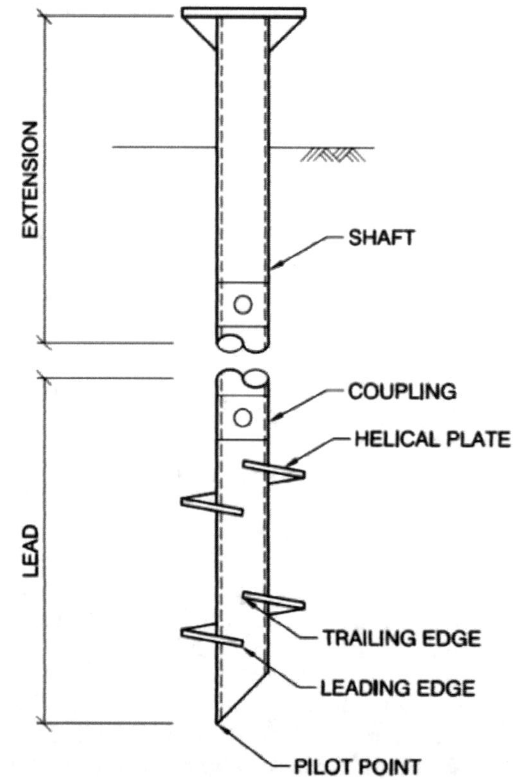

Figure 1. The basic components of the screw piles.

2. Carrying Capacity of Screw Piles

Different design methods determine the carrying capacity of screw piles in the literature. The most common methods are cylindrical shear method, individual bearing technique, and connection between torque for fixing of screw piles and carrying capacity of pile method. Using of the International Building Code encompasses these three design methods as well as different ways to find the carrying capacity of the screw piles [9,10]. The three hypothetical methods are based exclusively on estimating the maximum capacity of screw pile by equations, which differentiate whether the applied load is compression or tension and the pile is inserted in cohesive or cohesionless soil. However, the contact between the installation torque and carrying capacity is a formula based strictly on soil conditions as well as the characteristics of the screw piles. The carrying capacity of screw piles inserted in cohesionless soils can be calculated using Eqs. (1) to (4) [11]. The main variance in the analysis of cohesive and cohesionless soils under compressive loads is to take advantage of the internal friction of the soil as opposed to the undrained shear strength that is used in the analysis of cohesionless soil.

$$Q_c = Q_{helix} + Q_{bearing} + Q_{shaft} \tag{1}$$

$$Q_{helix} = 1/2\ \pi\ D_h\ \gamma'(H_b^2 - H_t^2)K_s \tan\varphi \tag{2}$$

$$Q_{bearing} = \gamma' H\ A_h\ N_q \tag{3}$$

$$Q_{shaft} = 1/2\ P_s\ H_{eff}\ \gamma' K_s \tan\varphi \tag{4}$$

Where

D_h: the average diameter of helix;

γ': the effective unit weight of soil;

H: the embedded depth of the pile;

H_b: the depth to the bottom helix;
H_t: the depth to top helix;
K_s: the coefficient of lateral earth pressure in compression loading;
A_h: the area of the bottom helix; and
P_s: the perimeter of the screw pile shaft as suggested by Meyerhof and Adams [12].

$$N_q = e^{\pi \tan\varphi} \tan(45 + \varphi/2) \tag{5}$$

When the ratio of H/D is under shallow conditions, the shaft friction is negligible [13], so Eq. 4 reduces to:

$$Q_c = 1/2\ \pi\ D_a\ \gamma'\left(H_b^2 - H_t^2\right) K_s \tan\varphi + \gamma' H\ A_H\ N_q \tag{6}$$

The pile load test is used to determine the axial carrying capacity and settlement of piles [14]. The axial load is applied in an increment by using a fixed penetration rate method and the corresponding settlement is observed. Failure occurs at the point where the settlement continues to move downward without an increase in load, leading to plunging failure in the pile [15]. An ultimate pile load is then determined as the vertical asymptote between the point where the restricting skin opposition is mobilized and approaches point limit is reached, as the load settlement curve draws near a vertical gradient [16]. Different techniques for determining the ultimate pile carrying capacity from a load test have existed in the literature. Some of these techniques such as the Davisson criterion, Brinch Hansen criterion, L1–L2 technique, Federal Highway Administration (FHWA) (5% of the helix diameter), and ISSMFE (10% of the helix diameter) are outlined in reference [17]. Through previous experience and research concerning screw piles, it was found that most studies included the behavior of screw piles in clayey and sandy soils without gypsum. Therefore, the behavior of screw piles inserted in sandy soil containing gypsum with high rate reached to (40%) will be studied. The study would be including the effect of static loads and periodic loads on screw pile in gypseous soil by manufacturing a laboratory physical model for testing and conclusions.

3. Soil Sampling and Material Properties

The gypseous soil is brought from the site of the University of Fallujah in Al-Anbar governorate. The soil has a gypsum content of 40%. The relative density of soil used in testing the model pile is 40% which considered medium to dense relative density based on DIN 1045 "verification of the safety of earthworks and foundations". The physical and engineering tests conducted on the soil in the laboratory. The properties of used soil are given in Table 1.

Table 1. Geotechnical properties of gypseous soil used in tests.

Property	Value	Property	Value
γ_{sat} (ASTM D1556-00)	17.6 kN/m^3	Specific gravity (B.S-13377: I976, No. 6)	2.4
$\gamma_{d,min}$ (ASTM D4254-00)	12.7 kN/m^3	Soil classification	SC
$\gamma_{d,max}$ (ASTM D4253-00)	16.9 kN/m^3	Passing sieve #200	14.5%
Relative density, Dr	40%	Cohesion, (ASTM D3080-72)	7.2 kPa
$\gamma_{d,\ used}$ according to Dr	14.1 kN/m^3	Angle of friction (ASTM D3080-72)	36°
void ratio, e	0.7	Permeability (ASTM D 2434-68)	8.04×10^{-4} cm/s
Water content, ω	25.1%	Gypsum content	40%

Screw piles consist of steel shaft and helix, which are distributed over the shaft with constant spacing. The helices are distributed along the steel shaft up to 400 mm and the rest of the shaft is 100 mm free head of pile above the soil surface. The dimensions of manufactured screw piles are given in Table 2 and shown in Fig. 2.

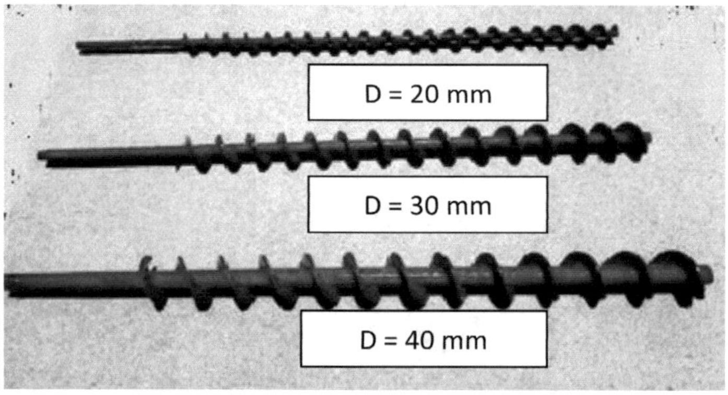

Figure 2. Models of used screw piles.

Table 2. Dimensions and configuration of used screw piles.

Screw pile of 20 mm helix diameter					
Shaft diameter	Helix diameter	Total length	Embedded depth in soil	No. of helix	Spacing between helix
8 mm	20 mm	500 mm	400 mm	22	18 mm
Screw pile of 30 mm helix diameter					
Shaft diameter	Helix diameter	Total length	Embedded depth in soil	No. of helix	Spacing between helix
12 mm	30 mm	500 mm	400 mm	15	26 mm
Screw pile of 40 mm helix diameter					
Shaft diameter	Helix diameter	Total length	Embedded depth in soil	No. of helix	Spacing between helix
16 mm	40 mm	500 mm	400 mm	13	30 mm

The pile cap is made of steel with dimensions of 120×120×6 mm. The screw pile is connected to the pile cap by making a hole in the center of the pile cap with the same diameter of the screw pile shaft to be by nut as shown in Fig. 3. The type of free pile cap was adopted in the tests used.

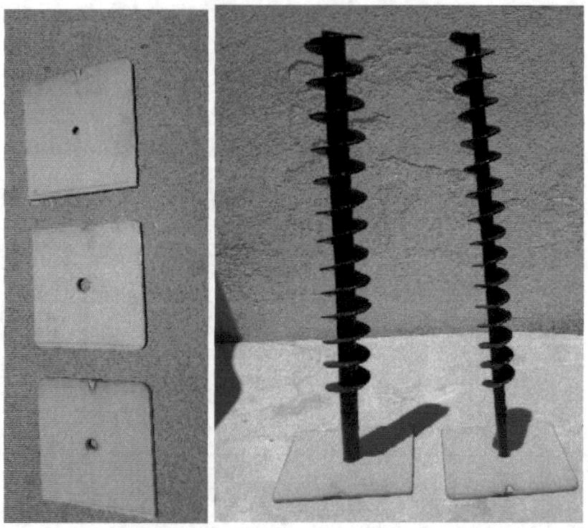

Figure 3. Screw pile with pile cap of dimensions 120×120 mm.

A steel box with dimensions of (700×700×600) mm was used for conducting the test where it was filled with gypseous soil to simulate field conditions as shown in Fig. 4.

Figure 4. Installation of screw pie in the steel box container.

4. Testing Procedure

The raining technique was used to pour the gypseous soil in the test box and satisfy the desired relative density. To achieve a uniform layer according to the desired density, a special raining device was designed for this purpose. The pouring height and sand discharge rate mainly affect the density of the gypseous soil layer in the raining method. The height of the box was divided into 5 layers, each layer of 120 mm height, the weight of soil required to fill the layer is 83 kg depending on the adopted density of soil of 14.1 kN/m^3 which satisfies relative density of 40%. The relationship between pouring height and the relative density of sand is shown in Fig. 5. The relative density of the gypsum soils used in all tests is assumed of medium density (40%), and to reach the specified relative density of 40%, the free fall height would be 720 mm.

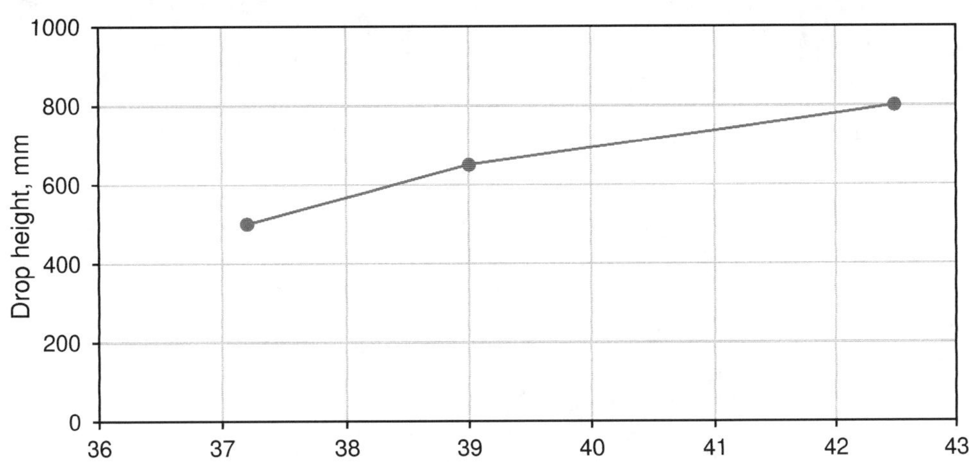

Figure 5. Calibration curve of raining technique.

The following steps were followed to install the screw piles after finishing the placement of the soil in the container and applying loads on the pile cap:

IOP Conf. Series: Materials Science and Engineering **901** (2020) 012006 doi:10.1088/1757-899X/901/1/012006

- On the top of the screw pile, apply the manual torque to install the screw pile in the soil to the required depth of 400 mm as shown in Fig. 6.
- The tests included inserting three screw piles of different helix diameters (20, 30, and 40) mm in the soil to the required embedded depth of 400 mm, there is a gap of 100 mm thickness between the pile cap and surface of the soil as shown in Fig. 7
- Then a square plate (pile cap) is fixed on the top of the screw pile. Subsequently, the dial gauge and LVDT were installed vertically to measure screw pile displacement as shown in Fig. 8.
- The hydraulic jack system is installed with a compressor to simulate axial static loads on the screw piles, which inserted in gypseous soil. The used hydraulic jack of full capacity equals to 5 tons.
- The load cell was fixed on the head of the loading shaft of hydraulic jack and connected with the indicator to measure the axial loads on the screw piles.
- The test was carried out after 24 hours from inserted the screw piles into the soil to give a sufficient time to rearrange the granules of the soil and remolding.
- Axial force is applied gradually. The applied load increased from 0.1 to 1.2 kN, and the average downward movement of the screw pile is measured under each increment of loading using a dial gauge and LVTD.
- The application of axial loading on the screw pile continued to reach the point of failure, which defined as the load corresponding to a settlement value of 20% of the diameter of the helix [18]

Figure 6. Insertion of screw pile in soil.

Figure 7. Gap between pile cap and soil surface.

Figure 8. Loading system of screw pile and gauges.

IOP Conf. Series: Materials Science and Engineering **901** (2020) 012006 doi:10.1088/1757-899X/901/1/012006

5. Results and Discussion

The laboratory tests were carried out for the carrying capacity of the screw piles under axial static loads. The test was conducted on screw piles of different diameters (20, 30, and 40) mm with an embedded depth of 400 mm penetrate in gypseous soil. The test continued to extremely reach the point of failure in the final displacement. The results obtained from tests can be summarized in Table 3, showing the diameters of the screw piles with displacement for each load increment of the screw piles.

Table 3. Displacement for each load increment applied on screw piles.

Axial load (kN)	Displacement of screw pile (mm) D = 20 mm	D = 30 mm	D = 40 mm
0	0.00	0.00	0.00
0.1	1.17	0.56	0.22
0.2	1.93	0.99	0.74
0.3	2.87	1.52	1.28
0.4	3.67	2.76	2.37
0.5	4.81	4.06	3.28
0.6	7.19	5.66	4.46
0.7	11.35	7.85	6.72
0.8	16.33	10.17	8.98
0.9	21.15	14.09	10.90
1.0	27.08	17.42	12.18
1.1	32.42	20.76	14.41
1.2	41.54	23.10	16.87

In Figs. 9, 10, and 11, there is a resistance of the screw pile to the displacement at the beginning of the tests. Since the failure occurred in screw piles when the displacement is reached more than 20% of the screw piles diameter. The failure load was 0.45 kN for the screw pile of 20 mm diameter which corresponds to a displacement of 4 mm, while the load that caused the failure of the screw pile of 30 mm diameter is 0.65 kN where the displacement reached more than 6 mm. The failure load of the screw pile of 40 mm diameter was 0.76 kN and the corresponding settlement is 8 mm. The failure mode in all cases is general shear failure that is one of the advantages of using screw piles to resist the axial loads in gypseous sandy soils. Therefore, the adopted criteria of failure may be extended to be 40% instead of 20%. This criterion can be approved by applying different theories to calculate the failure loads.

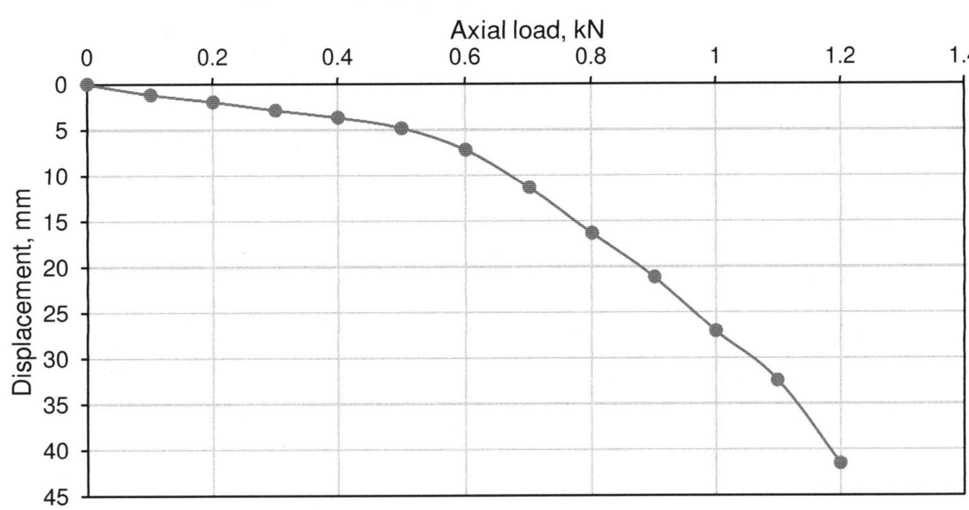

Figure 9. The load-displacement curve for a screw pile of 20 mm diameter.

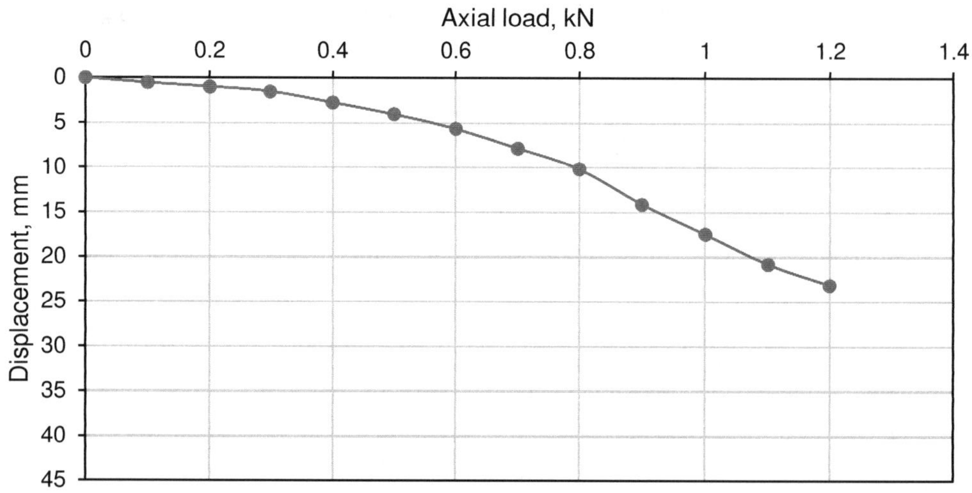

Figure 10. The load-displacement curve for a screw pile of 30 mm diameter.

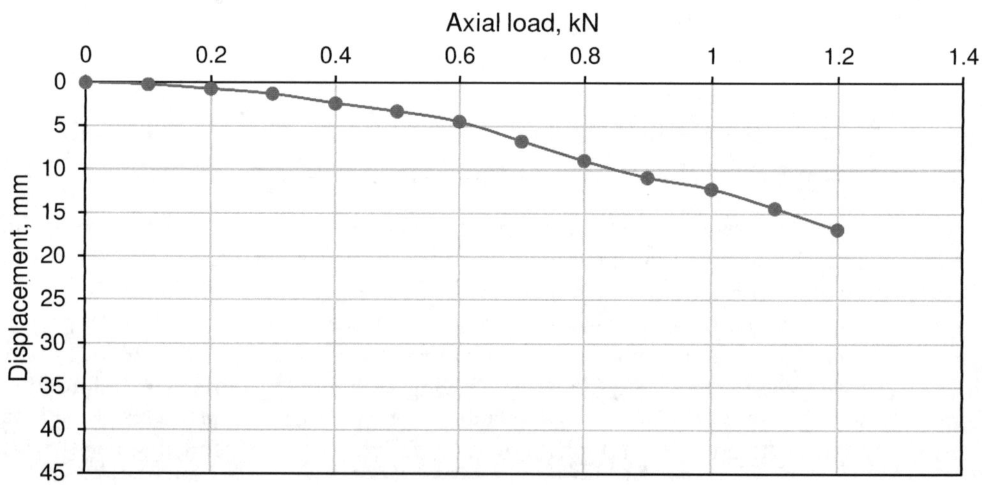

Figure 11. The load-displacement curve for a screw pile of 40 mm diameter.

6. Conclusions

In the current study the ultimate carrying capacity of screw piles of different diameters inserted in dry gypseous soil of medium relative density were investigated. Based on the results of the current study, the following conclusions can be drawn out:

- The ultimate carrying capacity of the screw piles increases with decreasing the L/D ratio.
- Increasing the diameter of screw piles inserted in gypseous soil gives high failure load and low displacement due to the high strength of dry gypseous sandy soil.
- Depending on the adopted failure, criteria, which specify the failure, load corresponding to 20% of screw diameter. The failure loads are (0.45, 0.65, and 0.76) kN and the corresponding displacements are 4, 6, and 8 mm for piles of 20, 30, and 40 mm diameter, respectively. The adopted failure criteria of 20% is very conservative and can be extended to 40%.

References

[1] N.S. Al-Dulaimi, Characteristics of gypseous soils treated with calcium chloride solution, M.Sc. Thesis, Civil Engineering Department, University of Baghdad, 2004.

[2] B.S.Z. Albusoda and R.S. Hessain, Bearing capacity of shallow footing on compacted dune sand underlain Iraqi collapsible soil, Engineering and Technology Journal, 31(19) Part (A) 2013 13-28.
[3] M.O. Karkush, Y.J. Al-Shakarchi, and A.N. Al-Jorany, Theoretical modeling and experimental investigation of leaching behavior of salty soils, In Conference on Construction and Building Technology, 2008 123-138.
[4] Arup Geotechnics, Design of screw piles: assessment of pile design methodology, Ove Arup & Partners Ltd, London, 2005.
[5] B. Livneh and M.H.M. Naggar, Axial testing and numerical modelling of square shaft helical piles under compressive and tensile loading, Canadian Geotechnical Journal, 45(8) 2008 1142–1155.
[6] Z. Guo and L. Deng, Field behaviour of screw micropiles subjected to axial loading in cohesive soils, Canadian Geotechnical Journal, 55(1) 2018 34-44.
[7] R. Schmidt and M. Nasr, Screw piles: use sand considerations, Struct. Mag., 2004 29–31.
[8] D. Xiao, C. Wu, and H. Wu, Experimental study on ultimate capacity of large screw piles in Beijing, In International Congress and Exhibition, Sustainable Civil Infrastructures: Innovative Infrastructure Geotechnology, Springer, Cham, 52-58.
[9] IBC, ICC., International building code, International Code Council, Inc.(formerly BOCA, ICBO and SBCCI), 2009, https://codes.iccsafe.org/content/chapter/4626/.
[10] M. Perlow Jr., Geo-Frontiers, 2011.AmericanSocietyofCivil Engineers, USA, 2011, pp.94–102.
[11] M.P. Mitsch and S.P. Clemence, Uplift capacity of helix anchors in sand, American Society of Civil Engineers, New York, 1985 26–47.
[12] G.G. Meyerhof, Bearing capacity and settlement of pile foundations, ASCE J.Geotech.Eng., 102 (GT3) 1976 197–224.
[13] M. Budhu, Soil Mechanics and Foundations, 3rd ed. John Wiley & Sons Inc, USA, 2011.
[14] Nasr, M.H., Large capacity screw piles, In Proceedings of the International Conference: Future Vision and Challenges for Urban Development. Cairo, Egypt, 2004 1–15.
[15] M.J. Tomlinson, Foundation Design and Construction, 7th ed. Pearson Prentice Hall, Malaysia, 2001.
[16] J.E. Bowles, Foundation Design and Analysis, 5th ed. McGraw-Hill, USA, 1996.
[17] M. Sakr, Installation and performance characteristics of high capacity helical piles in cohesionless soils, DFI Journal-The Journal of the Deep Foundations Institute, 5(1) 2011 39-57.
[18] [18] A.J. Lutenegger, Cylindrical shear or plate bearing? Uplift behavior of multi-helix screw anchors in clay Cylindrical shear or plate bearing?—Uplift behavior of multi-helix screw anchors in clay, In Contemporary Topics in Deep Foundations, 2009, pp. 456-463.

IOP Conf. Series: Materials Science and Engineering **901** (2020) 012007 doi:10.1088/1757-899X/901/1/012007

Probability Concepts of Transfer Load to the Foundation of Container Structure

Ayad Al-Rumaithi[1,a], Mushriq F. K. AL-Shamaa[1,b], and Mahdi O. Karkush[1,c]

[1]Civil Engineering Department, University of Baghdad, Baghdad, Iraq
[a]ayad.a@coeng.uobaghdad.edu.iq, [b]mushriqf@coeng.uobaghdad.edu.iq,
[c]mahdi_karkush@coeng.uobaghdad.edu.iq

Abstract: This paper presents stochastic analysis using the perturbation method to model the structure of a container to verify the distributions of probability of maximum and minimum axial forces reactions in piles. The proposed simulation of a container port terminal under 11 scenarios of load combinations was presented. The probability distributions for live loads are assigned according to the input parameters of simulation data. Part of the load itself is implicitly combined such as vertical live load which includes the weight of equipment and containers and wind load. The structural model was simulated in the software STAAD Pro., while the statistical analyses were performed with MATLAB. The results demonstrated that, the most significant external actions for the values of the axial forces reactions on the tips of the piles which considered appropriate to normal probability distributions.

Keywords: Probability, container foundation, stochastic analysis, perturbation method.

1. Introduction

With the technological development of the process of production and the increase of the world population occurred particularly in the last century, the requirement for products of the most varied categories increased considerably. This development is leading to the improvement of the techniques of shipbuilding, enabling larger ships, and containerization has made the port complexes become in supply sequences. Today almost all international trade and a considerable part of its domestic trade dependent on waterway transport as shown in Fig. 1. Structural systems in general are responsible for the stability of constructions, port structures, as well as, have additional importance, due to their functioning. Supply chains whose interruptions would result in huge costs to society. Chaining of containers, which allow the carrying out of operations loading and unloading. As for their structural kind, a structure can be classified as continuous or discrete. Due to the type of cargo that moves, has as its main peculiarity the actions arising from its equipment, specific for the transportation of cargo containerized.

Investigations of the behavior of the container foundation has often been studied by several researchers. In 2016, Vivek and Durga analyzed and designed marine berthing construction with several loading cases, they used information of Visakhapatnam Port to model the structure based on different scenarios of loading combinations. They studied members with variable section dimensions to satisfy all marine safety requirements [2]. In 2017, Kanakeswararao and Ganesh studied the behavior of pile under vertical and lateral load based on different approaches. P-Y curves and Vesic's methods are used for calculation of vertical load in pile. They used linear elements to model pile and assumed Winkler soil spring in both horizontal and vertical directions. They calculated modulus of soil based on code 2911 and depended on Brom's method to compare the results [3].

Figure 1. Typical Container terminal [1].

In 2018, Ćosić et al. presented a method to evaluate soil-pile-bridge pier interaction under seismic state. Nonlinear dynamic analysis was used individually for the analysis of pile and pier. The fragility curves were made depending on the probability theory of log-normal distribution and regression considerations. They used a difference of seismic state in a relation of scale for the intensity measure parameter, or Pin Grid Array according to the incremental nonlinear dynamic analysis [4]. In 2019, Su et al. used seismic fragility curves to assess large-scale pile in wharf with soil pile interaction. they found that soil pile interaction has a considerable effect on the behavior under seismic state. Piles that are supporting Port of Los Angeles Berth 100, USA have been studied with or without taking into consideration soil pile interaction. They concluded that a reliable estimation of seismic fragility in wharf structures is so important so that soil pile interaction can be either higher or lesser than the one without this consideration [5]. Other studies were investigated such as in references [6 and 7]. The current study is an attempt to present the uncertainty of reactions in the pile foundation of a container terminal under several scenarios of loading combinations. The next section describes the structure analyzed in this study.

2. Description of The Structure

The structure is a segment of a container port of a length of 42 m and a width of 20 m. It is supported by 60 piles. 24 of these piles are inclined in both directions to achieve resistance against lateral loads. The dead load in this structure is self-weight with a superimposed dead load. Four live loads are considered to be applied to this structure. Figure 2 shows a diagram of the structure with the loads. The first live load (LL1) is two vertical line loads on the rails which represent equipment, container, and wind loading. The second (LL2) and third loading (LL3) are ship collision scenarios at different locations. The fourth live load (LL4) is the pressure of the sheet pile adjacent to the structure.

The live loads are assumed to uncertain nature, so it requires stochastic analysis to get the variability of piles of axial forces with different load combination scenarios. The next section describes the methodology of the analysis used in this paper.

3. Methodology

Stochastic finite element analysis was conducted using the perturbation method. This method is based on calculating the response from each realization of input forces. Then the response from all realizations is collected to get the statistics including minimum and maximum piles of axial forces. In this study stochastic analysis with million realizations was conducted. By applying the superposition method of linear systems, finite element analysis using STAAD Pro. needed only to be done by applying unit loads at the location of live loads instead of repeating the analysis one million times. Then response from these unit loads are multiplied by live loads magnitudes to get real responses. The process is summarized in Fig. 3.

IOP Conf. Series: Materials Science and Engineering **901** (2020) 012007 doi:10.1088/1757-899X/901/1/012007

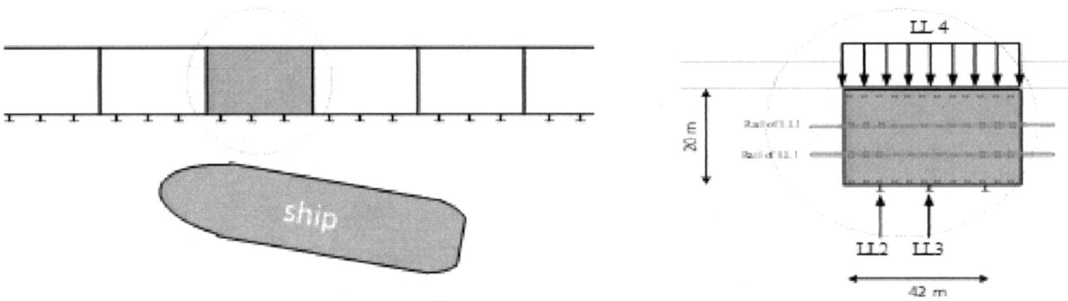

a) Top view with the ship b) Top view with live loads

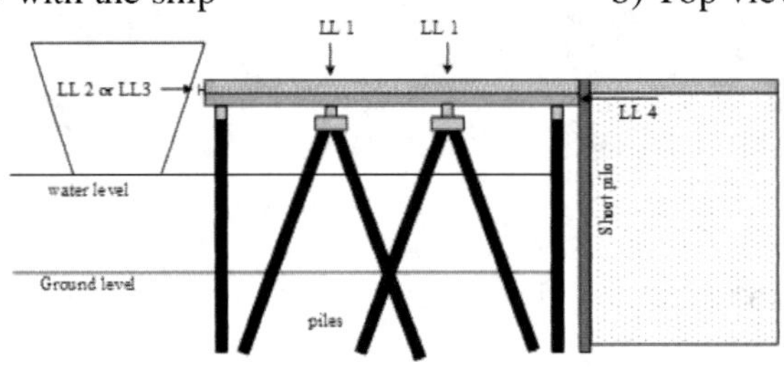

c) Side view

Figure 2. Schematic diagram of the structure with live loads.

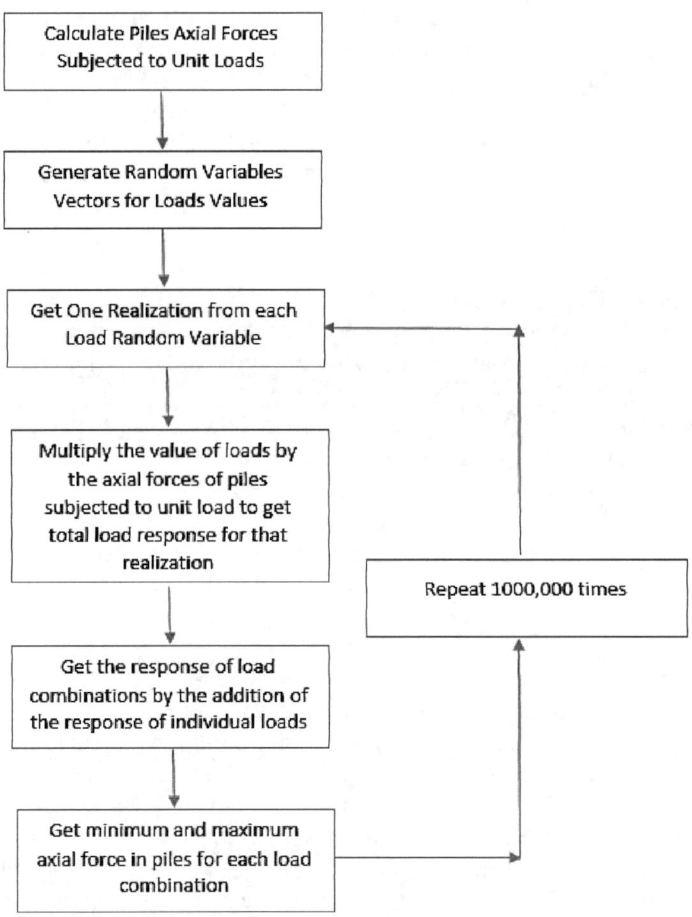

Figure 3. The method of calculation of stochastic linear response by superposition method.

IOP Conf. Series: Materials Science and Engineering **901** (2020) 012007 doi:10.1088/1757-899X/901/1/012007

The finite element model, live loads, and live loads random variables definitions, statistics, and generation are discussed in the next sections.

4. Finite Element Model

The structure beams and piles are modeled using beam elements, while the slab is modeled using plate elements. Piles tip are assumed to hinge. The slab dimensions are 20 × 42 m with a depth of 0.225 m. The beams heights are 0.625 m and widths are 0.7 m width. Piles section is tubes with an outer diameter of 0.7 m and a thickness of 0.15 m. The number of piles is 60 and the length of vertical piles is 20 m. The above dimensions are arbitrary and chosen based on British standards [8-11]. The material for the structure is concrete with a modulus of elasticity of 21.7 GPa and a density of 23.5 kN/m^3. Fig. 4 shows the analytical model and rendered view of STAAD Pro model.

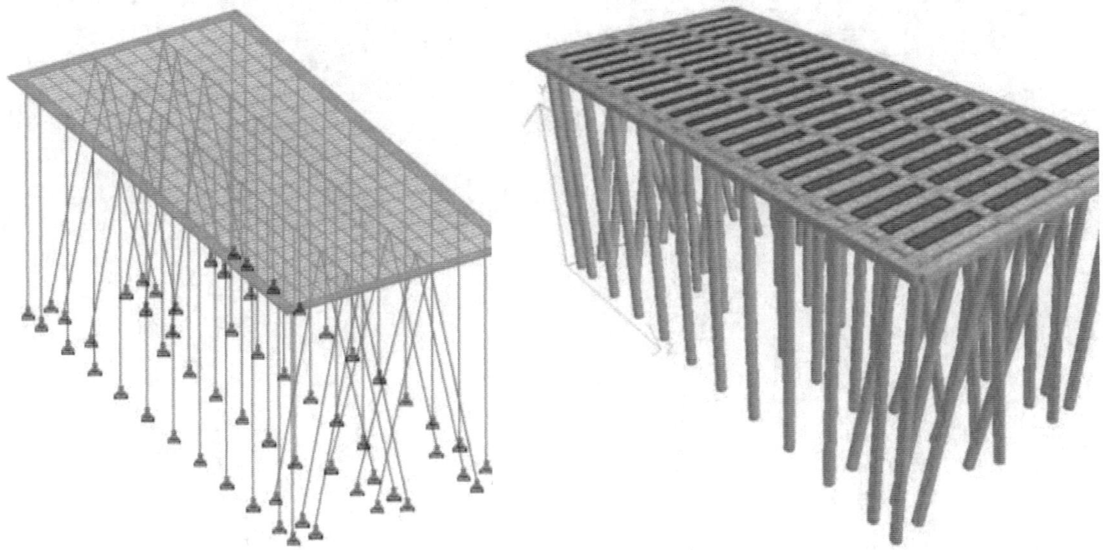

a) Analytical model b) Rendered view

Figure 4. STAAD Pro finite element model.

Some of the piles are inclined in both directions to achieve lateral loading resistance which can be seen in the front and lateral view of the analytical model in Fig. 5.

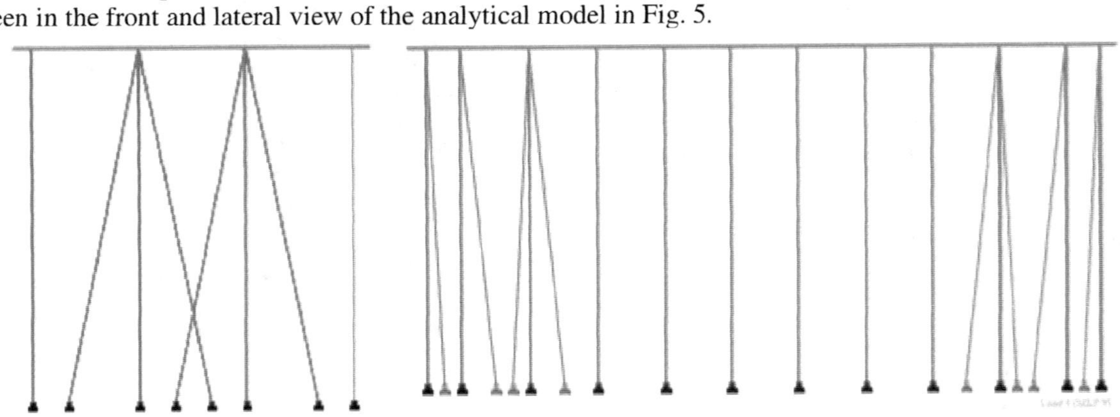

a) Front view b) Lateral view

Figure 5. Front and lateral views of the analytical model.

Dead load is assumed to be self-weight combined with a uniformly distributed superimposed dead load of 5.3 kN/m^2 exposed to the slab. Four-unit loads are applied at live loads locations. The locations of the loads are the two uniformly distributed vertical loads representing the combined effect of equipment, containers, and wind at central longitudinal beams for load 1, lateral concentrated force near the corner at the edge facing the water for load 2 which represents ship collision, lateral concentrated force at the center of that edge for load 3 that represent ship collision at a different location, and lateral uniform distributed load represents sheet pile pressure. The loads locations in analytical models are shown in Fig. 6.

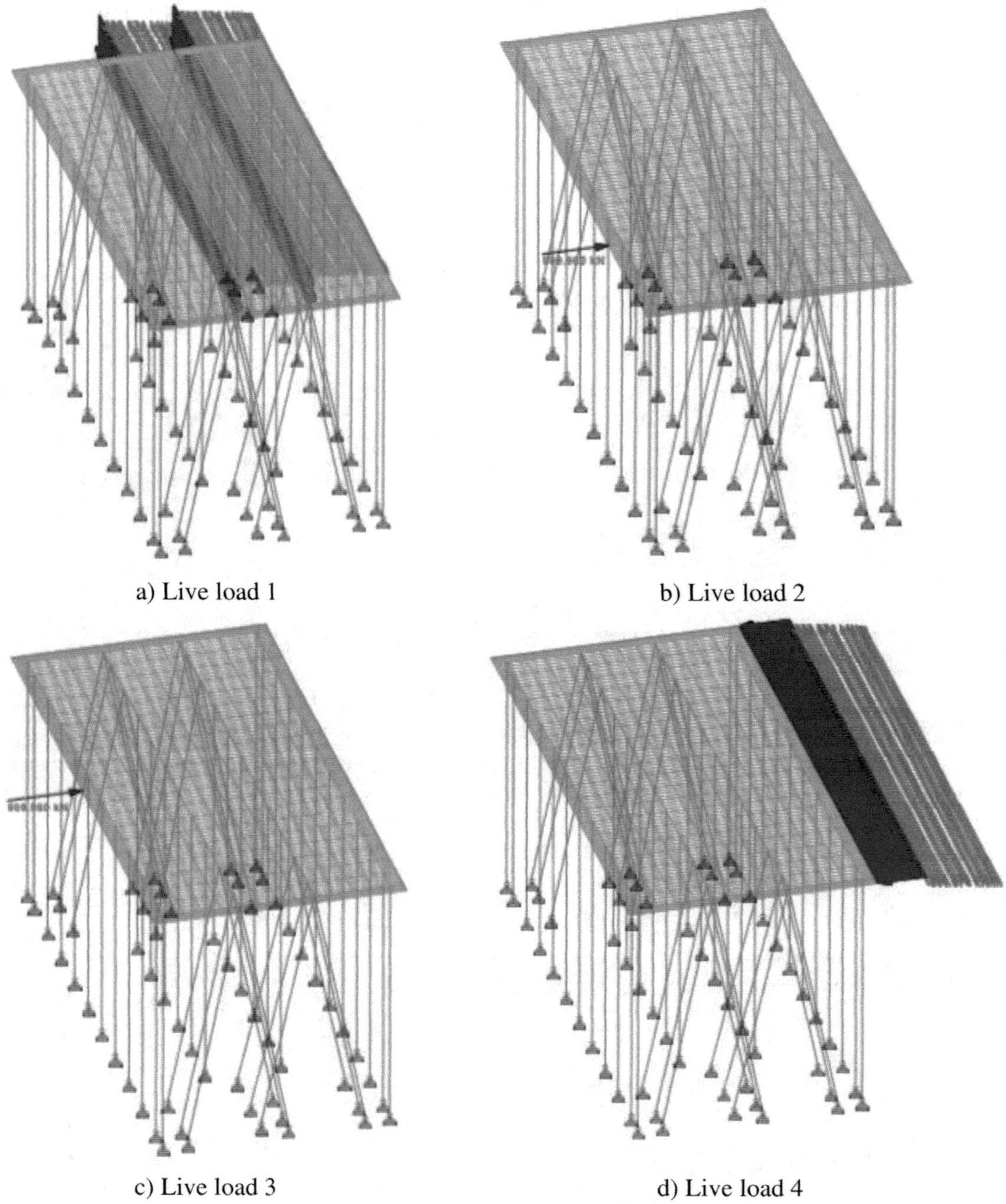

a) Live load 1

b) Live load 2

c) Live load 3

d) Live load 4

Figure 6. Live loads locations.

5. Live Loads Distributions and Generation

Live load values are assumed to be close to design values. Their distribution type and standard deviation are assumed to achieve uncertainty in their values. Table 1 shows live loads distribution types, mean and standard deviation.

Table 1. Live load properties.

Load Case	Unit	Distribution Type	Mean	Standard Deviation
Live Load 1	kN/m	Weibull	80	15
Live Load 2	kN	Normal	3000	600
Live Load 3	kN	Normal	3000	600
Live Load 4	kN/m	Normal	150	15

The distribution type used for loads 2, 3, and 4 is a normal distribution which probability density function (PDF) is expressed in the following relationship [12]:

$$f(x) = \frac{1}{\sigma\sqrt{2\pi}} \exp\left[-\frac{1}{2}\left(\frac{x-\mu}{\sigma}\right)^2\right] \quad (1)$$

where μ and σ are mean and standard deviation, respectively. Load 1 follows Weibull with the following PDF:

$$f(x) = \left(\frac{k}{\lambda}\right)\left(\frac{x}{\lambda}\right)^{k-1} \exp\left[-\left(\frac{x}{\lambda}\right)^k\right] \quad (2)$$

where λ and k are the distribution parameters. These above distributions can be generated from uniform random variables using the inverse transform technique:

$$X = F^{-1}(u) \quad (3)$$

where u is a realization of uniform random variable ranging from 0 to 1. $F(x)$ is the cumulative density of function of the distribution to be generated which is related to PDF by:

$$F(x) = \int_{-\infty}^{x} f(x)dx \quad (4)$$

The mean for any random variable can be calculated from:

$$\mu = E(x) = \int_{-\infty}^{\infty} x\, f(x)\, dx \quad (5)$$

and variance:

$$\sigma^2 = \int_{-\infty}^{\infty} (x-\mu)^2 f(x)\, dx \quad (6)$$

where the standard deviation is the square root of variance. The generated loads of PDFs are illustrated in Fig. 7.

Using the method explained in Fig. 3, the response from each unit live loads are multiplied by their values in all one million realizations. Then, the responses are summed into combinations. All combinations scenarios for live loads are taken into account except those with live loads 2 and 3 together because both of them represent ship collisions at different locations. The self-weight and superimposed dead load are added to all combinations. The total number of load combinations is 11. After the load combination response values are calculated for all piles in the realizations, minimum and maximum axial load of the 60 piles is found for each load combination with each realization. The total number of minimum and maximum piles of axial forces is 11 million each, representing 11 minimum and 11 maximum axial forces for all the million realizations.

IOP Conf. Series: Materials Science and Engineering **901** (2020) 012007 doi:10.1088/1757-899X/901/1/012007

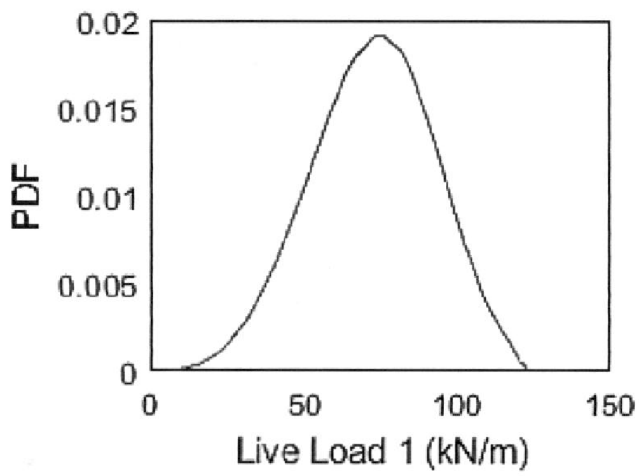

a) Live load 1

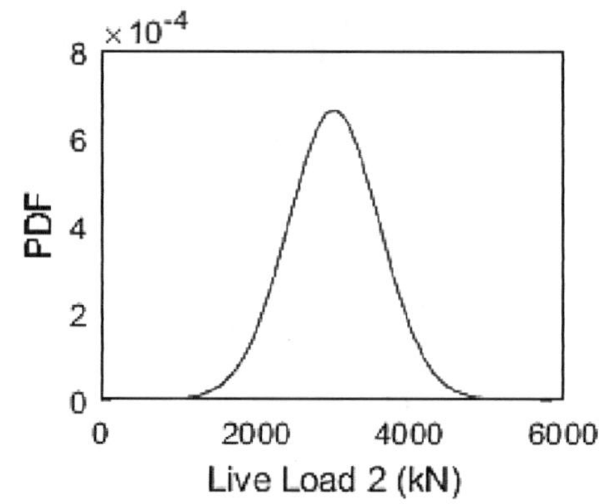

b) Live load 2

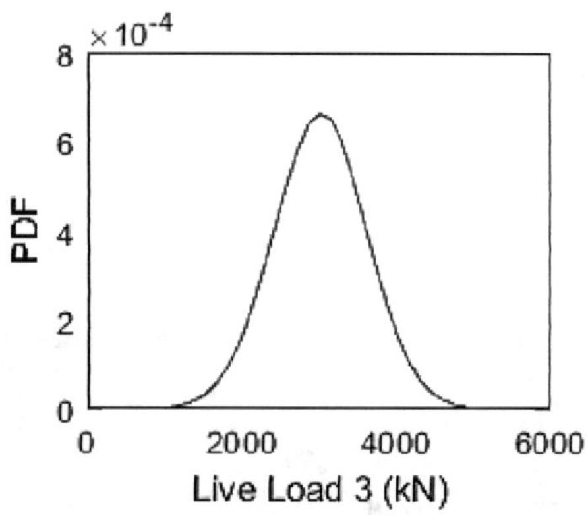

c) Live load 3

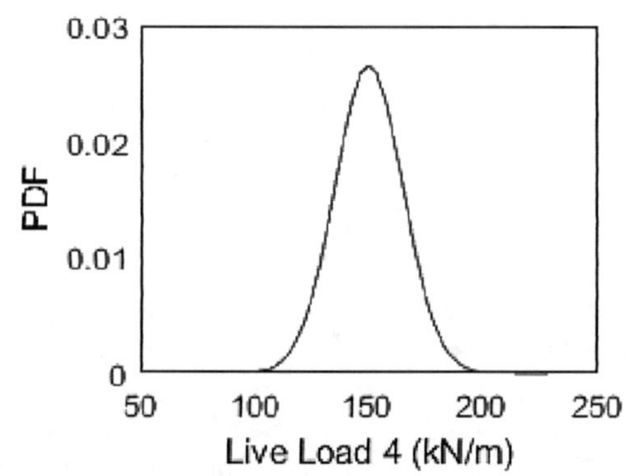

d) Live load 4

Figure 7. Distributions of live load types.

6. Stochastic Analysis Results

Figure 8 shows the PDFs of minimum axial forces (tension) of piles for selected important load cases. Table 2 gives the mean and standard deviation of those PDFs. It can be seen from the results that lateral loads are the most dangerous case in tension. The effect of the lateral impact of the ship at the center is less than that near the corner because the impact location is more distant to the inclined piles that absorb lateral forces. The combination of sheet pile pressure and ship collision has less effect than sheet pile alone because of the opposite force direction. Vertical load decreases the effect of other loading cases because it adds compressive forces to the piles. The case of only vertical load was excluded from these results because it increases the axial load in all piles.

Table 2. Minimum axial force in piles (Tension).

Load Combination	Mean (kN)	Standard Deviation (kN)
2	-726.535	243.743
3	-137.824	124.722
4	-828.762	131.236
1+2	-655.445	244.529
1+3	-66.334	125.67
1+4	-754.215	132.028
2+4	-739.802	131.637
3+4	-206.133	180.079
1+2+4	-669.509	133.083
1+3+4	-135.504	181.028

IOP Conf. Series: Materials Science and Engineering **901** (2020) 012007 doi:10.1088/1757-899X/901/1/012007

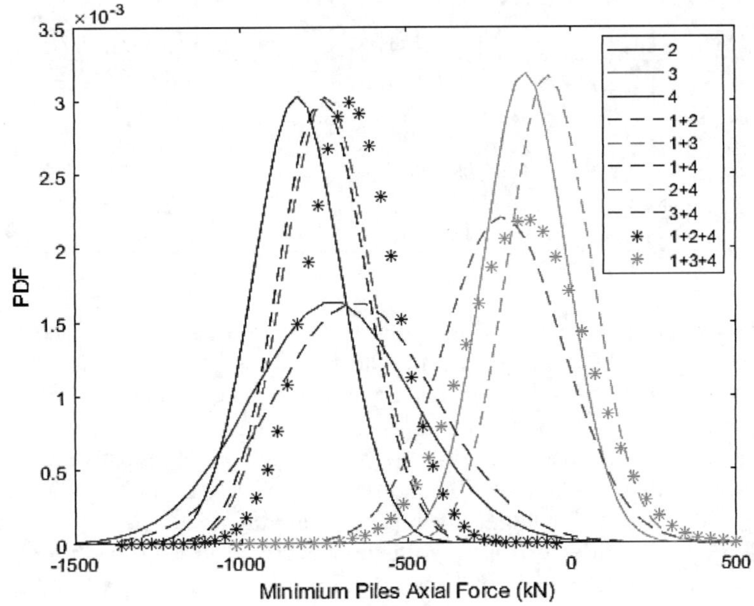

Figure 8. Minimum axial forces in piles (Tension) probability density functions.

Figure 9 shows the PDFs of maximum axial forces (compression) of piles for selected important load cases. Table 3 gives the mean and standard deviation of the corresponding PDFs. Vertical loading combined with sheet pile pressure is the most dangerous case. Lateral loading of the ship collision when combined with that case will have a decreasing effect because of the opposite direction. Other load combinations of only lateral loads are not shown because they had a decreasing effect instead of increasing maximum axial loads of piles.

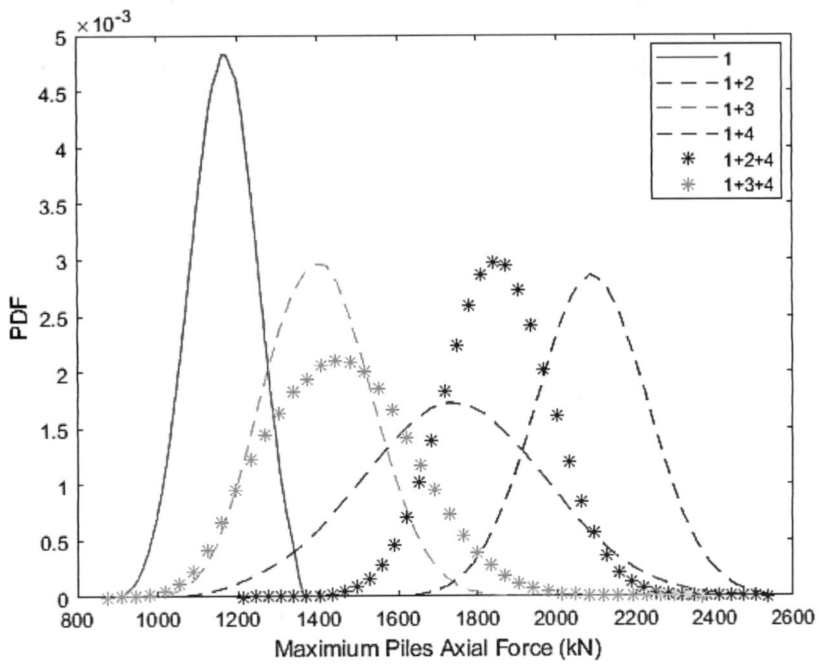

Figure 9. Maximum axial force in piles (Compression) probability density functions.

Table 3. Maximum axial force in piles (Compression).

Load Combination	Mean (kN)	Standard Deviation (kN)
1	1162.233	78.933
1+2	1740.305	229.968
1+3	1400.829	130.68
1+4	2087.865	139.427
1+2+4	1849.463	133.495
1+3+4	1456.997	179.948

7. Conclusions

This paper presented a stochastic analysis to model a structure of a container subject to uncertain loads using STAAD Pro and MATLAB. The study aimed to find the probability distributions of maximum and minimum axial forces in the piles. Eleven load combinations were used, representing dead load and four types of live loads. Live loads had uncertainty expressed by probability density functions. The results of minimum axial forces probabilities showed that lateral loads are the most dangerous in tension. The location of the ship collision had a significant effect on axial forces where collision on the corner has more effect than the center of the edge. Sheet pile loading also had a significant effect. However, combining the loading of sheet piles with ship collision or vertical loading had a reduction effect. The results of maximum axial forces showed that the combined effect of vertical loading with sheet pile pressure is the most dangerous case. That was reduced when combined with ship collision because of the opposite direction.

References

[1] Information on https://www.ft.com/
[2] T. Vivek and R. D. Prasad, Analysis and design of marine berthing structure, Int. Journal of Engineering Research and Application, 6(12) (Part -3) 2016 46-56.
[3] T. Kanakeswararao and B. Ganesh, Analysis of pile foundation subjected to lateral and vertical loads, Int. J. Eng. Trends Technol, (2) 2017 113-127.
[4] M. Ćosić, R. Folić, and B. Folić, Fragility and reliability analyses of soil-pile-bridge pier interaction, Facta Universitatis, Series: Architecture and Civil Engineering, 16(1) 2018 093-111.
[5] L. Su, H.P. Wan, Y. Dong, D.M. Frangopol, and X.Z. Ling, Seismic fragility assessment of large-scale pile-supported wharf structures considering soil-pile interaction, Engineering Structures, 186 2019 270-281.
[6] L.M. Palmer, Development of a simplified analysis approach for predicting pile deflections of piers subjected to lateral spread displacements and application to a pier damaged during the 2010 Maule, Chile, M8. 8 Earthquake (2018).
[7] D.P. Schuetz, D.B. Robbins, P.M. Schuman, and D.J. Kelly, D. J. (2019). Investigation and initial stability analysis of a wharf on severely deteriorated steel h-piles, In Ports 2019: Port Engineering, 2019 273-283, Reston, VA: American Society of Civil Engineers.
[8] British Standard Institute, 2000. BS 6349-1: Maritime structures – Part 1: Code of practice for general, London (2000).
[9] British Standard Institute, 2004. BS 6349-4: Maritime structures – Part 4: Part 4: Code of practice for design of fendering and mooring systems, London (2004).
[10] C.A. Thoresen, Port's Design Handbook, 3ed. London: ICE Publishing (2014).
[11] G. Tsinker, Handbook of Port and Harbor Engineering: Geotechnical and Structural Aspects, New York: Chapman & Hall (1997).
[12] L.G. Alberto, Probability and Random Processes for Electrical Engineering, Pearson Education India, (1994).

IOP Conf. Series: Materials Science and Engineering **901** (2020) 012008 doi:10.1088/1757-899X/901/1/012008

Comparison among Different Methods to Estimate Ultimate Capacity of Bored Pile

Bushra S. Albusoda[1,a], Mohammed S. M[1,b] and Mohammad F. Abbas[2,c]

[1]Department of Civil Engineering, University of Baghdad, Baghdad, Iraq
[2]Department of Civil Engineering, University of Al-Muthanna, Al-Muthanna, Iraq
[a]dr.bushra_albusoda@coeng.uobaghdad.edu.iq, [b]mohammed.bridge@gmail.com,
[c]engmoh863@gmail.com

Abstract This study investigates the ultimate load capacity of concrete 1.5 m diameter bored pile based on three methods: field-loading test, field Standard Penetration Test results, and numerical modeling. A comparison of the ultimate load capacities from these methods was made (the measured large-scale field-loading test, the correlated SPT *N* values applying Novo SPT software and numerical modeling where Plaxis3D software with Mohr-Coulomb model was utilized). The result of the pile field-loading test was more accurate than the other methods and used as a reference for comparison among their results. In general, the value of ultimate load capacity calculated from Novo SPT software and that obtained from numerical analysis were compatible in comparison with field load results.

Keywords: Pile loading, Novo SPT software, bored pile, Plaxis 3D.

1. Introduction

Concerning the doubts regarding the design and analysis of pile foundations, it has become a practice, and mostly obligatory, to conduct large-scale pile loading tests. These field-loading tests are costly, requiring much time, and for limited projects, the expenses are often tough to justify. Piles loading tests are usually executed to measure the ultimate capacity of piles of a foundation system at the project site and to estimate the quality of all piles. Pile loading also has a significant role to play as a mean in the estimation of settlements of foundations, despite the doubts expressed by many investigators that the settlement of a single pile has little or no link to that of a pile group [1]. Preferably, to allow an estimate of the distribution of base and shaft resistance during loading stages, testing of piles should use transducers. Analysis of results of such tests should consider the presence of residual stresses brought during pile construction, if it does not consider these stresses a confusing calculation of the shaft and base resistances can be made [2].

After assessment and validation in the site, the designer approves the design load that already calculated based on the site investigations report or modified accordingly for the safer design for project piles. The possible variation of piles quality having the same configuration is relatively narrow down by conducting pile integrity tests according to ASTM - D5882 for all piles [3].

To study the behavior of large pile (1.5 m) diameter, a static vertical loading and unloading tests were carried out at experimental non-working pile in Al-sodoor Bridge project site according to ASTM-D1143. Both of pile loading test data and the Novo SPT software ultimate loads components were compared with Plaxis3D software analysis based on input data from site investigations report to attain more insight into the behavior of the large pile diameter subjected to vertical loading and unloading.

The Mohr-Coulomb soil model was employed for modeling the granular soil behavior and its interaction with field pile in this software.

2. Study Area

The pile load-settlement data and SPT test results of the study were obtained from Al-sodoor Bridge site in Diyala Governorate. The 1.5 m large diameter pile with 20 m length was tested using a 750-ton working load applied on it by axial compression as shown in Fig. 1. The reaction force is transferred by hydraulic jacks which conveying the field load test from a frame of steel girders loaded by an array of concrete blocks. Accurate certified calibrated instrumentations were installed onto the top of the pile test so that an accurate measurement of the test pile displacement can be obtained. Figure 2 illustrates configuration for hydraulic jacks' group and the controls for measuring pile loading and settlement, this test was carried out according to ASTM-D1143.

Figure 1. Pile loading test.

Figure 2. Hydraulic jack group.

3. Pile Design Based on Soil Correlations

In general, the design of piles is based on empirical correlations specified experimentally on different soils by several researchers by conducting laboratory and field tests thus, this situation often confusing the geotechnical designers. Several site investigations boreholes may not be enough to explore the sub-soil conditions of the whole pilling project. So, conducting pile loading tests are the most reliable way to confirm the integrity of as-constructed piles at the site. It is important to conduct a pile-loading test to verify the actual pile capacity [4]. In the prediction of pile load capacity from the load-settlements curve within the given criteria, it is necessary to use pile loading test data specified that the pile load tests should be supplemented by detailed site investigation to define the entire soil profile accurately [5].

4. Analysis Method and Material Properties

4.1. Pile Ultimate Load Capacity Analysis Using Static Load Test

To ensure the geotechnical and the structural reliability of the pile also to predict settlement of other piles, the pile load test is frequently carried out. This field loading was performed using a reaction method. The test procedure involves applying an axial compression load on the top of the experimental large (1.5 m) diameter pile test (with 20 m length from the ground surface level) by four hydraulic jacks on the top of the test pile as shown in Fig. 1. The static pile load test caused a mobilization along the soil-pile interface where this mobilization generally occurs well before the ultimate structural capacity of the pile is reached. Once the test is complete, the working pile is returned to an unload condition and the rebound of the pile can be measured.

According to ASTM-D1143, the non-working pile test was loaded to twice its design-working load that already estimated based upon the soil investigation report for site pilling and then unload leaving the load off until the pile rebound substantially ceases. Figure 3 illustrates pile test settlement and rebound vs. loading and unloading while Figs. 4 and 5 illustrate the time versus load and (settlement and rebound) respectively.

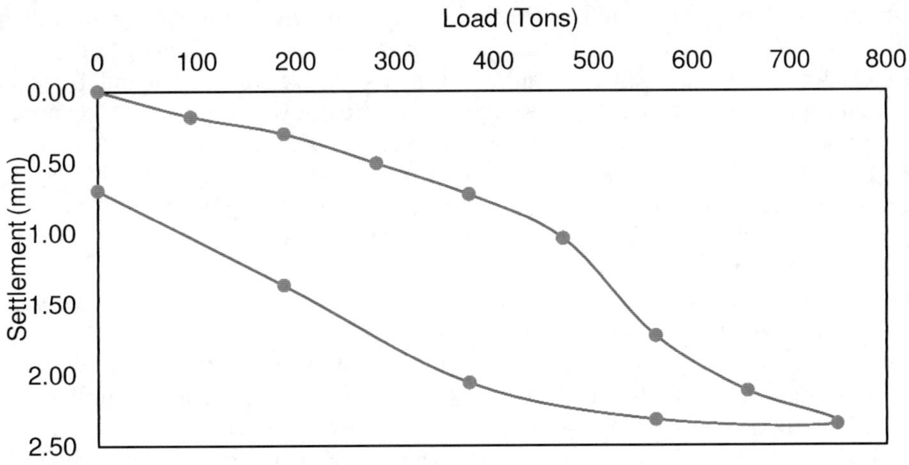

Figure 3. Relation between pile settlement and rebound vs. loading and unloading.

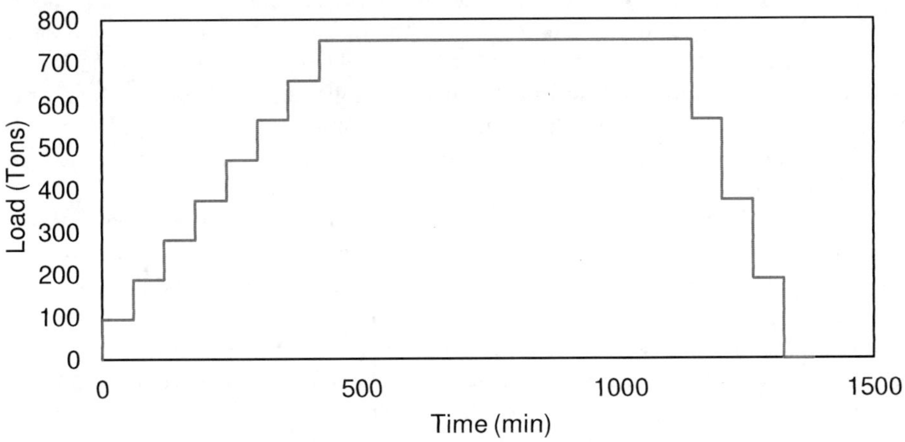

Figure 4. Loading and unloading of pile test with time.

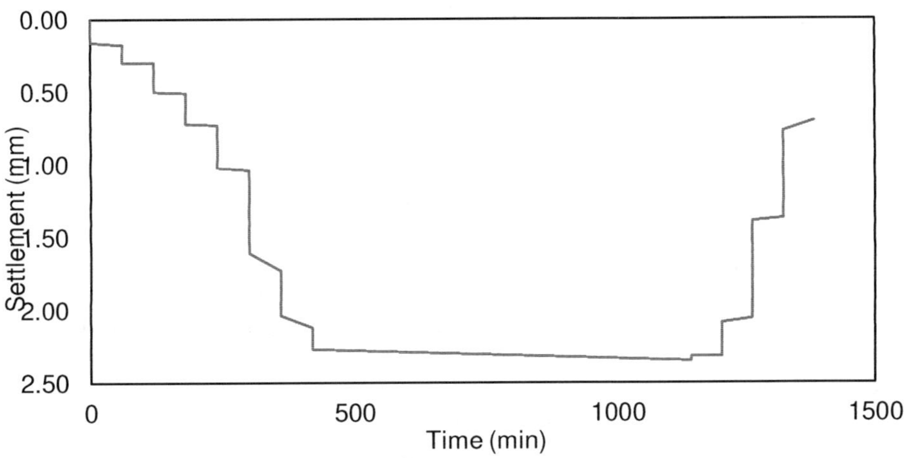

Figure 5. Settlement and rebound of pile test with time.

4.2. Ultimate Load Capacity Analysis from Field SPT N Values

It is known that, the prediction of the piles ultimate load capacity as being the region of highest uncertainty in the design of foundation. For predicting piles ultimate load capacity, several approaches and methods have been proposed to conquer their uncertainties. Some simplified postulations or empirical approaches concerning soil stratigraphy, soil-pile structure interaction, and distribution of soil resistance along the pile were included in these approaches. So, they do not provide quantitative values that are directly useful in foundation design [6]. In recent years, as a supplement of the static and dynamic load capacity analysis is the determination of piles load capacity from field loading data that has been used by geotechnical engineers. This is due to the fast progress of field-testing devices, a better appreciation of the soil's behavior, and the subsequent comprehension of insufficiencies of traditional laboratory tests and some of their restrictions [7].

The Standard Penetration Test is yet the most generally used field test to characterize soil materials and for the determination of pile capacity, which is one of the earliest purposes of this field test. Nevertheless, some limits and disadvantages are involved with the SPT regarding the clarification and repeatability of test data. These problems are due to the uncertainty of the energy provided by different SPT hammers to the anvil system and regarding the field test procedures (the borehole diameter, the rod length, the split spoon sampler type, the blow rate, and the anvil type) [8]. The direct pile capacity evaluations are conducted by applying SPT N values with some modification factors to empirical correlations. Yet, there is a great deal of suspicion about averaging and filtering data that deals with pile ultimate capacity, the failure area in the vicinity of pile tip, utilizing a total stress method, and the piles capacity of narrow base penetration in compacted layers [9]. Depending on many theories, the indirect SPT approaches utilizing friction and undrained shear strength parameters predicted from measured data. In the indirect approaches, only soil parameters are attained from Standard Penetration Test and the procedure of the pile bearing capacity calculation is matching the static method, and thus contains a similar basis of faults [10].

In foundation design codes as well as in the literature there are several prescribed SPT-based design methods, most of them being empirical. The field tests are faster and cheaper than those carried out in the laboratory and do not necessitate any sampling. SPT test is suitable to estimate the resistance and density of granular media and is largely practiced within the scope of foundation projects. It should be however mentioned that the diversity of empirical interpretations of SPT and the dependence of this field test on the test procedure and the device features are the source of uncertainties and discrepancies between methods aimed to evaluate bearing capacity and settlement of foundations. These uncertainties in geotechnical pile design, notably within the scope of big sized projects, often lead to carry out static pile loading tests to experimentally determine pile load capacity and settlements as well. Although the SPT-based pile design literature is wealthy, one needs to be aware of local geotechnical conditions from which several empirical formulas were derived. Caution is then necessary when using any SPT-based method. Several studies on pile test databases were undertaken to empirically derive methods of pile design [11] or to assess existing methods [12].

4.3. Ultimate Load Capacity Analysis from Novo SPT Software

Novo SPT version 2.79.2014 is a software that deals with the analysis of Standard Penetration Test data and connecting between soil properties and the number of blows depended on more than 310 formulae. These formulae were quoted from more than 70 academic researches and papers are implemented in Novo SPT program along with powerful features for organizing the correlations such as statistical charts, reports, import, export data, and more. The SPT raw data are corrected/normalized for field procedures (Energy level, Borehole diameter, Samper type and Rod length) to obtain N_{60} and applying overburden correction resulting $(N_1)_{60}$ values (based on several types of research chosen by the user) where this parameter is widely used in engineering analysis including soil liquefaction among other correlations.

This software has a wide ability of calculating soil (Static and Dynamic) parameters and representing them graphically with borehole depth based on several researchers selected by the program's user [13]. The soil parameters that can be calculated by this program are: Overburden Correction factor (C_N), (E),

IOP Conf. Series: Materials Science and Engineering **901** (2020) 012008 doi:10.1088/1757-899X/901/1/012008

(Ø), (D_r), Undrained Shear Strength (S_u), Shear Wave Velocity (V_S), (G), Liquefaction Potential (CCR), Allowable Bearing Capacity (q_{all}), Settlement (S) for a shallow foundation, Pile Bearing Capacity, CBR and other soil parameters, each equation is derived for specific soil type and may have limitation for use. Table 1 summarizes the calculated load resistance components for (1.5 m) diameter pile based on field SPT N values by utilizing Novo SPT software.

Table 1. Pile-load capacity values by utilizing Novo SPT software.

Method	f_s (Ton)	q_b (Ton)	Comments
GEO,1996 and Yau,2000	515	-	For bored pile
Hassan and O'Neill, 1994	1805	-	For drilled piles
Meyerhof, 1976	313	809	For small displacement piles (bored)
Quiros and Reese, 1977	780	-	For drilled piles
Reese and O'Neill, 1988	1805	556	For drilled piles
Reese and Wright, 1977	1393	592	For drilled piles
Yves Robert, 1997	940	776	For bored pile in granular soils

4.4. Ultimate Load Capacity Analysis by Plaxis3D Software

Numerical analysis is increasingly accepted in geotechnical problems consideration where finite element evaluations are progressively employed in foundations design. In this study, many methods for evaluating the ultimate load of piles from field SPT data were used and matched with large-scale field loading test results. In this part, the Plaxis3D, which is a software, based on a three-dimensional finite element program specially developed for the analysis of foundations. This software described the behavior of the model by graphical procedures and some input parameters, which permit the user to generated models automatically, so induced good output facilities and detailed evaluation procedures.

To interpret the behavior of 1.5 m diameter single pile with 20 m length with the input parameters listed in Table 2, a model is made (with a 5 × 5 m) working area using Mohr-Coulomb model (which is an elastic perfectly-plastic model) where the model is isotropic and does not account for soils stress-dependency where soils tend to stiffen with increased pressure [14]. It is recommended using this model as an initial estimation of soil because of a relatively fairly accurate and fast for analysis. To generate the model of the pile in the program by employing a solid member from the standard option list and it is placed inside the interface layer at the center coordinates of the mesh. The generated mesh and the load-settlement curve of the system (for pile: L = 20 m, D = 1.5 m) by this software are shown in Figs. 6 and 7 respectively. Figure 8 illustrates the pile skin resistance-settlement curve. The borehole log for the study site is shown in Figure 9.

Table 2. Material input properties from site used for Mohr-Coulomb Model.

Property	Values	Information Reference
γ_{sat} (kN/m^3)	20	Project site investigation report
Drainage type	Undrained	Quick loading
E (MPa)	58	AASHTO, 1996
ν	0.49	Undrained loading
Cohesion Cu (kPa)	40,50,80	Site investigation report for depths 0-8, 8-10, 15-20 m respectively
Friction angle (Ø°)	36,46	Site investigation report for depths 6-15, 19-20 m respectively
Dilatancy angle (ψ°)	6,16	Plaxis-3D Material Manual
γ_c (kN/m^3)	24	Pile materials
Poisson's ratio (μ)	0.2	Pile materials
Modulus of Elasticity (MPa)	200	Pile materials

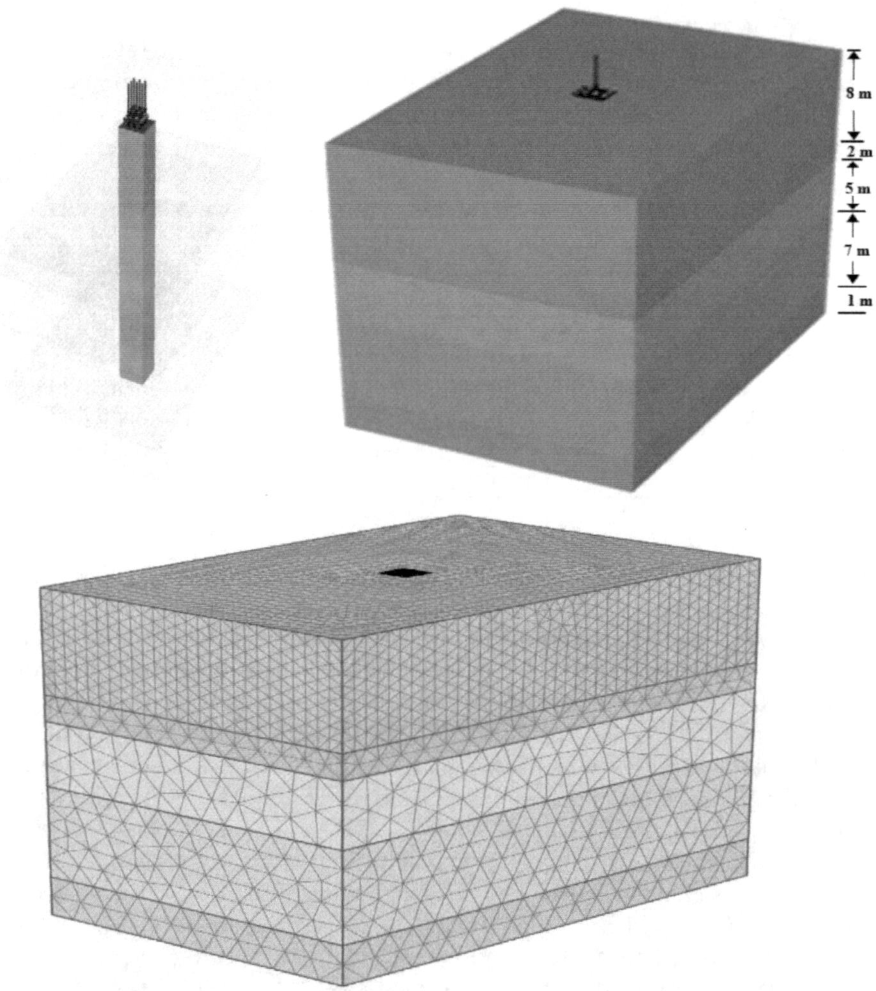

Figure 6. Generated mesh of the system (for pile: L=20m, D=1.5m) as executed by Plaxis-3D.

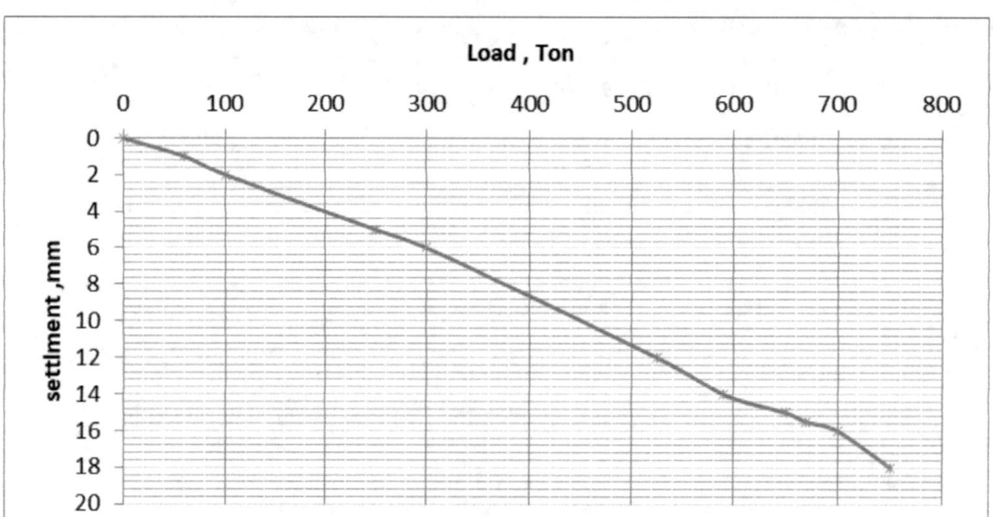

Figure 7. Load-Settlement curve (for pile: L = 20 m, D = 1.5 m) as executed by Plaxis-3D.

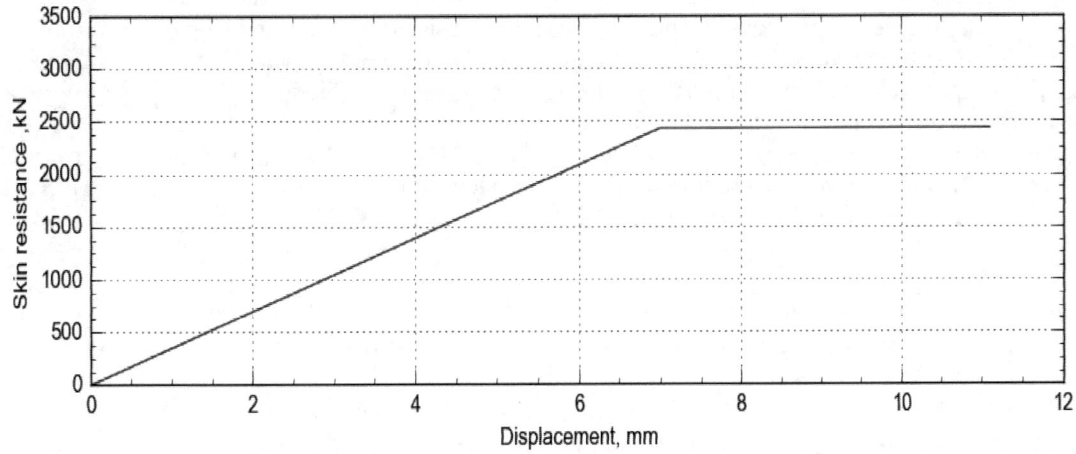

Figure 8. Skin resistance-Displacement curve (for pile: L = 20 m, D = 1.5 m) as executed by Plaxis-3D

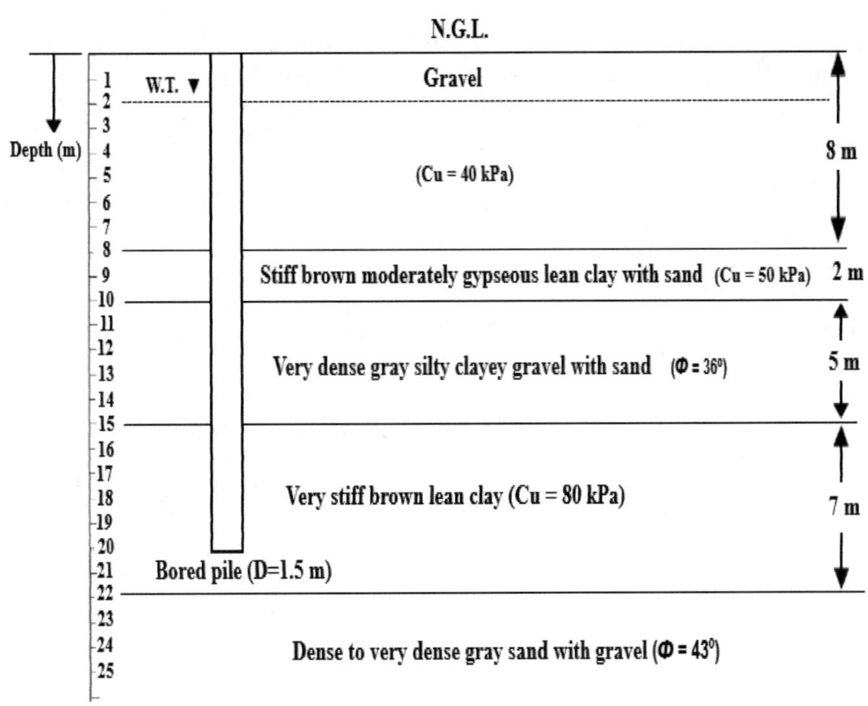

Figure 9. Borehole log for the study site.

5. Discussion of Results

Concerning pile ultimate load capacity values based on correlation with SPT N values, it is clear that these correlations were originally derived empirically based on different soil properties from soil property of the study site, which reflect the wide range of these bearing capacity values. Excluding high load capacity values from N values, there is a fair agreement with other value from field load test as well as with load capacity value obtained from Plaxis3D software analysis.

Evaluating the ultimate load capacities of piles is an importing approach in geotechnical engineering. Find a suitable method for predicting carrying load capacity of the pile is the trend of the investigators to peruse additional researches on this subject. In recent years, conducting field-testing such as SPT, which can be used to predict the ultimate load capacity of the pile, because of its low cost, its simplicity, and being able to give a numerical parameter, that may be related to crude but straightforward design rule. The analysis of pile ultimate load capacity by finite element method using Plaxis3D software and comparing it with field loading methods, tried to find a reasonable prediction for its load capacity. The results indicate that, in this site, the field load capacity measured value (750 Ton) was fair close to the one obtained by the finite element method (FEM) (about 1000 Ton) based on input data from site investigation report, which is also close to pile load capacity components calculated based on SPT N values by utilizing Novo SPT software (1,071 Ton).

6. Conclusion

- Many factors are affecting SPT and produce a wide scattering in N values within test site where density, structure, water table, or lateral stress within the test medium will lead to a high range in calculated pile load capacities.
- The considerable difference in SPT apparatus and field procedures, a significant variability of measured penetration resistance can occur.
- The Mohr-Coulomb model gave a good prediction of the ultimate pile load as compared with empirical correlations based on field SPT N values.
- Conducting a pile load test is an essential matter in geotechnical foundation engineering concerning heterogeneity of site soil properties, being several boreholes for site investigation cannot give clear insight of soil substrata.
- It is a suitable and fast approach for simulation of soil behavior by using little input parameters of the Mohr-Coulomb model for analyzing geotechnical problems.

Reference

[1] H.G. Poulos, Pile testing and settlement prediction, Conference Paper in Geotechnical Special Publication, March 2012.
[2] A. Eslami and B.H. Fellenius, Pile capacity by direct CPT and CPTu methods applied to 102 case histories, Canadian Geotechnical Journal, 34 1997 886-904.
[3] N.T. Singh, Verification of pile load capacity using static pile load test, Conference Paper in Geotechnical Special Publication, December 2016.
[4] T. Matsumoto, K. Matsuzawa, and P. Kitiyodom, A role of pile load test–Pile load test as element test for design of foundation system, In Proc. 8th Inter. Conference on the Application of Stress Wave Theory to Pile: Science, Technology, and Practice, 2008 39-58.
[5] H.G. Poulos, Pile behavior-theory and application, Geotechnique 39 1989 365-415.
[6] A. Eslami and B.H. Fellenius, CPT and CPTu data for soil profile interpretation: review of methods and a proposed new approach, Iranian Journal of Science and Technology, Transaction, 28(B1) 2004 69-86.
[7] T. Lunne, P.K. Robertson, and J.J.M. Powell, Cone Penetration Test in Geotechnical Practice, Blackie Academic & Professional (1997).
[8] N.V. Nayak, Foundation Design Manual, Dhanpat Rai & Sons pub (1985).
[9] H.G. Poulos and E.H. Davis, Pile Foundation Analysis and Design, JohnWiley, New York (1980).
[10] B.H. Fellenius, D.E. Harris, and D.A. Anderson, Static loading test on a 45 m long pipe pile in Sandpoint, Idaho, Canadian Geotechnical Journal, 41(4) 2004 613-628.
[11] Robert, Y., 1997. A few comments on pile design. Can. Geotech. J. Vol.34. pp.: 560-567.
[12] M. Bustamante, R. Frank, and S. Christoulas, Evaluation de quelques méthodes de calcul des pieux forés, Revue française de géotechnique, (54) 1991 39-52.
[13] NovoSPT User's Manual, Design and Programming by Alireza Afkhami, 4188 Hoskins Road North Vancouver, BC Canada [Online]. Available at http://www.novotechsoftware.com.
[14] Plaxis-3D, Plaxis Material Manual, Version 1, 2013.

Effects of Soil-structure Interaction and Different Simulation on the Response of Concrete Box Culverts

Salah R. Al-Zaidee[1,a], Ashraf Hameed Alsalmani[2, b*] and Alhusain Salam[1, c]

[1]Civil Engineering Department, University of Baghdad, Baghdad, Iraq
[2]Mechanism and Equipment Department, University of Baghdad, Baghdad, Iraq
[a]salah.r.al.zaidee@coeng.uobaghdad.edu.iq, [b]ashraf.hameed@coagri.uobaghdad.edu.iq
[c]alhusainsalam2018@gmail.com

Abstract. This paper shows how different simulation and soil-structure interaction can affect the response of concrete culverts embedded in dry cohesionless soils. Three simulations have been considered. In the first, second, and third simulations, the applied loads have been determined based on empirical relations of PCA without including the soil-structure interaction. The differences between these simulations are in using the beam element for the first one while using the frame element for the second one and the plane strain element for the third one. In addition to neglect the soil-structure interaction, the axial and shear deformations have been neglected in the first simulation. The axial deformations have been included in the second simulation; finally, all of the axial and shear deformations are included in the third simulation.

Keywords: Culverts, Traditional Analysis, Finite Element Analysis, Soil-structure interaction.

1. Introduction

The civil engineer routinely encounters analysis and design of box culverts in work. In traditional analysis and design, loads acting on culvert are determined based on analytical solutions of the theory of elasticity. In these solutions, the soil is simulated as a semi-infinite elastic isotropic media [1]. The main drawback of these solutions is they are derived from the absence of the embedded structure; therefore, they cannot simulate soil-structure interaction. Although the culverts are rather simple structures, the loadings applied to these structures during their construction and subsequent service life can be complex. These structures are subjected to substantial vertical and lateral earth pressure and are often subjected to significant temporary loadings during the construction of the embankment [2]. Both AASHTO Standard Specifications for Highway Bridges and AASHTO LRFD Bridge Design Specifications stipulate procedures for calculating the design loads on culverts based on the Marston-Spangler theory [2]. This theory reflects how the construction process may affect the loads transfer to the buried culvert. In the trench installation, some of the weight and applied load on the more deformable central prim transmits to the stiffer adjacent soil. In the embankment installation, forces transmit for the softer side soils to the stiffer central prism [3]. An analytical solution of the differential equilibrium equation is used in the Marston-Spangler theory to quantify these effects. Solution results have been presented in a nondimensional form to be suitable in the design practice [4].

In 1975, the Portland Cement Association adopted the Marston-Spangler theory to prepare design charts for analysis of the concrete box culverts. In these charts, the walls, slabs, and foundations of the culverts are considered a beam with neglecting the effect of concrete Poisson's ratio, axial deformation, and soil-structure interaction [5]. Spite the powerful of analytical solutions, they are limited in nature and applied to regular geometry and typical load conditions. In contrast, the approximate finite element solutions are general and can be applied for different geometry and loads conditions, and they can deal with nonhomogenous and nonlinear soil nature [6]. Neglecting the soil-structure interaction remains the main drawback in the traditional analytical solutions [7]. This paper aims to show how different modeling of the culverts with and without soil-structure interaction may affect the resulting design forces. Three different simulations have been considered. These simulations have been arranged in a hierarchy form. In the first simulation, traditional analysis of PCA has been used as the reference case. In the second modeling, the loads acting on the culvert have been determined using the PCA approach. At the same time, structural analysis has been achieved using frame elements to include the effect of axial deformation. Finally, the contribution of the shear deformations of the deep culvert members have been mimicked in the third simulation by using the plane strain finite element model. This paper aims to use plane strain finite element models to simulate the interaction between the culvert and the surrounding soil and to show how design forces and soil pressures are affected due to the interaction. Finite element models have been prepared with Abaqus commercial finite element packages.

2. Loads on the Culvert

According to [8], loads acting on the culverts include the weight embankment and the wheel loads due to highway traffic. How the wheel load is distributed through soil media to act on the culvert depends on the nature of pavement, i.e., rigid or flexible, the soil nature and stiffness, and the embankment depth [5]. It is relatively difficult to have a mathematical expression for the embankment load transmitted to the culvert as it depends on different factors, including the angle of internal friction, the density, the homogeneity, and the moisture content of the soil. Until now, the Marston-Spangler theory stills the most accurate analytical basis to determine the weight of the embankment on the culvert. According to this theory, the acting load would equal to the weight of the prism located directly above the culvert and the upward or downward shear forces depending on the construction method. As the stresses due to lateral earth pressure generally counteract the stresses due to the vertical loads acting on the culvert, therefore they can be neglected conservatively [5]. When they have to be computed, the lateral pressures are estimated based on the at-rest condition as there is no significant soil yielding or movement due to lateral bracing of the walls [9].

3. Finite Element Analysis

For the time being, the finite element method represents the most powerful computation approach that can be used to simulate complex structures with complicated boundary conditions. It has been developed in the later 1950s. It uses a piece-wise polynomial to approximate the field variation in the considered media. In the stress analysis discipline, the displacements are considered the main unknow that are approximated with the shape functions. Currently, an isoparametric formulation is used where the same polynomial is adopted to approximate the field and the geometry [10]. From the geotechnical point of view, the finite element technique can simulate both deformations and mobilization in one model rather than in two, like it is done in classical soil mechanics. It also can deal with the coupled analysis of deformation and flow [11]. Katona pioneered using the finite element technique to model the buried pipes and culverts. This work developed the well-known public software CANDE (Culvert Analysis and Design) [12]. [13], and [14] also achieved contributions to adopt the finite element method for the analysis of buried structures.

4. Case Studies

In this paper, three case studies have been considered to include a square one-cell box culvert, rectangular one-cell culvert, and a two-cell culvert. Geometry and dimensions for the case studies are presented in Fig. 1 below. To be comparable with the classical solution of PCA, the units have been maintained in their original imperial form. The soil mass has been taken large enough to simulate the deformation vanishing at the constrained boundaries. All considered soils have been assumed dry with 35° angle of internal friction. A corresponding Poisson's ratio of 0.333 and at rest condition, k_o, have been assumed based [15].

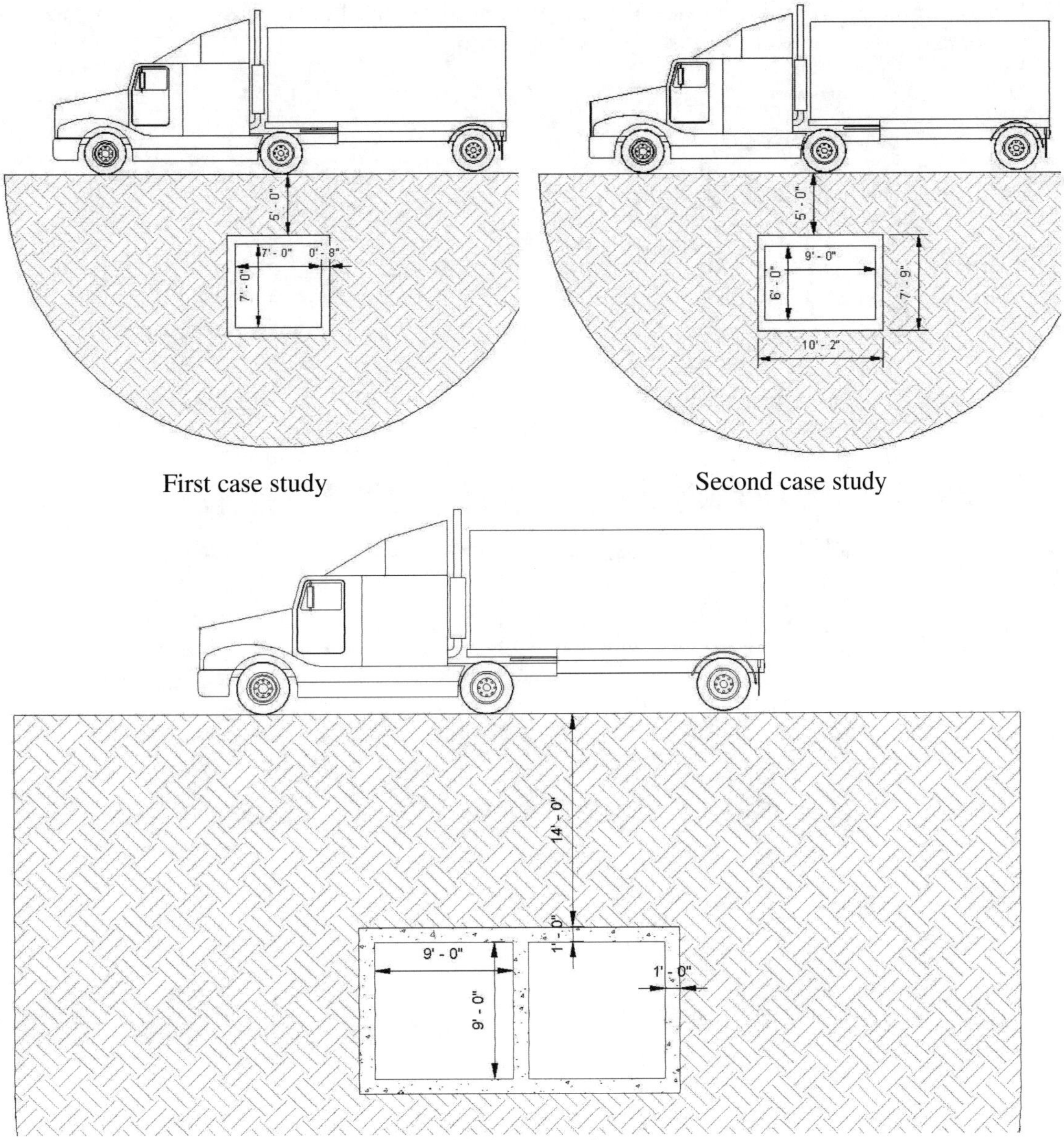

Third case study

Figure 1. Geometry and dimensions and of the case studies.

For each case study, three simulations have been considered. In the first simulation, the loads have been computed according to [5], see Fig. 2, and the culvert has been simulated as a beam element with neglected axial deformation. Points, where internal forces have been determined, are presented in Fig. 3 for the case study. For the second simulation, the frame elements have been used to include the axial deformation, as indicated in Fig. 4. Finally, as indicated in Fig.5, the plane strain elements, with two degrees of freedom per node, have been used to meshed the culvert for the third simulation. For the first, second, and third simulations, the underneath soils have been replaced by Winkler foundations with a subgrade coefficient determined based on Eq. 1.

$$k_s = 0.65 \sqrt[12]{\frac{E_s B^4}{E_f I_f}} \frac{E_s}{B(1-\mu^2)} \quad (1)$$

IOP Conf. Series: Materials Science and Engineering **901** (2020) 012009 doi:10.1088/1757-899X/901/1/012009

While, where E_s, E_f= modulus of soil and footing respectively; B, I_f= footing width and its moment of inertia based on cross-section. μ=soil's Poisson's ratio.

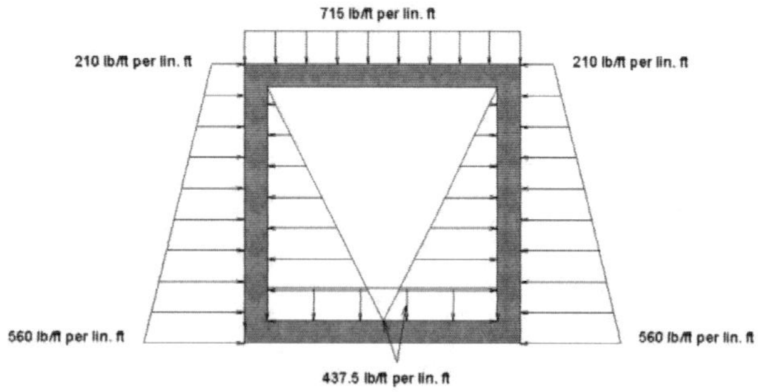

First case study.

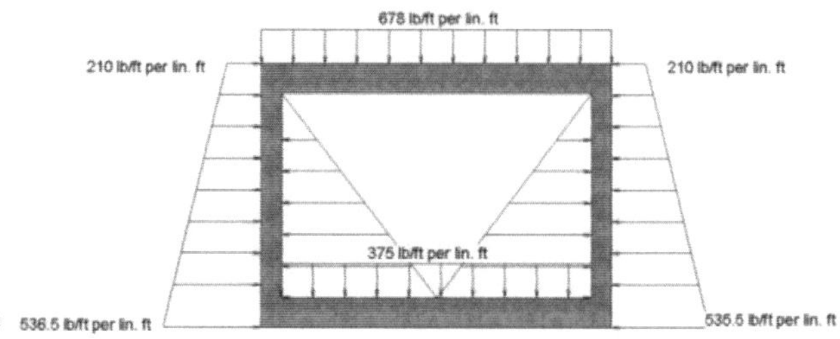

Second case study.

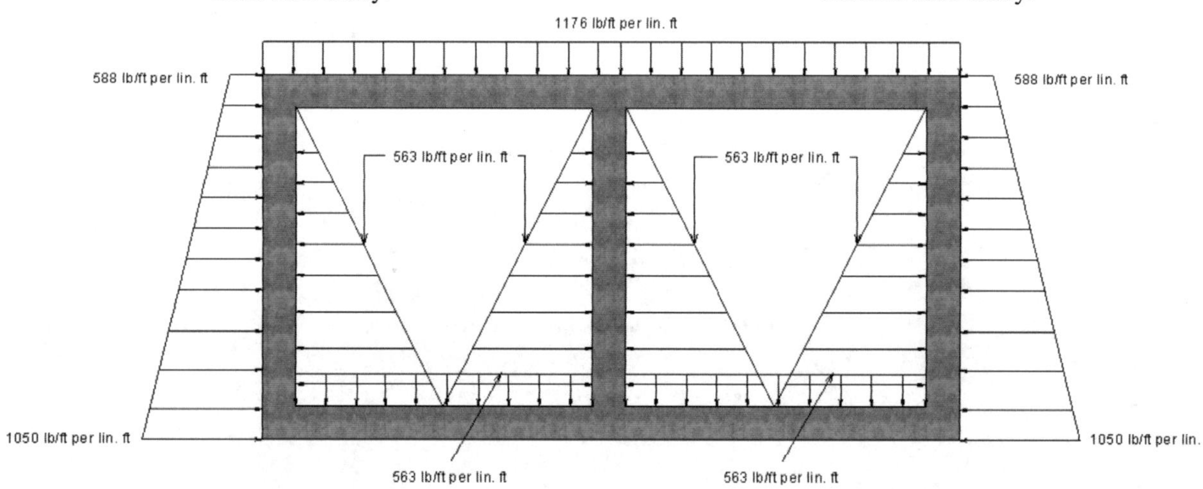

Figure 2. Loads acting on the case study culverts determined according to [5].

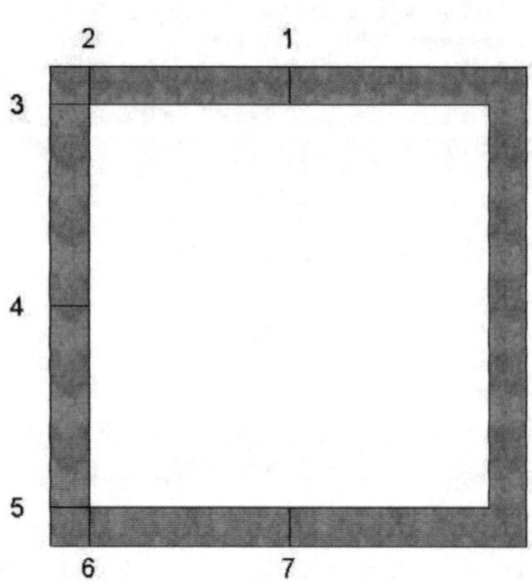

First case study.

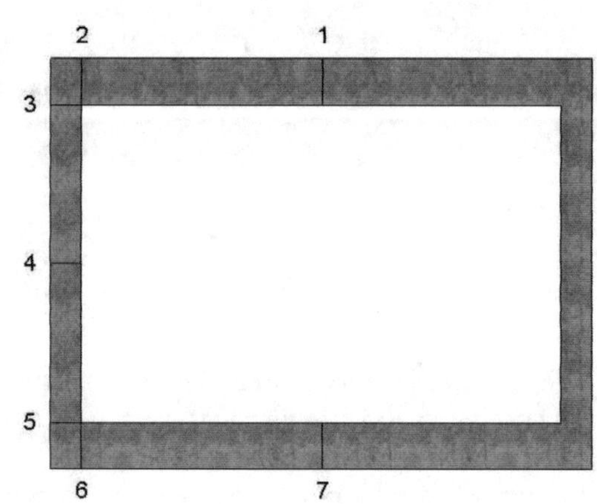

Second case study.

Figure 3. Points where internal forces have been determined for the first case study.

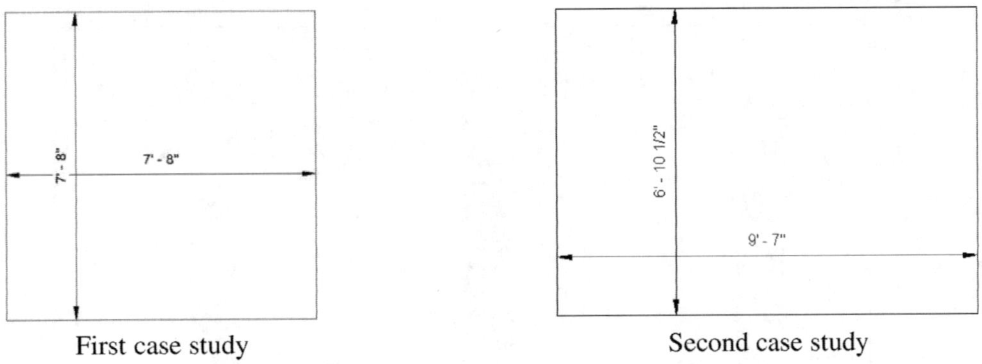

Figure 4. Frame element simulation for the culverts.

Figure 5. Plane-strain element simulation for the culverts.

5. Results and Discussion

To simplify the result's visualization and interpretation, they have been presented in a bar chart to compare the variation between the three models for each case study. The presentation has been by referring to the points presented in Fig. 3 and to the load case of Load Case 1, Load Case 2, and Load Case 3 to refer respectively to the self-weight, hydrostatic pressure, and the combination of soil and truck pressures. The results for the three case studies have been presented in Fig. 7 through Fig. 14.

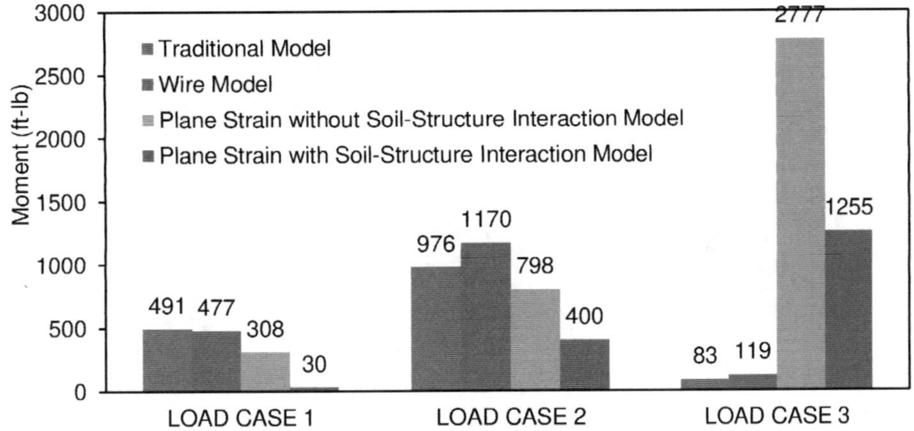

Figure 6. Moments variation at Section 4 of case study 1.

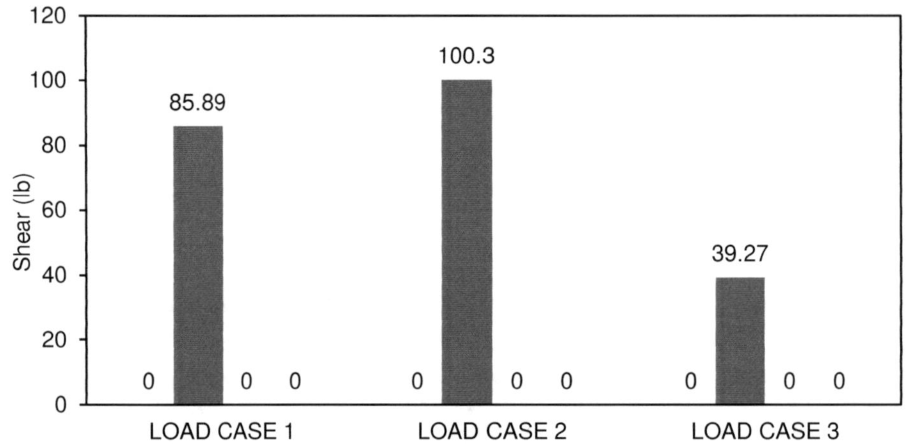

Figure 7. Shear variation in Section 4 of case study 1.

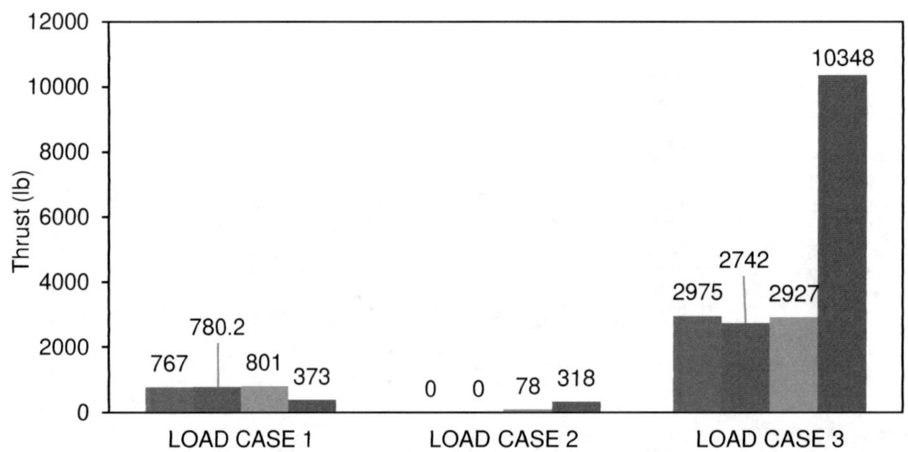

Figure 8. Thrust variation in section 4 of case study 1.

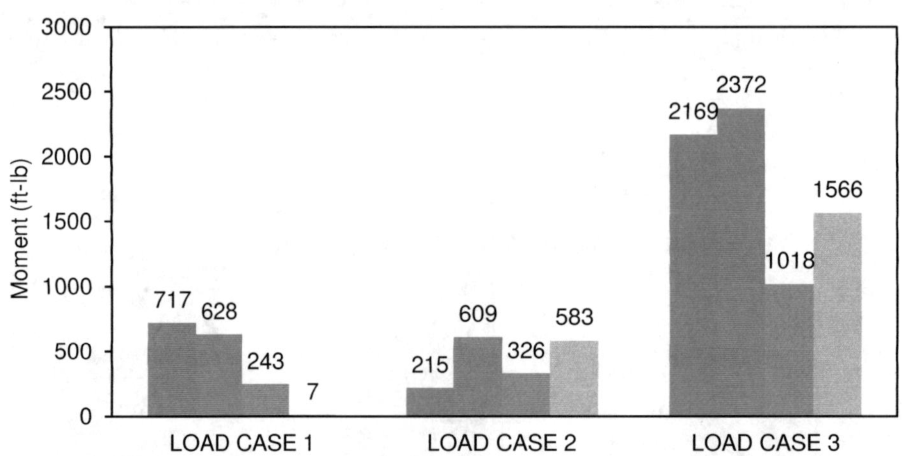

Figure 9. Moments variation in section 5 of case study 2.

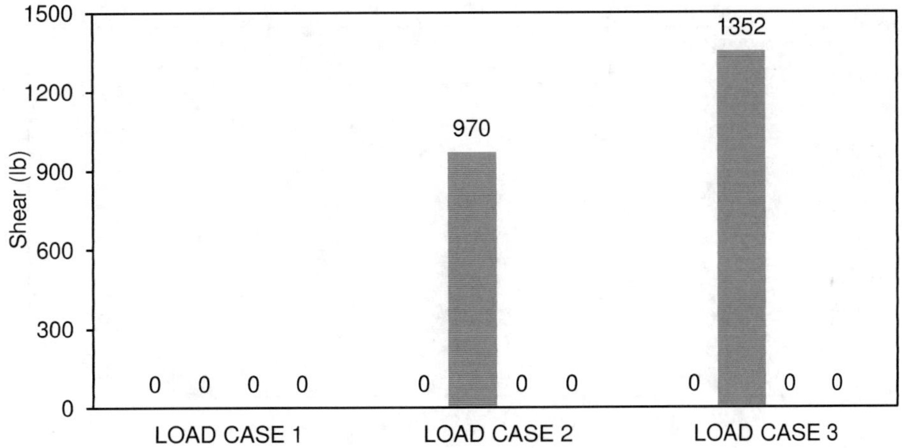

Figure 10. Shear variation in section 5 of case study 2.

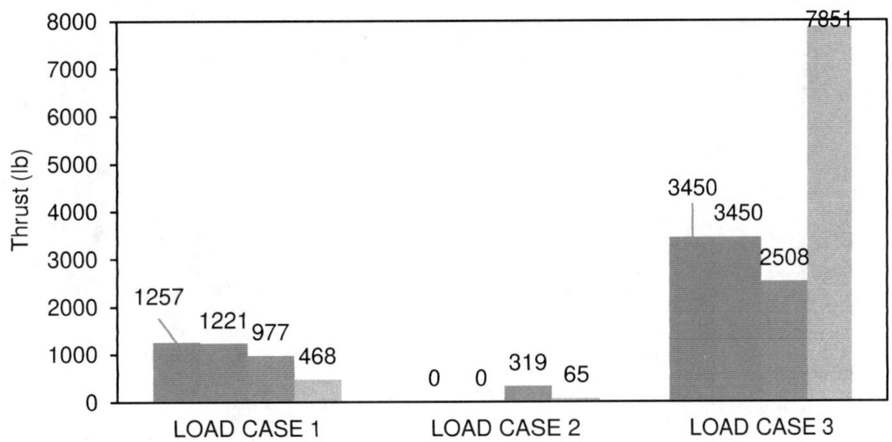

Figure 11. Thrust variation at Section 5 of case study 2.

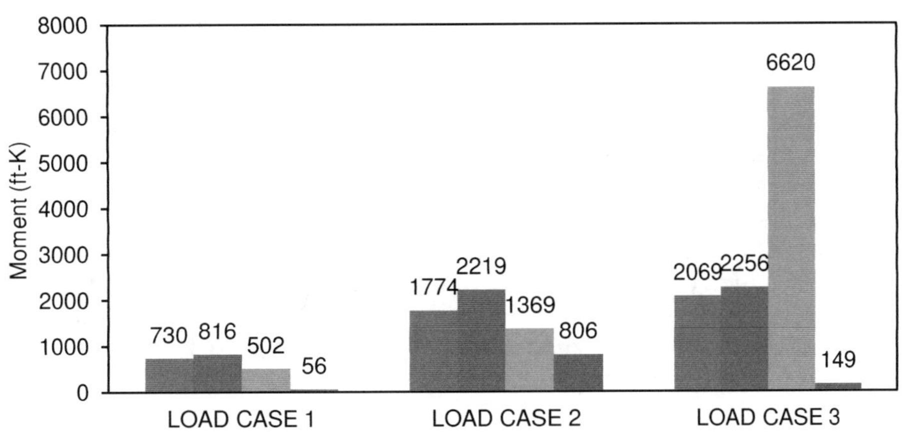

Figure 12. Moments variation at Section 5 of case study 3.

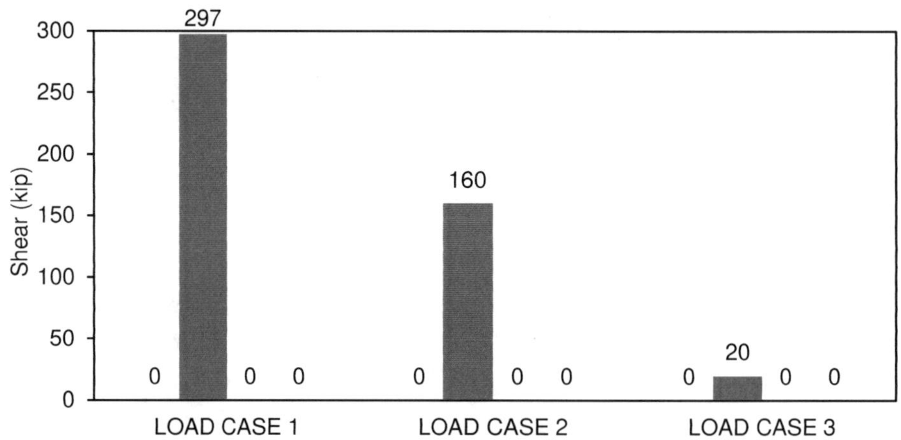

Figure 13. Shear variation in Section 5 of case study 3.

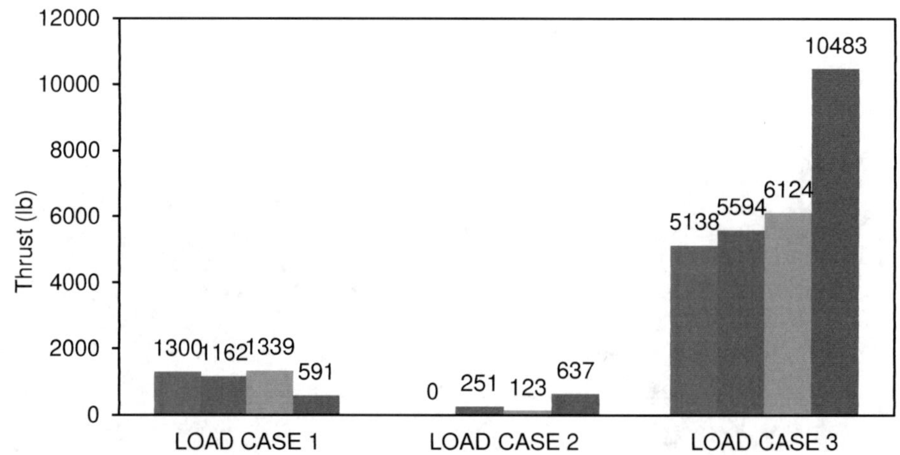

Figure 14. Thrust variation in section 5 of case study 3.

6. Conclusions

The figures above indicate significant differences between different analysis models due to the differences in adopted assumptions for each model. Consider the first case study for an instant. From Fig. 4, one can see that for the first load case, which represents the culverts self-weight, the results for the first three models are close to each other. Insignificant axial and shear deformations in this section make the adopted assumptions more representative and the traditional analysis more valid. Moreover, for the more sophisticated soil-interaction model, a significant difference can be seen compared with the other models due to the significant effect of the soil-structure interaction. In load case three, which represents the combined effect of soil weight, truckload, and the lateral pressure from the earth and surcharge load, a more profound difference can be observed between the first two models and the second two models. The first two models with the traditional analysis and frame element models neglecting the shear deformations affect the calculated stress as the finite element method determines the stresses from the strains. These stresses are, in turn, affecting the moment in those sections.

References

[1] M.E. Harr, Foundations of Theoretical Soil Mechanics., McGraw-Hill Book Company, 1966.

IOP Conf. Series: Materials Science and Engineering **901** (2020) 012009 doi:10.1088/1757-899X/901/1/012009

[2] K. Kim and C. Yoo, Design Loading for Deeply Buried Box Culverts, Alabama: Highway Research Center Auburn University, 2002.
[3] G.F. Sowers, Introductory Soil Mechanics and Foundation: Geotechnical Engineering, Fourth Edition., Macmillan Publishing Co., Inc., 1976.
[4] M.G. Spangler and R.L. Handy, Soil Engineering, Fourth Edition., Harper&Row, Publishers, New York, 1982.
[5] PCA, Concrete Culverts and Conduits., Illinois: Portland Cement Association, 1975.
[6] F. Azizi, Applied Analysis in Geotechnics., E&FN Spon, 2000.
[7] R.E. Hunt, Geotechnical Engineering Techniques and Practice., McGraw Hill., 1986.
[8] AASHTO, AASHTO LRFD Bridge: Design Specification., American Association of State Highway and Transportation Officials., 2012.
[9] J.E. Bowles, Foundation analysis and design, 5th Edition., McGraw-Hill Companies, Inc., 1996.
[10] K.J. Bathe, Finite Element Procedures., PHI Learning Private Limited., 1996.
[11] G.S. Vegard and N. Steinar, Development and Implementation of Effective Stress Soil Models, Norwegian University of Science and Technology Department of Civil and Transport Engineering, 2014.
[12] M.G. Katona, A simple contact–friction interface element with applications to buried culverts, International Journal for Numerical and Analytical in Geomechanics, 7(3) 1983 371-384.
[13] J. Duane, R. Robinson, and C.A. Moore, Culvert-soil interaction finite element analysis, Journal of Transportation Engineering, 112(3) 1986 250-263.
[14] G.A. Leonards, C.H. Juang, T.H. Wu, and R.E. Stetkar, Predicting performance of buried metal culverts, Transportation Research Record, TBR., 1985 42-52.
[15] B.M. Das, Principles of Foundation Engineering, 7th Edition, CENGAGE Learning, 2010.

IOP Conf. Series: Materials Science and Engineering **901** (2020) 012011 doi:10.1088/1757-899X/901/1/012011

Two-Way Bubbled Slabs Strengthened by Prestressing Strands

Durai H. Al-Rashedi[1,a*] and Ali H. Ali Al-Ahmed[1,b]

[1]Department of Civil Engineering, University of Baghdad, Baghdad, Iraq
[a]durainasser@gmail.com, [b]ali_hussein_alahmed@yahoo.co.uk

Abstract. Nine proposed two-way simply supported slabs were numerically tested by finite element method using ABAQUS commercial software to investigate the behavior of two-way slabs strengthened by prestressing strands, one-third of them were solid and the other two-third were bubbled. All slabs have the same dimensions of (1600 ×1600 ×140) mm. Some of these specimens were strengthened by using two or three strands with nominal diameter (6.4) mm in both directions. The slabs were divided into three groups depending on the type of slab (solid or bubbled) and the bubble diameter to slab thickness ratios (D/H). From the numerical tests, it was found that the ultimate load capacity for the bubbled slab at D/H ratio 0.64 and 0.80 was decreased to 12.6 and 14% respectively, while, the maximum deflection for bubbled slab at D/H ratio 0.64 and 0.8 was increased to 3-4 times as compared with the conventional solid slab.

Keywords: Two-way slabs, bubbled slabs, prestressed strands, flexural failure.

1. Introduction

Implementation of plastic balls in reinforced concrete (RC) two-way slabs is considered one of the latest ways to reduce slabs weight. This type of slabs is called bubbled slab system which is widely used in Europe, that's due to a reduction in self-weight up to 35%, increasing the span of slabs and reducing the total cost of the building [1-5]. When long spans are required, post-tensioned prestressed strands could be used to control the deflection and to increase slabs capacity [6, 7, 8, 10, 11]. Many studies and tests were performed to investigate the flexural and shear behavior of bubbled slabs. It was found that the Bubble Deck slab for the same strength of solid slab had (87, 66) % bending stiffness and concrete volume respectively [12]. Due to the presence of the plastic bubbles which affect negatively on shear strength as a result of removing a significant amount of concrete, the reduction factor for the bubbled slab shear capacity is taken (0.6) of the solid slab [13]. For this reason, it is recommended to remove the bubbles around the columns, as shown in Fig. 1 [7].

2. Proposed Slab Models

Nine two-way RC slab proposed models were numerically tested in this research. These slabs are with proposed dimensions of (1600×1600×140) mm and are divided into three groups depending on their type (solid or bubbled). Group one includes three solid slabs. While, Group two and Group three consist of three slabs having a bubble diameter of 90 mm and 112 mm, So the bubble diameter to depth ratio (D/H) are 0.64 and 0.8 respectively, for each group, one slab is considered without strengthening and the others are supposed to be strengthened with 2×2 or 3×3 unbonded prestressed strands. Table (1) describes the proposed slabs designations and details.

Figure 1. Bubbled slab system.

Table 1. Characteristic of the proposed tested slabs

Group	Design specimen	Type of Specimen	Bubble Diameter (mm)	(D/H)*	Number of Bubbles	Weight reduction (%)**	Number of strands in both directions
I	SD1	solid	N.A	0	N.A	0	N.A
	SD2	solid	N.A	0	N.A	0	2
	SD3	solid	N.A	0	N.A	0	3
II	BD1	bubbled	90	0.64	169	27.25	N.A
	BD2	bubbled	90	0.64	169	27.25	2
	BD3	bubbled	90	0.64	169	27.25	3
III	BD4	bubbled	112	0.80	100	33.61	N.A
	BD5	bubbled	112	0.80	100	33.61	2
	BD6	bubbled	112	0.80	100	33.61	3

* refer to a bubble diameter to slab thickness ratio
** % reduction in weight = (w. solid slab-w. bubbled slab) / w. solid slab x100, [9]

3. Proposed Specimens Details

All slab models are proposed to be designed with rebar Ø 4 @ 50 mm for the main reinforcement in both directions. This reinforcement produced a steel ratio of 0.0022 which is more than the minimum reinforcement as recommended by the ACI Code 318M-2019 [14]. The concrete compressive strength fc' and steel yielding stress fy are considered to be 25 and 420 MPa respectively. For strengthening slabs, while, the prestressing strands grad 250 low relaxation with nominal diameter (6.4) mm respectively.

4. Modeling of Specimens in ABAQUS Computer Program

Three proposed groups of two-way RC simply supported slabs are implemented in the finite element analysis (FEA) using ABAQUS computer program. As mentioned before, each group consists of three specimen models, one is without strengthening and the others are strengthened with 2×2 or 3×3 unbonded prestressed strands. Depending on bubble to depth ratio (D/H), Group one (G1) having B/H=0 (solid slabs) and group two (G2) having B/H=0.64 (slabs with 90 mm bubbles), while group three (G3) having B/H=0.80 (slabs with 112 mm bubbles). Figures 2, 3, and 4 show the details and finite element modeling for the proposed slabs. Moreover, these slabs are analytically subjected to five concentrated loads up to failure, one of these loads is subjected at the center of the models and the other four loads are subjected at the middle quarter of the model slabs. To prevent local failure, steel plates of dimensions (100 ×100 ×10) mm are located under the applied loads. While, steel plates with dimensions (140 ×70 ×10) mm are located at the ends of each strand and contacted the model side face.

IOP Conf. Series: Materials Science and Engineering **901** (2020) 012011 doi:10.1088/1757-899X/901/1/012011

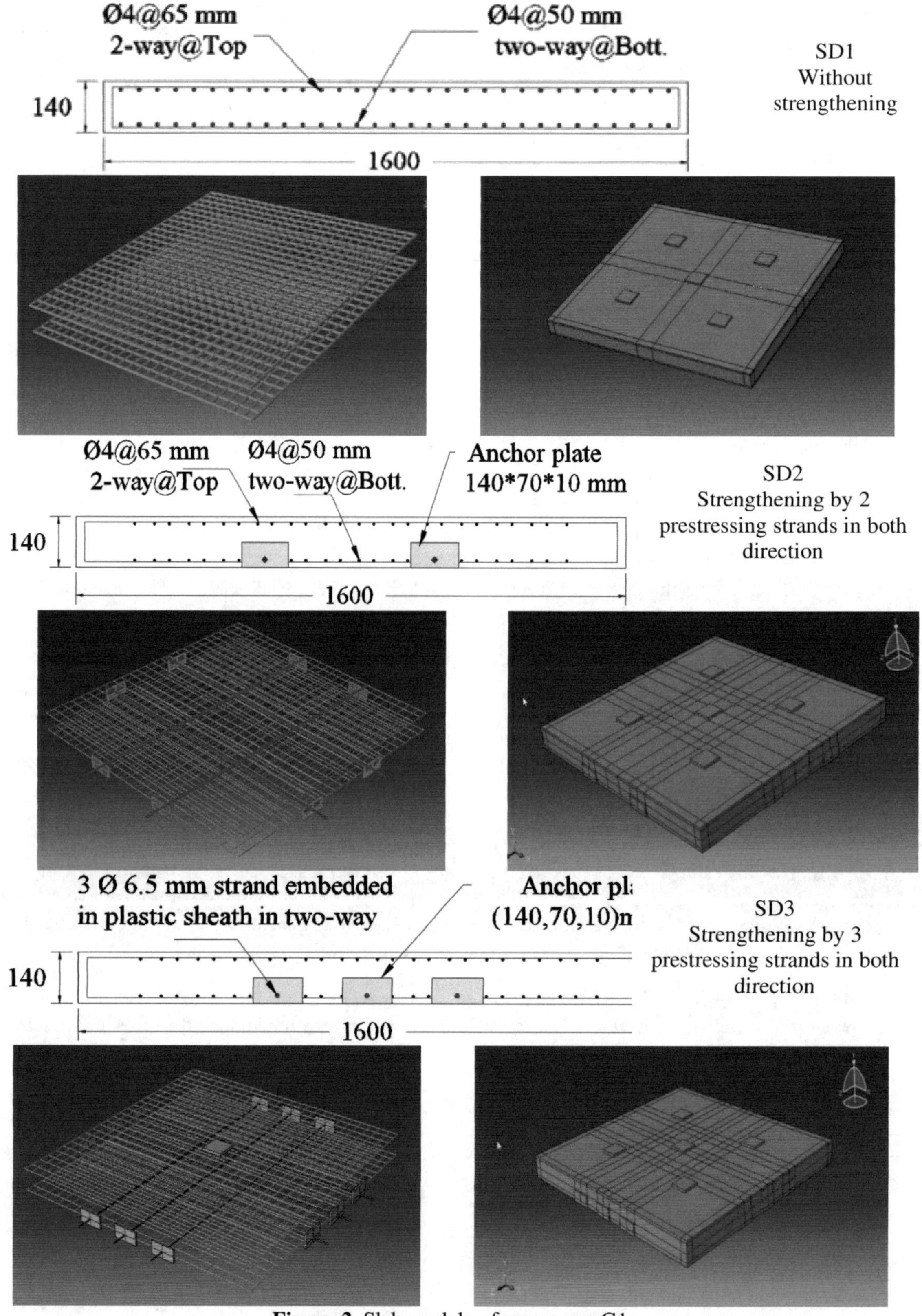

Figure 2. Slab models of group one G1.

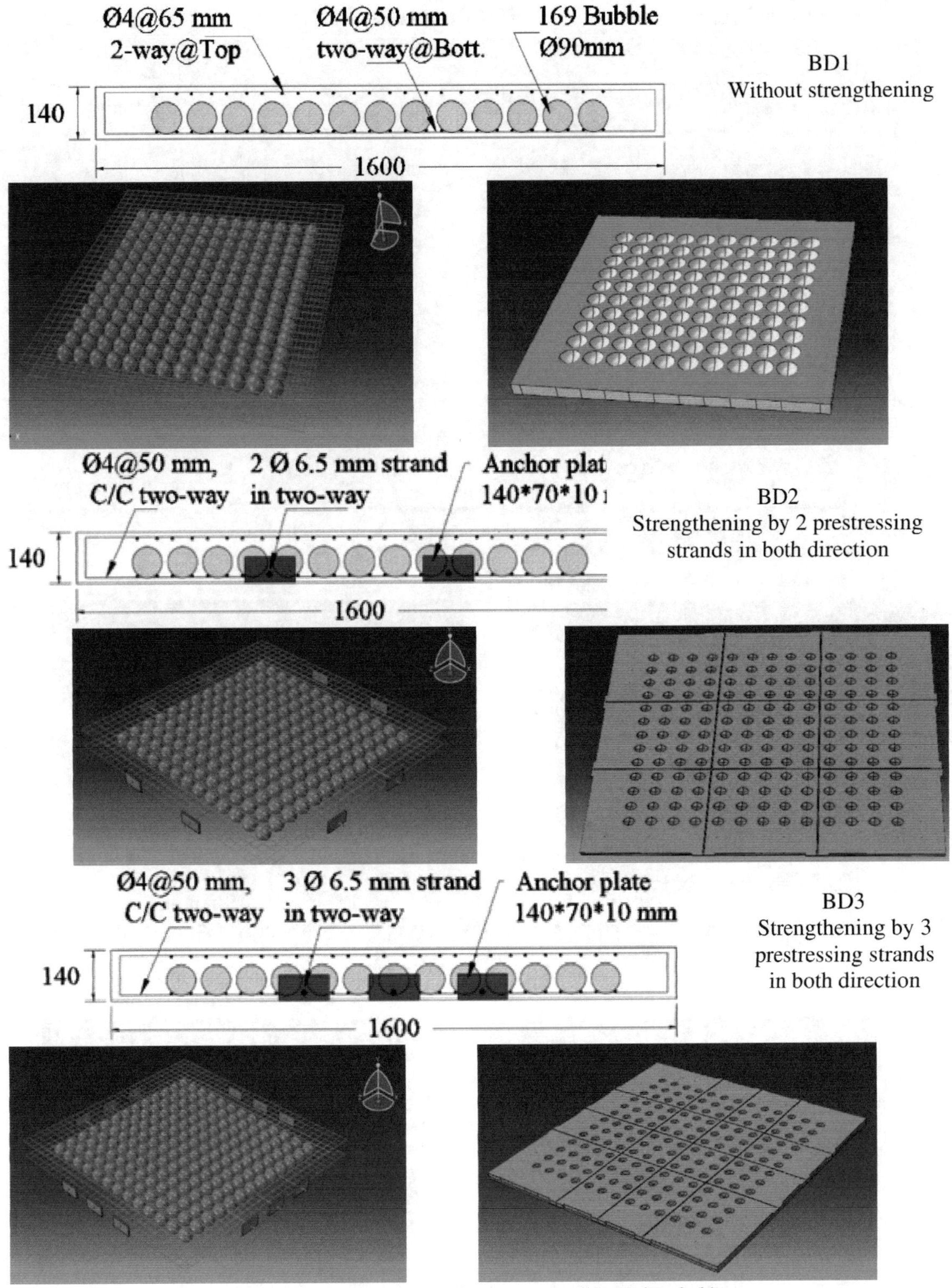

Figure 3. Slab models of group two (G2) (D/H=0.64).

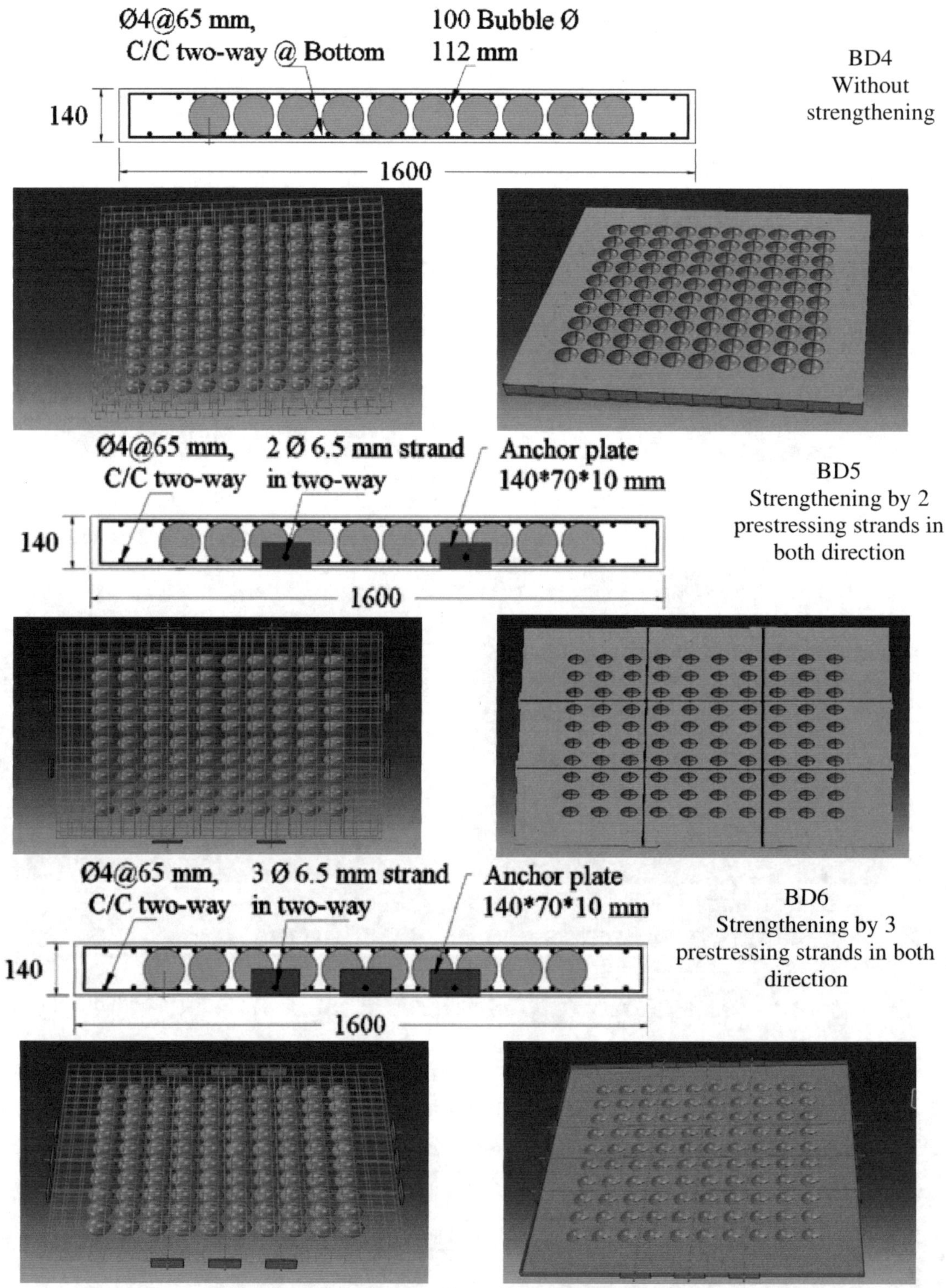

Figure 4. Slab models of group two (G2) (D/H = 0.80).

Steel strands are modeled as shown in Fig. 5. The strand is modeled through a cylindrical sheath having a diameter 2 mm larger than the diameter of the proposed strand to simulate the actual behavior of the unbounded prestressed strands. It is worth to mention that each proposed strand is stressed at a level of 65% from its yield stress.

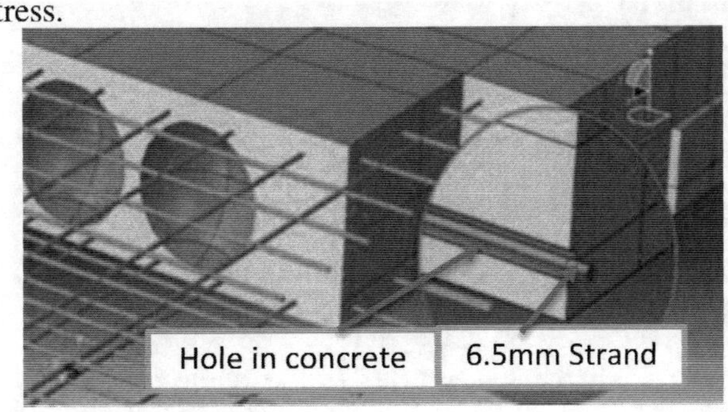

Figure 5. Proposed strands modeling in ABAQUS computer program.

5. FEA Results

5.1. Ultimate Load Capacity

Table 2 illustrated the results of ultimate load capacity and deflection for slabs numerically tested by finite elements analysis, FEA. Where group one B/H (0), showed a higher ultimate load than the other two groups (G2, G3), B/H (0.64, 0.80) by (14.5-16) % respectively. And group three showed dropping in ultimate load than the group two by 2% when increasing the ratio of bubble diameter to slab thickness from (0.64 to 0.80). From the other hand, there is increasing in ultimate load for slabs in the same group when increasing the number of prestressing strands, from (0 to 3) in both directions, this increase can be evaluated as below: The group one G1, there is an increase by (88-98)% for SD2 and SD3, respectively. Group two G2, the increase is (56- 73) % for Bubbled Deck (BD2, BD3), respectively, and group three G3, the increase is (40- 55) % for Bubbled Deck (BD5, BD6) Sequentially.

Table 2. Ultimate load capacity and ultimate deflection for slabs models.

Group	Slab designation	B/H	Ultimate load (kN)	Ultimate deflection (mm)	Camber (mm)	Net deflection (mm)
	SD1	0	285	8.3	---	8.30
1	SD2	0	538	6.5	0.0524	6.45
	SD3	0	564	5.4	0.0918	5.30
	BD1	0.64	249	6.5	---	6.50
2	BD2	0.64	389	8.5	0.0808	8.42
	BD3	0.64	430	9.0	0.123	8.88
	BD4	0.80	245	7.3	---	7.30
3	BD5	0.80	345	7.9	0.0885	7.81
	BD6	0.80	380	7.32	0.135	7.20

5.2. Load-Deflection Response

Deflections are read at the center of the slabs model. The load-deflection curves are plotted to show the behavior of the slabs when using prestressing strands as shown in Fig. 6. Also, the results showed there is a decrease in deflections for all groups when increasing the number of strands from (0 to 2×2 or 3×3) as following; G1 by (55-65) %, G2 by (50-70) % and G3 by (60-63) %. On the other hand, an increasing number of strands from (2×2 to 3×3), cause an increase the camber of the prestressed solid slab by (75) % and by (52) % for both types of Bubbled slabs of ratios of B/H (0.64, 0.80). While Fig. 7 introduces the effect of D/H ratio. The results showed an increase in the deflections of the Bubbled slabs in comparison with solid slabs by (3-4) times the deflection of solid slab, and an increase deflection with increasing the ratio of D/H from (0.64 to 0.80) by 28%, when reading deflections at 60% of the ultimate load.

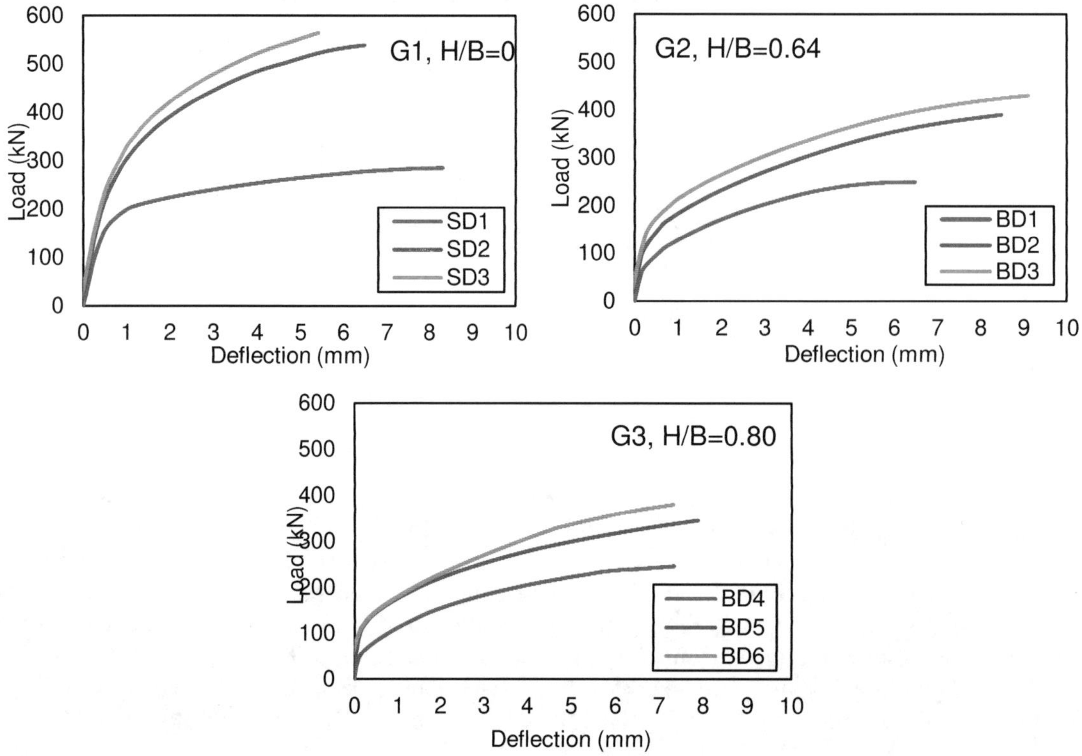

Figure 6. Load-deflection curves for slabs (Effect number of strands).

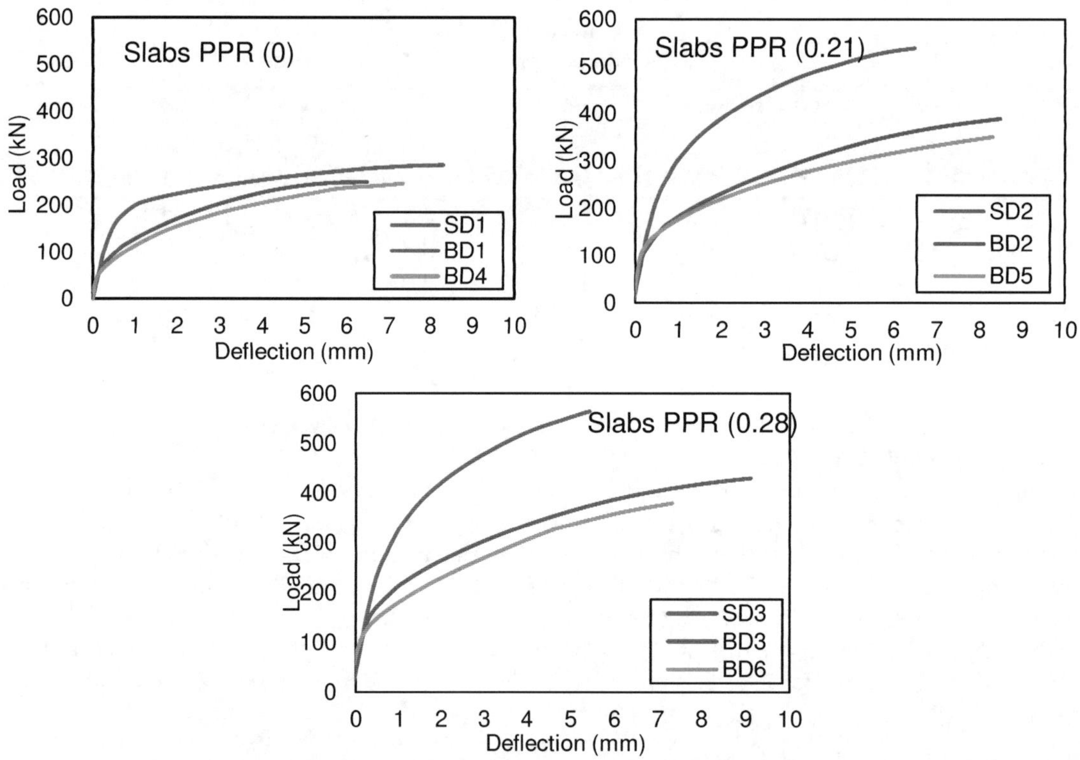

Figure 7. Load-deflection curves for slabs (Effect of B/H).

5.3. Cracks Pattern

After reviewing the failure patterns of the models shown in Fig. 8, the following conclusions are noted;

- The mode of failure is almost the same for all slab's models will be flexure failure. The cracks appear as a green region start from the center and extend diagonally toward the corner to make (x) shape.
- For the slabs models without strengthening, (SD1, BD1, BD4), the cracks manifest clearly and with great intensity as wide strip forming (x) shape. These cracks decrease and diverge among themselves with an increase in strands number from (0 to 3), this because of the prestressing tendon effect significantly on increasing the flexure stiffness of slabs and decreasing the cracks.
- The number of cracks were increased with increasing of the bubble diameter to slab thickness ratio (B/H) as compared with the conventional solid slab.

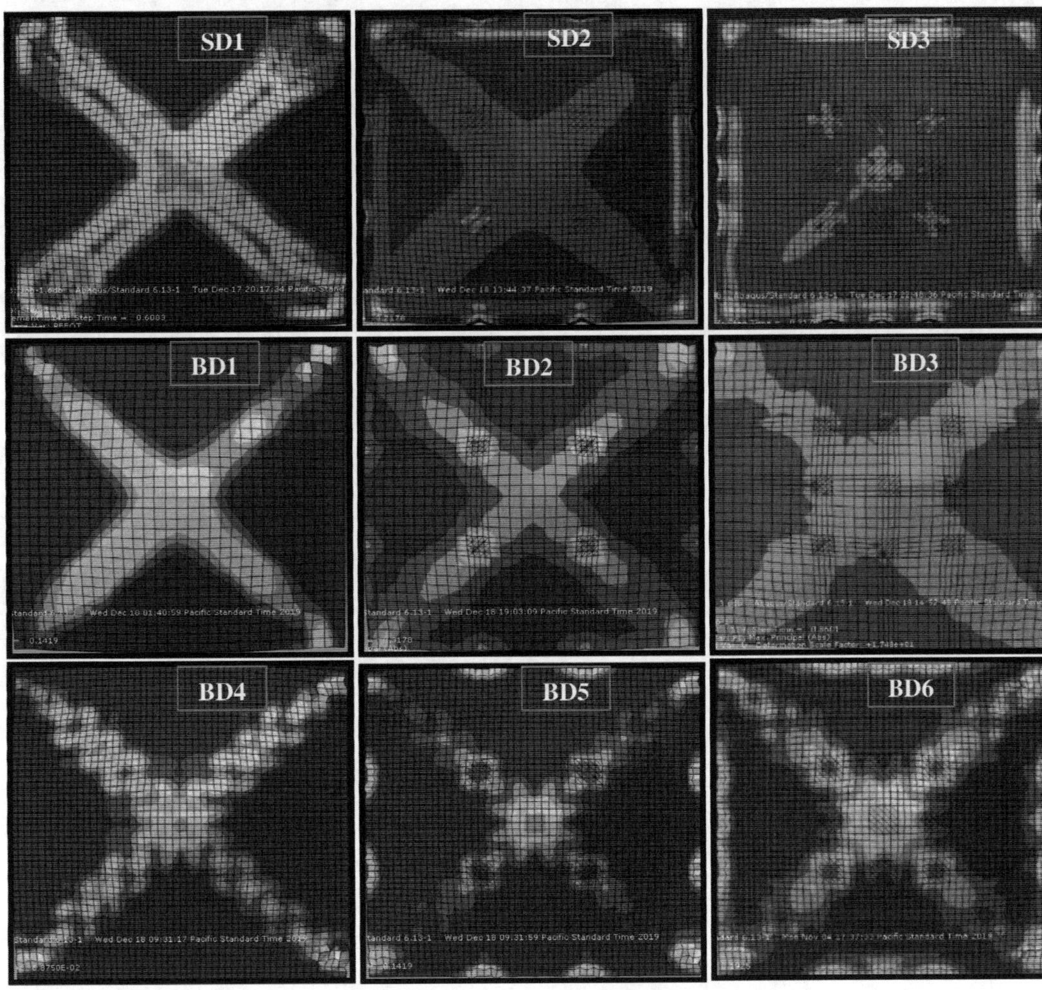

Figure 8. Cracks pattern of slabs models.

6. Summary and Conclusions

From the analytical results obtained in this study the following conclusions can be drawn:

- The weight of bubbled slabs with bubble diameter to slab thickness ratio (D/H), 0.64 and 0.80 can be reduced to 27% and 34% respectively, as compared with conventional sold slab.

- The Bubbled slabs that have ratios of D/H (0.64, 0.80) % can carry (87, 85) % respectively of the ultimate load capacity compared with the reference solid slab.
- Strengthening the Bubbled slabs have ratios of B/H (0.64, 0.80), by increasing the number of prestressing strands from (0 to3) enhance the ultimate load capacity by (40-73) % and reducing the deflection by (50-70) %. The same applies to the solid slabs when increasing the strands from (0 to 3) in both directions leads to increasing the ultimate load by (88-98) % and reducing the deflection by (67-70) %.
- The mode of failure for all slabs is the flexural mode and the pattern of the cracks takes the form of a letter (x) that extends to the corners of the models.

References

[1] A. Churakov, Biaxial hollow slab with innovative types of voids, submitted to Journal of Construction of Unique Building and Structure (2014).
[2] T. Lai, Structural Behavior of Bubble Deck Slabs and Their Application to Lightweight Bridge Decks. M.Sc. thesis, Dept. Civil. Eng. MIT, Cambridge, Jun (2010).
[3] A.V. Joseph, Structural Behavior of Bubble Deck Slab. M.Sc. thesis, Dept. Civil. Eng. Mahatma Gandhi Univ., Indian, Nov. 2015.
[4] S. Calin and C. Asavoaie, Method for Bubble Deck Concrete Slab with Gaps, published by Technical University "Gheorghe Asachi", Iaşi, Romania, (2009).
[5] Kyng Consulting, Bubble Deck Design Guide for compliance with BCA using AS3600 and EC2, Western Australia, DG_v.1.2, 31 Oct. (2008).
[6] N. Oukaili, H. Yaseen, Effect of Prestressing Force on Performance of Polymer Bubbled Deck Slabs, submitted to the 7th International Conference on FRP Composites in Civil Engineering, International Institute for FRP in Construction, Vancouver, Canada, (2014).
[7] C.J. Midkiff, Plastic Voided Slab System Application and Design. M.Sc. thesis, Dept. Arch. Eng., Kansas State Univ., Kansas State, USA, (2013).
[8] P. Sokal, Design of Two-Way Pre-stressed Slabs: submitted to 14th International Conference of Ph.D. Students JUNIORSTAV, At Brno, Czech Republic, (2012).
[9] M.S. Mushfiq, S. Saini, N. Rajoria, Experimental Study on Bubble Deck Slab: submitted to International Research Journal of Engineering and Technology (IRJET), (2017).
[10] A.E. Naaman, Prestressed Concrete Analyses and Design, 2nd edition, Ann Arbor, Michigan 48105, USA, (2004).
[11] BubbleDeck UK (2008, Sep.) Ltd, Bubble Deck Structure Solutions, Information on: http://www. BubbleDeck-UK.com
[12] G. Björnson, BubbleDeck Two-Way Hollow Deck, Sep. 2003. Information on: http://www.bubbledeck- uk.com/pdf/BDOverview9-03.pdf
[13] BubbleDeck UK (2006, Apr.). BubbleDeck Slab Properties, Information on: http://www.BubbleDeck-UK.com
[14] Building Code Requirements for Structural Concrete, 318, 2019.

Simulation of Sediment Transport in the Upper Reach of Al-Gharraf River

Mariam Hadi Daham[1,a*], Dr. Basim Sh. Abed[1,b]

[1]Department of Water Resources Engineering, University of Baghdad, Baghdad, Iraq.
[a]maryam.hadi95@gmail.com, [b]drbsa62@ucoeng.uobaghdad.edu.iq

Abstract. For the past few years, the sediment began to accumulate in Al-Gharraf River which reduces the flow capacity of the River. In the present research, a numerical model was developed using Hec-Ras software, version 5.0.4. to simulate the flow and sediment transport in the upper reach of the river. The hydrological and cross-section data measured by the Ministry of Water Resources, for the reach located between Kut and Hai cities and having a length of 58200 m, was used to perform calibration and verification of the model. Moreover, field sampling of suspended and bed loads was gathered for five months starting from 7/2/2019, and laboratory tests of samples were conducted to be used as input data in the model. Actual flowrates for the year 2018 were used to predict the sediment accumulation in the river, and the results show that the accumulated deposition was ranged between 2.5 cm to 1m, while the erosion depth ranged between -4cm to -2cm. Furthermore, the estimated amounts of sediment discharges were ranged between 70 and 1308 Tones/day. Then four scenarios of flowrates ranged between 100-300 m^3/s were used to analyze the possible sedimentation problems and to propose the treatment needed to preserve the morphology of the river The results show that the invert changes of bed depth were ranged between -2 to 9 cm, and the sediment discharge ranged between, 33 to 970 Tones/day. A lining of specified sub-reaches using the concrete or by the grouted riprap was proposed to avoid sedimentation or erosion respectively.

Keywords: Al-Gharraf river, sediment transport modeling. Hec-Ras 5.0.4.

1. Introduction

The sediment accumulation in Al-Gharraf River became one of the important problems that concern the Ministry of Water resources due to its significant influence on the River abilities of convenience and the benefits of it. There were no previous studies concerning the sedimentation in the upper reach of Al-Gharraf River. Kayyun and Dagher [1] formulated a model using Hec-Ras software to simulate the sediment transport within the Tigris River located between Al- Muthana Bridge and ends at the Sarai Baghdad gauging station located in the center of Baghdad City, the model simulated from 2012 to 2017. The results indicated that erosion was found in the upstream and the downstream of the River reach with a depth equal -0.51m, And the deposition located in the center of the stream with depth equals 0.39 m. Safi et al. [2] simulated a quasi-unsteady analysis model to simulate the sedimentation and erosion in Kunar River which located in Afghanistan. The results showed that Kunar River erosion occurs in upstream of the reach with maximum erosion depth about -2.449 m and sedimentation happens in downstream of reach with maximum deposition depth about 1.495 m.

Khassaf and Abbass [3] studied and estimated the transporting amount of sediment using Hec-Ras software, in upstream of Al-Yaa'o regulator, located on Euphrates River within Al-Qadissya City. The conducted maximum amount of total sediment load from the model was (248419) Tons/year. Mohammed et al. [4] studied sediment transport capacity in Euphrates river at Al-Abbasia reach using

Hec-Ras model, the result of simulating the model for one year shows that the average sediment discharge was 50 Tones/day. Mustafa et al. [5] constructed an HEC-RAS model to study the sediment transport for a specified reach in the Euphrates River, the study reach located between Haditha dam and Heet station which located in Heet city. The results of the simulated model for one year showed that the total sediment discharge was 237.38 Tons/day. (Dysarz et al. [6] created a model using Hec-Ras to study the deposit accumulation during 10 years in Ner River located in Poland, the result showed the average depth of deposition accumulation was 0.5 m.

There are several solutions to treat the deposition in the streams like statistical methods, or numerical modeling for the sediment analysis, either way, the sediment samples must be gathered to identify the bed type and measure the concentration of sediment in the River. AL-Gharraf River is the largest tributaries of Tigris River there for it's the main resource of water for the cities lies in the middle and south of Iraq. The selected reach for this study extends from Kut city, downstream of Kut Barrage and it ends near Hai city with reach length equal to 58200 m. The reach of the River includes important structures like Gharraf Head Regulators and Hai Regulator and also includes few subdivisions. In this study, only subdivisions that have a flowrate value greater than 1 m^3/s were selected, which are Janabyat Al Hai, Gamela, and Um al Tubaikh, Fig.1. This research aims to create a numerical model to simulate the sediment transport in the upper reach of Al-Gharraf River and find out the suitable treatment for sedimentation problems in that reach.

Fig.1 Satellite image of the upper part of Al-Gharraf River, Google earth

2. Basic Equations Concern Sediment Modeling

There are basic equations concerning sediment transport in the Rivers. Firstly, the sediment continuity equation, from [7]:

$$-\frac{\partial QS}{\partial X} = (1-\lambda P)\, b\, \frac{\partial \eta}{\partial T} \qquad (1)$$

Where; b= channel thickness in m, η= channel altitude in m, λP= porosity, QS= sediment discharge in Tones/day. The Following equation was chosen in sediment transport function called the Laursen-Copeland transport function equation, from [7]:

$$C_M = 0.01\gamma(\frac{ds}{D})^{1.167}\left(\frac{\tau}{\tau C} - 1\right) f(\frac{u*}{\omega}) \quad (2)$$

Where: γ = the water-specific weight in kg/m^3, ds= mean diameter of sediment in mm, D = Effective depth for the flow in m, τ = the shear tension of channel in pa bottom, τ_C = Critical shear stress of channel bed in pa. ω= fall velocity of deposit in m/s, u $*$=Shear velocity m/s, and, C_M = concentration of Deposit flowrate Tones/day.

3. Inputs and Boundary Conditions

In this study, the Hec-Ras, version 5.0.4, the software was used to simulate the sediment transport in Al-Gharraf River. This software was adapted in the United State in, February 2016, by the US Corp of Engineers to analyze the schemes of the Rivers.

3.1. One-dimensional steady flow modeling

The one dimensional model in the steady-state must be created first and calibrated before simulating the sediment transport model, A schematic of the River was entered and extends from the upstream of the Gharraf Head Regulator, at station 58200m and it ends near Hai city at station 100m.that having a total number of 24 stations through an interval space ranges from 100 to 1000m between each cross-section. All the survey data were measured by MOWR [8]. Even though the hydrological data for the mainstream, outlets, and Regulators were from MOWR [8,9]. The boundary condition for the upstream was the discharges and for the downstream, the normal depth was selected with an energy slope value of 8 cm/km. While for the gate openings, actual data were taken for both Regulators that equal to 260 cm for Al-Gharraf Head Regulator and 210 cm for Hai Regulator,

3.2. Inputs of Morphological model

Hec-Ras present Quasi unsteady flow method to analyze sediment transport in the Rivers. The selection of the hydraulic model must be appropriate and the model should be calibrated using various manning roughness coefficient values. Several required data that quasi unsteady flow needed, such as the flowrates, the entered flowrates in this research were actual measured values of the Al-Gharraf River by the Ministry of Water Resources, MOWR. These data were considered as the upstream boundary condition, located at the station No.58200 m, which represents the flow time series. While The selected boundary condition for Hydraulic structures was Gate Time Series, Both of The Regulators gates openings data were taken from actual operation, for Head Regulator it ranged between 0.60 m to fully opened, and for Hai Regulator it ranged between 1m to fully opened. The adopted boundary condition at the last station of the study, station No. 100, was the Stage time series, which were the measured values also conducted by MOWR. The selected boundary conditions for subdivisions are the Normal depths, as well as the selected value of the energy slope was 8 cm/km. The mean of the measured temperature values during sediment sampling was 20°C, which was considered as input data into the quasi-unsteady window.

3.3. Sediment Transport modeling

The sediment transport Model was conducted by entering data into the following files: file of geometry, file of quasi unsteady flow, and the file of sedimentation information. The sediment data file contains the transport functions, these functions are sensitive and must be chosen carefully. In this study the bed type of the River bed was silty clay, Therefor, Laursen-Copeland was the selected function to characterize the sediment transport in Al-Gharraf River. The selected boundary condition for the upstream of the study was the rating curve which calculates the sediment flowrate and it requires the discharge of the water, the sediment load, and the grain size distribution for the upstream. Moreover, the sediment file requires the grain size distribution data for each cross-section, these inputs are the result of analyzing the bed material samples in the laboratory. A total number of six-bed gradation templates were inputted into the specified sections were the bed material samples gathered, and nine samples of suspended load were gathered, including three sets of suspended sediment samples at the upper reach of Al-Gharraf River which governed by the present study at cross-section No.1, at station

58200, to be as inputs of concentration of sediment discharge. Figure 2 illustrates the grain size distribution for the upstream of the study reach.

4. Results and Analysis

In the following articles, the calibration and the verification of the flow and sediment transport models will be illustrated. Moreover, the results of the real and assumed flow conditions will be analyzed.

4.1. Calibration and Verification of the steady flow model

The calibration process was conducted to know whether the model simulation accurate or not, several manning roughness values were tested to determine the optimal manning roughness coefficient value for the steady-state of Al-Gharraf River, The calibration process occurs by comparing the observed values measured by MOWR at Al-Hai Regulator with simulated values of the model. The measured discharge value ranged between 160-180m^3/s with a discharge interval of 5m^3/s, and water surface elevation (WSE)ranged between 12.04 to 12.7 m**,** Table 1. The tested values of manning roughness were ranged between 0.024-0.028, and the simulation time was one day. Figure 3 shows the optimal manning roughness value for steady-state was 0.026 with Root Mean square, RMSE value 0.0914, and the coefficient of determination value equal to 0.995. While the verification process was conducted by using the measurement data for the year 2018, using 0.026 the optimal value of manning roughness for the steady-state. The coefficient of determination value, R^2 was 0.99, Fig.4.

Table 1. The observed data of Hai Regulator for the year 2017.

Month	Q (m^3/s)	WSE (m)
June	160	12.04
July	165	12.1
August	170	12.2
Septemper	175	12.5
October	180	12.7

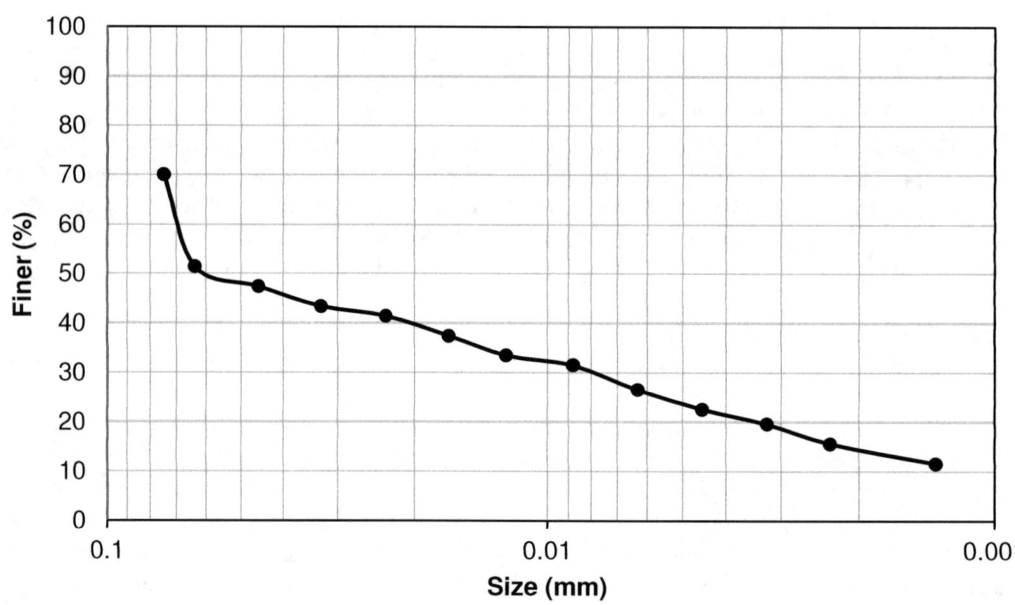

Figure 2. Particle distribution curve for bed material at the upstream reach of the river.

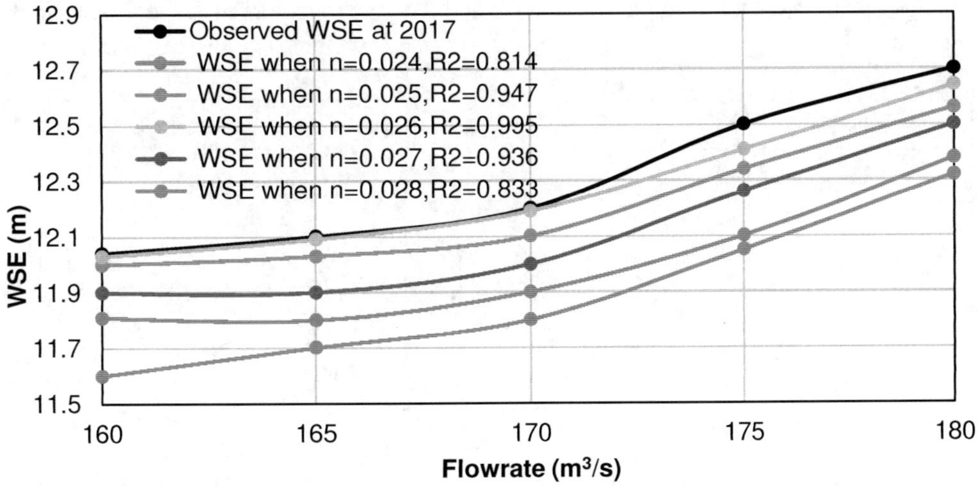

Figure 3. Calibration of the steady-state flow model for the year 2017.

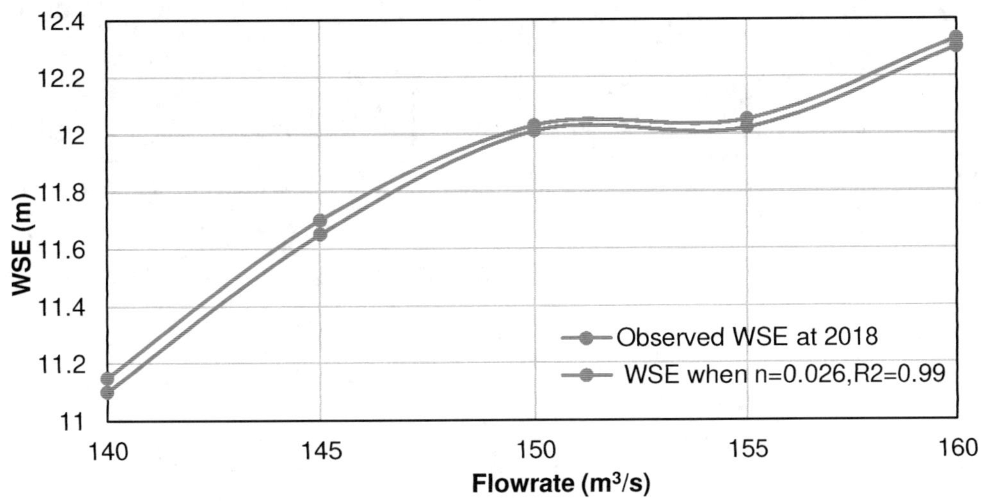

Figure 4. Verification of steady-state flow model for the year 2018.

4.2. Calibration and Verification of the Sediment transport model

The Calibration of the sediment model is a significant process to check the model validation to simulate the actual condition of Al-Gharraf River and solve the sedimentation problem. In the present study, the model was calibrated using observation measurements of the cross-section conducted by the MOWR [8] and the measurements conducted by [10]. Eight sections were selected where the significant deposition and erosion occur, to calibrate the sediment transport model, four of them were illustrated in Fig.5. In the calibration of the simulation model, the comparison will be presented for the four years periods extended from 2012 to 2016 for selected sections to show the influence of sediment accumulation or erosion occurs in those cross-sections with the measured cross-sections by Arrar [10] using a maximum accumulation depth during the simulation time equal to 48 cm, these results achieve high values of R^2 which ranges from 0.99 to 0.975. The obtained value of RMSE was 0.0353. While the verification process was conducted by performing for two different cross-sections where significant

Sedimentation and erosion also occur. The value of the R^2 value was ranged between 0.996 to 0.998 and RMSE value equal to 0.01, which was obtained when comparing the simulation results with the measured data in the other two sections, Fig.6. That indicates good agreements between these results, as well as the good ability of the simulation model to estimate the transport of the sediments in the reach of the river.

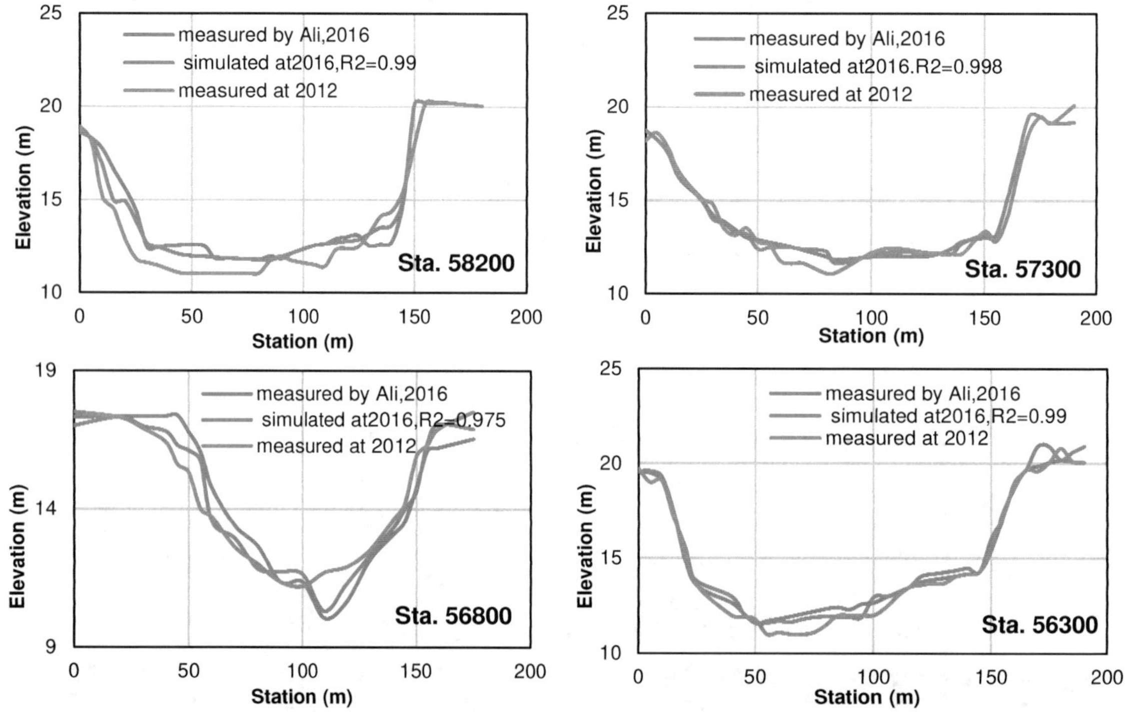

Figure 5. The calibration process of the sediment transport model in selected cross-sections.

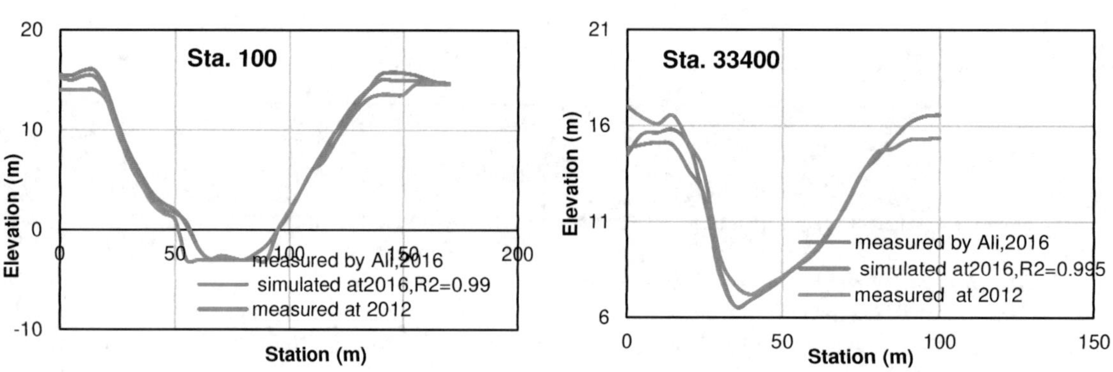

Figure 6. The verification process of the sediment transport model in selected cross-sections.

4.3. Result of sediment simulation

The sediment transport model was simulated for one year using all the necessary flow data that measured during 2018, to visualize the River sections and to conduct the invert change depth and the sediment discharge values. Invert change is the difference between the initial bed level and the predicted bed from the model simulation, where, all the negative signs in figures and tables mean erosion while the positive signs mean sedimentation. The results of simulating the model for one year showed that the maximum sediment accumulation occurs upstream of Gharraf Head Regulator. At station 58200 m, cross-section No. 1, with depth equal to 1 m for the year 2018, and the sediment discharge of it was 1308 Tones/day, while the minimum sediment accumulation depth was near Hai city, at station 100, cross-section No.

24, with depth equal to 2.5 cm and sediment discharge 70 Tones/day. The maximum erosion occurs in station 17300, cross-section No. 14, with erosion depth equal to -4 cm. Fig.7.

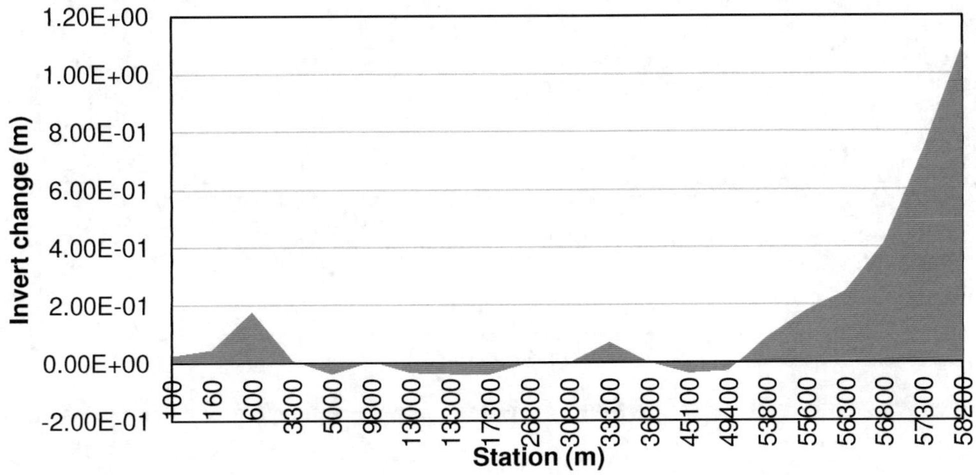

Figure 7. Accumulative changes of invert bed depth along the reach of River for the year 2018.

Another simulation of the sediment transport model was conducted at different assumed values of discharge within the capacity of Al-Gharraf River; 90-100, 190-200, 290-300 m^3/s for one month, to Handle the sedimentation problem in Al-Gharraf River. The results show that the value of invert change depth was ranged from – 2 to 9 cm, which means the maximum value of sedimentation depth, was 9 cm and the maximum erosion depth was -2 cm, and the sediment discharge ranged between 32 and 527 tones/day when 90-100 m^3/s was released Fig. 8. For the case of discharge equals 190-200 m^3/s, the invert change depth ranged from -2 cm to 9 cm also, and the sediment discharge ranged from 77 to 970 Tones/day Fig. 9. While the values of invert change and the sediment discharge in the cases of Q equals 290-300 m^3/s were -2 to 1 cm and 91 to 350 Tones/day respectively, Fig.10.

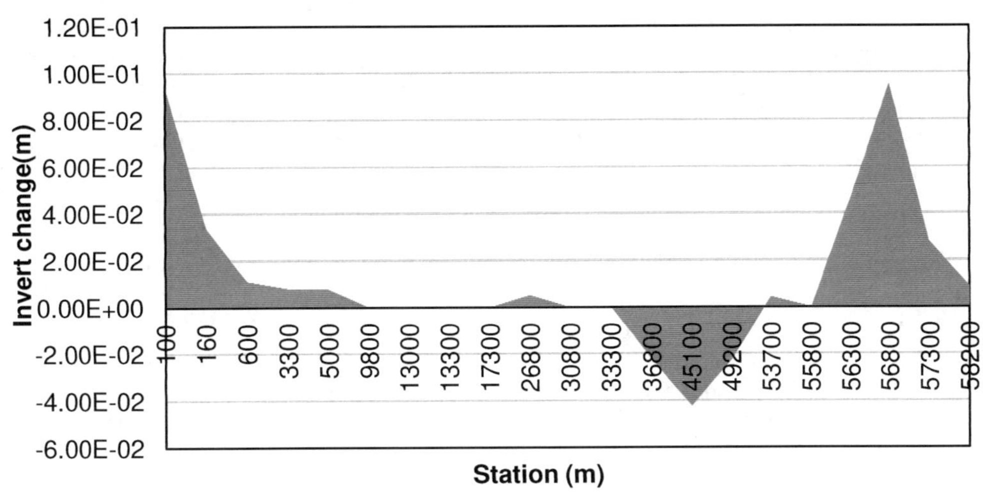

Figure 8. Accumulative changes of invert bed depth along the reach of River when Q=90-100 m^3/s.

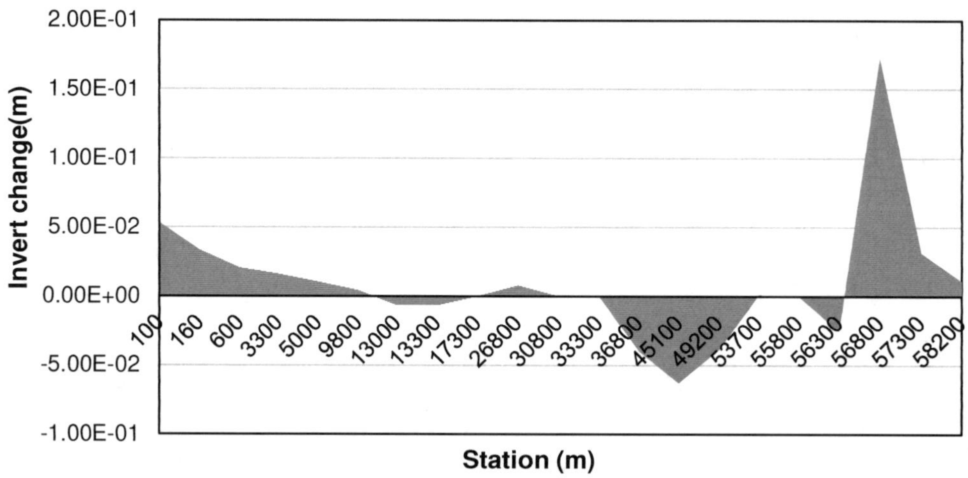

Figure 9. Accumulative changes of invert bed depth along the reach of River when Q=190-200 m^3/s.

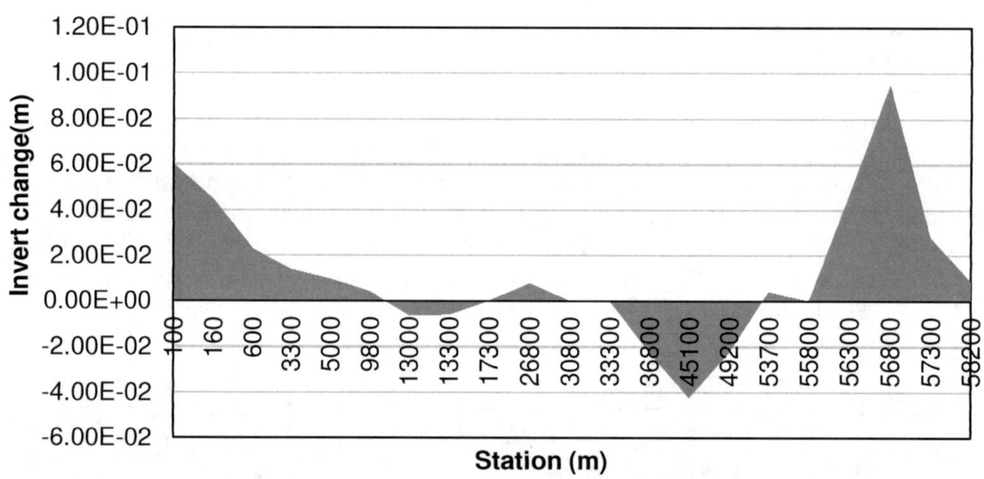

Figure 10. Accumulative changes of invert bed depth along the reach of River when Q=290-300 m^3/s.

4.4. Suggested treatments for Sedimentation problems

The Simulation during the period of field sampling extended from 7/2/2019 to 7/7/2019 using the values of the discharges that measured during that period are ranged between 250-210 m^3/s. The simulation results show the cross-sections in which the maximum deposition occurs was at RS 58200, RS 57300, RS 56800, and RS 600. And the maximum erosion occurs at RS 13000, and RS 5000, Fig.11. The values of accumulative sediment depths were ranged between 0.15 and 0.37 m, and the scouring depth values were ranged between -0.02 and -0.04 m. To minimize the sediment accumulation, the Manning Roughness coefficient (n) value must be reduced for a distance preceding the sections that sediment accumulated in. That will maximize the velocity and reduce the deposition of sediment. While in the erosion sub reaches the Manning Roughness value must be increased to reduce the velocity values which results in minimizing the scour in these sub reaches.

The actual value of Manning (n) used in the present study was 0.026. Several values of manning roughness coefficients were tested and the results obtained indicate that the suitable Manning (n) values that minimize the deposition and erosion of sediment are 0.015 and 0.028 respectively. The concrete lining with n equals 0.015 was proposed to reduce the roughness of River with a percent of 42%. While the Grouted Ripraps was proposed for the lining the erosion sub reaches with n equals 0.028 that increases the roughness of River with a percent of 8%. The obtained results from the simulation model point out that the deposit depths when the proposed lining used were reduced from 0.15 - 0.37m to 0.012-0.21 m. The resulted erosion depths were reduced from -4cm to 2 mm when the proposed lining was used.

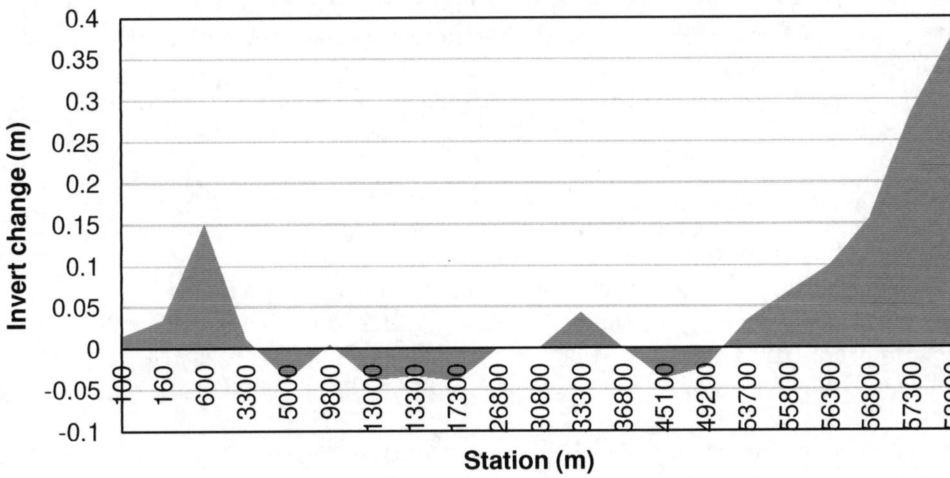

Figure 11. Accumulative changes of invert bed depth for the period (7/2/2019 to 7/7/2019).

Table 2. The treatment results of the selected cross-sections along the reach of Al-Gharraf River

Station	Invert change before treatment (m)	Invert change after treatment (m)	Percent of change
58200	0.377	0.21	44%
57300	0.283	0.101	64%
56800	0.156	0.06	62%
13000	-0.039	2.34E-03	94%
5000	-0.038	1.45E-03	96%
600	0.151	0.012	92%

5. Conclusions

The following conclusions were extracted from the results obtained from the simulation and sedimentation model:

- The developed numerical model was well simulated the flow and the hydraulics of sediment transport in the reach of the study due to good agreement between the simulated and measured results of water surface elevation and the morphology of the river.
- The suitable manning roughness coefficient that produces a good simulation of Al-Gharraf River for the steady-state was 0.026.
- To minimize the sediment accumulating occurs at specified sites along the reach of the river, lining with concrete for these sites will reduce the accumulation of sediment, and the results show that the invert changes of the governing bed dropped by a range between 44-92 %.
- To minimize the scouring occurs at some site along the reach of the river, lining with grouted riprap was proposed to reduce the erosion of the bed and banks by 92-96 %.
- The simulation results of the sediment transport model show that when the flowrate value ranged between 290-300 m^3/s, a minimum accumulation of sediment resulted in the river, and the estimated amount of sediment transported in the reach was ranged between 91 and 350 Tones/day. So, it may be considered as the optimum flowrate passes through the river.

References

[1] T.S. Kayyun and D.H. Dagher. Potential sediment within a reach in Tigris River. International Journal of Hydraulic Engineering, 7(2) 2018 22-32.

[2] W.H. Safi, V. Prasad, and R. Khare. Simulation for sedimentation assessment a case study of Kunar River in Afghanistan. International Journal of Advanced Research, 5(2) 2017 406-416.

[3] S.I. Khassaf and M.J. Abbass. Modeling of sediment transport upstream of Al-Yaa'o regulator in Al-Najaf city. International Journal of Engineering Research & Technology (IJERT), 2(11) 2013 1367-1381.

[4] H.S. Mohammed, U.A. Alturfi, and M.A. Shlash. Sediment transport capacity in Euphrates river at Al-Abbasia reach using Hec-Ras model. International Journal of Civil Engineering and Technology, 9(5) 2018 919-929.

[5] A.S. Mustafa, S.O. Sulaiman, and K.M. Al-Alwani. Application of HEC-RAS model to predict sediment transport for Euphrates River from Haditha to Heet. Al-Nahrain Journal for Engineering Sciences, 20(3) 2017 570-577.

[6] T. Dysarz, E. Szałkiewicz, and J. Wicher-Dysarz. Long-term impact of sediment deposition and erosion on water surface profiles in the Ner River: Water Journal, 9(3) 2017 168.

[7] G.W. Brunner. HEC-RAS River Analysis System Modeling User's Manual US Army Corps of Engineers Hydrologic Engineering Center. Information on http://www.hec.usace.army.mil/. (2016).

[8] Ministry of Water Resources, MOWR, The General Authority for Surveying of the Ministry of Water Resources. Data on the cross-sections and coordinate of the upper reach of Gharraf River (2012).

[9] Ministry of Water Resources, MOWR. Data on the water levels and hydraulic information of Gharraf River (2012-2019).

[10] A. Arrar. Hydraulic simulation of Gharraf River between Kut and Hay cities. M.Sc. Thesis. Department of Water Resources Engineering, University of Baghdad, Iraq (2018).

Effects of Sandy Soil-structure Interaction on the Natural Period of RC Building Frames

Aqeel T. Fadhil[*1, a]**, Salah R. Al-Zaidee** [1, b] **and Mustafa M. Jasim**[1, c]

[1]Department of Civil Engineering, University of Baghdad, Baghdad, Iraq
[a]aqeel.fadhil@uobaghdad.edu.iq, [b]salah.r.al.zaidee@coeng.uobaghdad.edu.iq,
[c]mustafa.m.j@coeng.uobaghdad.edu.iq

Abstract. Building natural period, T, is a key character in building response for wind and seismic induced forces. In design practice, the period, T, is either estimated from empirical relations proposed by the design codes or determined from analytical or numerical models. The effect of the soil-structure interaction is usually neglected in the design practice and analysis models. This paper uses a sophisticated finite element simulation to investigate the effect of soil-structure modeling on the fundamental period of RC buildings subjected to wind and seismic induced forces. A typical interior building frame has been imitated using the frame element for beams and columns with constrains to model the almost-rigid diaphragm nature for floors and roof while a three-dimensional element has been adopted for the raft foundation and the soil mass. Different sandy soils have been considered. Standard Penetration Test, SPT value, has been taken as an identification index for soil nature. All other soil properties have been estimated based on well-accepted correlations. After applying the gravity dead and live loads, and eigenanalysis has been achieved to show how soil flexibility affects the effective stiffness of the structure and the corresponding natural period. Finally, the results are used to introduce a correction factor to modify the natural period estimated from a linear analysis with ideal supports to reflect the aforementioned effects.

Keywords: Soil-structure interaction, Finite Element, Natural Period, RC Buildings, Frames.

1. Introduction

According to Clough and Penzien [1], in earthquake engineering, soil nature has two main effects. Firstly, it can affect the nature of vibratory wave that transfers from rock basement to the natural surface. Secondly, the soil can alter the natural period, T, of the structure and hence changes its dynamic response. In the Uniform Building Code (UBC) [2], the first effect has been simulated by the soil profiles S_A through S_F and the corresponding acceleration-based coefficient, C_a, and velocity based coefficient, C_v [3]. This study deals mainly with the second aspect that has essential aspects not only in seismic loads but also for other dynamic loads, including machine vibration and wind flow.

Physically, the natural period, T, for lateral vibration of building frames represents the time required for the structure to complete one cycle when vibrates freely under initial displacement and/or initial velocity. It is an important factor in building response for wind, seismic, and other lateral forces [3]. In routine design works, the period value, T, is usually estimated from empirical equations like Eq. 1 that has been adopted for decades by American standard [4].

$$T = C_t(h_n)^{\frac{3}{4}} \tag{1}$$

where:

C_t, in the metric system, has a value of 0.0853 for steel moment-resisting frames, and value of 0.0731 for reinforced concrete moment-resisting frames and eccentrically braced frames, and a general value of 0.0488 for all other buildings.
h_n is building height in meters.

With the valuable assistance of digital computers, more elaborate free vibration analyses have been executed by designers to estimate the natural period, T. Unfortunately, most of these computer analyses neglect the soil-structure interaction and adopt ideal supports. Different researchers have considered this effect in an attempt to define a consistent correlation to be considered in the design practice. Anand and Kumar [5] published a review on the academic research related to the seismic soil-structure interaction. The researchers gathered the work of 113 references and summarized the historical development in this area and the different approaches used to solve the problems related to seismic soil-structure interaction. The researchers also presented brief articles about the guidelines included in different regional codes including the United States, Europe, Japan, New Zealand, and India.

Hokabadi and Fatahi [6] conducted a 3D finite element model to study the effect of modeling different foundations on the seismic characteristics of the model including the natural frequency of the frame. The researchers modeled a fifteen-story building and concluded that different foundation models can change the seismic behavior of the model dramatically. Bhojegowda and Subramanya [7] performed finite elements multistory models with different heights; five-levels models, ten-levels models, and fifteen-levels models. These models had various base-fixity types and also different soil types ranging from soft to hard soil to examine the effects of these parameters on the response of the structure including base shear, bending moments, and natural period of the model. The study implied that models with pile foundations can be treated with a fixed base since the difference in response was almost negligible. The natural periods for flexible-footing models were higher than those with fixed ones while the bending moments and displacements were lower.

Guerdouh and Khalfallah [8] carried out a numerical analysis on the effect on soil modeling and interaction on the seismic behavior of a single-story framed structure with a height of 3m and a span of 5m. The concrete frame with a typical cross-section of (0.4m by 0.4m) rested on soil consisting of three layers: top soft soil, middle medium soil, and bottom hard soil, with depths of 10m, 20m, 50m respectively. The frame was subjected to low, medium, and strong earthquake excitations. The study intended to study multiple response aspects: the effect of soil-structure modeling on both the soil and the frame, the change in the earthquake lateral displacement of the frame, and the earthquake intensity. Different outcomes were drawn from the study emphasizing that the type of soil has a great impact on the seismic response and frames on soft soils showed higher lateral displacement than on harder soils. Some other researchers also introduced valuable work in attempts to cover many parameters relating to the soil-structure interaction with seismic response. These researchers include Belletti et al. [9], Venkatesh & Deshpande [10], Sharma et al. [11], Al Rumaithi [12], Mekki et al. [13], Behnamfar and Alibabaei [14], and Barnaure and Manoli [15]. This study adopts relatively sophisticated finite element models to show how different soil simulations and different soil properties can affect the estimated natural period, T, of building frames.

2. Modeling of Building Frame and Underneath Soil

2.1.Physical Modeling Aspects

To simplify the analysis process, the three-dimensional building frame can be broken down into a series of two-dimensional frames, as indicated in Fig. 1. This approximation is used for decades in analysis and design processes [16]. The superimposed dead and live loads that act in the form of load per unit area can be transformed into line loads that act on the supporting beams, as presented in Fig. 1. For a two-way slab system, triangular and trapezoid loads are a more accurate simulation [17].

2.2.Units

For dynamic and free vibration problems where inertia forces are included, one should be careful regarding the unit system and should be aware that he/she implicitly adopts a specific unit system for ground acceleration. In this study, a consistent unit system of m, N, kg, second has been respectively adopted for length, force, mass, and time.

2.3.Finite Elements for Frame

For the equivalent frame, the frame finite element indicated in Fig. 2 is adopted in the simulation of beams and columns [18]. ABAQUS Finite element software is utilized to simulate the model and run the analysis and extract the results. The software has a library that contains a variety of elements for the user to choose from based on the required degrees of freedom of the modeled element. In this research, an 8-node linear brick element is employed. The element, recognized as C3D8R, is a reduced integration, hourglass control. As it is based on a kinematic assumption of plane section before loading remains a plane and normal to the member axis after loading, it can generally reduce the three-dimensional behavior of beams and columns into an equivalent one-dimensional simple behavior, [19]. It is adequate for solid sections where there is no possibility of local bending effects [20]. It implicitly neglects shear strain. This approximation is generally accepted for relatively thin members [21]. A cubic Hermite shape function indicated in Eq. 2 is adopted to simulate displacement along element length.

$$N_i = a_i + b_i\xi + c_i\xi^2 + d_i\xi^3 \tag{2}$$

where a_i, b_i, c_i, and d_i are coefficients that can be determined in terms of nodal displacement q_i. This formulation is suitable when the element is subjected to distributed forces or inertia forces according to d'Alembert's principle [22]. The consistent mass matrix has been formulated using Eq. 3 below. Three Gaussian points have been adopted to execute the integration numerically. Finally, rotary inertial components have been neglected to avoid possible computational difficulties [23].

$$M = \rho A \int_0^{l_e} N_i^T N_i dx = \frac{\rho A l_e}{2} \int_{-1}^{1} N_i^T N_i d\xi \tag{3}$$

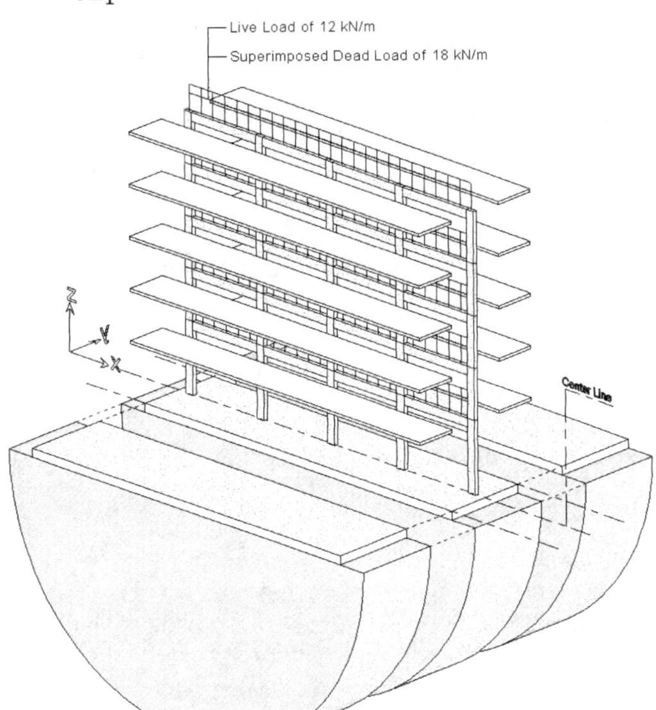

Figure 1. Building frame model.

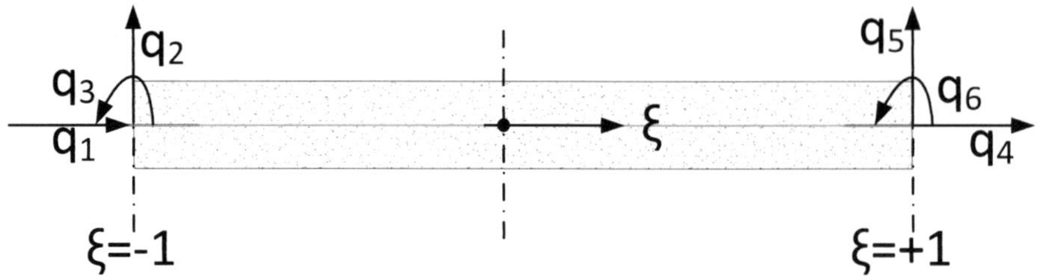

Figure 2. Frame element adopted in the simulation of beams and columns.

2.4.Simulation of Rigid Floors Diaphragm

Even though they are flexible out of their plane, floor systems are almost rigid in their plane [24]. Rigid body constraints are adopted to enhance the proposed model by imposing an infinitely rigid behavior [25]. With this technique, a master node has been selected at mid of each floor; all other nodes within the floor are considered as slave nodes where their horizontal displacements are assumed equal to that of the master node, see Fig. 3. For the computational purpose, this kinematic assumption is rewritten in the form of the transformation matrix, **T**, that is used in turn to reduce the stiffness matrix of girder as indicated in Eq. 4 below:

$$K_{Reduced} = T^T K_{Original} T \tag{4}$$

From Eq. 4 above, one concludes that rigid body constraints not only simulate the rigid diaphragm in a more accurate form but also reduces the corresponding computational efforts.

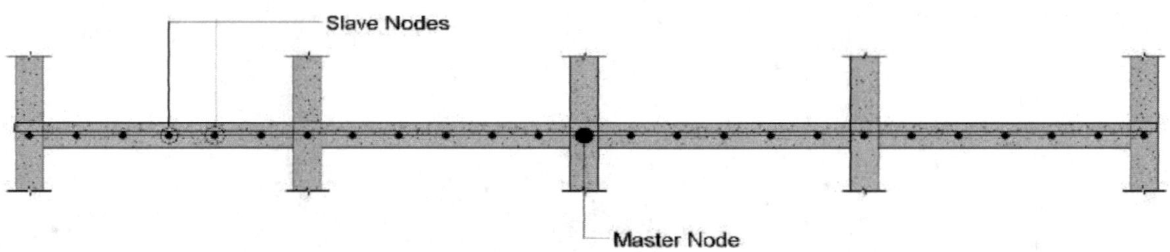

Figure 3. Master and slave nodes to simulate the rigid floor diaphragm.

3. Finite Element for Foundation

Foundation support for the typical equivalent frame indicated in Fig. 1 has been assumed to extend from mid-distance between two consecutive frames. Three-dimension elements have been used in the discretization process. According to ACI-318M Code [26], the elastic modulus for concrete can be correlated to its compressive strength, $\mathbf{f_c}'$, based on Eq. 5 below. According to Darwin et al. [17], a value of 0.2 is representative of concrete Poisson's ratio in the elastic range.

$$E_c = 4700\sqrt{f_c'} \tag{5}$$

When shear strain in included in a relatively thin member, as in the case of the foundation, the shear-locking phenomenon may occur due to overemphasis of shear contribution with the full Gaussian integration that is adopted to determine the integration of stiffness matrix [27]. Therefore, a reduced integration with a single Gaussian point located at the center of the element has been adopted in this research [28]. Unfortunately, recently, it has been found that the reduced integration schemes may lead to what is called the spurious zero-energy modes. These modes are deformed into an hourglass shape. In this research, this problem has been overcome through an hourglass control technique where additional stiffness term is adopted to prevent the formulation of the hourglass modes [29]. As indicated in Fig. 1a, two elements have been used along the raft depth to model its possible curvature with the linear brick element [30].

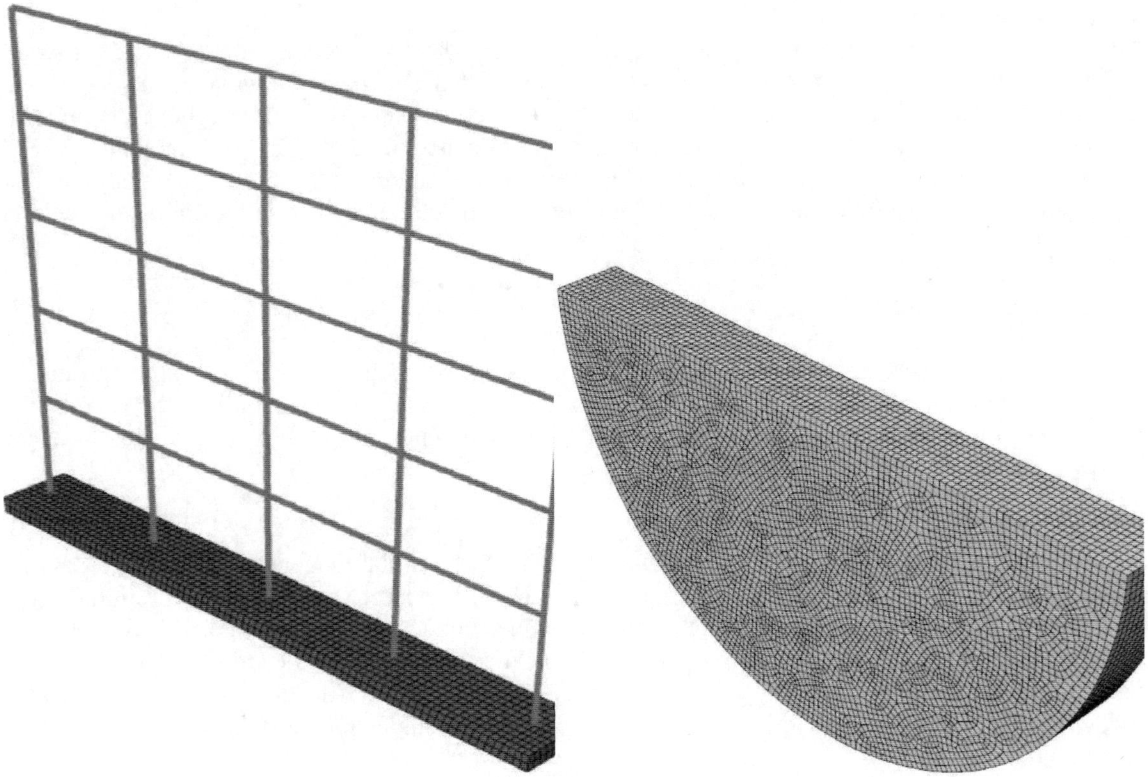

a. Raft foundation. b. Soil mass.

Figure 4. Finite element mesh.

3.1. Finite Element Mesh for Soil

As indicated in Fig. 4b, a linear brick element has been adopted in soil simulation. According to Cook [30], the linear element should be adopted only with low strain gradient and fine mesh. Both conditions are satisfied in the mesh presented in Fig. 4b.

3.2. Final Equations of Motion and Solution Algorithm

After modeling the frame, foundation, and soil mass, they have been assembled to behave as a complete system with the following equation of motion.

$$\begin{bmatrix} m_{ff} & m_{fs} \\ m_{sf} & m_{ss} \end{bmatrix} \begin{Bmatrix} \ddot{v}_f \\ \ddot{v}_s \end{Bmatrix} + \begin{bmatrix} c_{ff} & c_{fs} \\ c_{sf} & c_{ss} \end{bmatrix} \begin{Bmatrix} \dot{v}_f \\ \dot{v}_s \end{Bmatrix} + \begin{bmatrix} k_{ff} - k_{ffg} & k_{fs} \\ k_{sf} & k_{ss} \end{bmatrix} \begin{Bmatrix} v_f \\ v_s \end{Bmatrix} = \begin{Bmatrix} p_f \\ p_s \end{Bmatrix} \quad (6)$$

where the subscript f and s refer to the frame and soil respectively. The matrices **m**, **c**, and **k** respectively indicate the mass, damping, and stiffness of the system. The stiffness $\mathbf{k}_{ffg}$ shows the effect of axial forces on the stiffness and softening of the frame. It can be determined in an exact or approximate scheme. The approximate approach with fine mesh has been adopted here according to the recommendation of Al Zaidee et al. [31]. As the paper concerns with the free vibration, the load vector **p** has been dropped from Eq. 6. Damping ratios, ξs, have been used instead of the damping matrix **c** to simulate the energy dissipation in the system. This simulation is approximated in nature when applied to a nonproportional system with soil-structure interaction [24]. For the sake of reducing the program run time and avoiding unpreferred numerical computation divergence, tie contact is chosen the simulate the contact between the superstructure frames and foundation.

IOP Conf. Series: Materials Science and Engineering **901** (2020) 012013 doi:10.1088/1757-899X/901/1/012013

3.3. Soil Physical Parameters

According to Prakash [32], in a soil dynamic problem, the soil elastic models are applicable for strain range no higher than 10^{-4} mm/mm. Although the strain may extend beyond this range in some cases, the elastic model has been adopted to formulate and solve the problem as an Eigenvalue problem [28]. The elastic model is in terms of three basic parameters of elastic modulus, E, shear modulus, G, and Poisson's ratio, ν. In the mechanics of material and theory of elasticity, it has been found that only two of these parameters are independent, and the third one can be determined from the following relation [33]:

$$G = \frac{E}{2(1+\nu)} \tag{7}$$

Based on regression analyses of extensive test data, dynamic shear modulus, G, damping ratio, ξ, density, ρ, and Poisson's ratio, ν, have been correlated to standard penetration test, SPT, N value. According to Das and Luo [34], the initial dynamic shear modulus, G_{max}, can be estimated from the following relation for round-grained sand:

$$G_{max} = 35 \times 161.5\ N^{0.34} (\sigma_0)^{0.4} \tag{8}$$

where G_{max} and confinement pressure, σ_0, are in kPa. The initial tangent modulus, G_{max} can be adopted for shear strain in the range not higher than 10^{-4}. For larger strain values, the initial shear modulus, G_{max}, can be reduced for a suitable tangent value. For adopted strain range, the dry sandy soil is lightly damped with damping ratio, ξ, of 1% [34]. Other soil parameters can be correlated to SPT value based on relations adopted in static analysis. According to Das [35], soil Poisson ratio is related to the internal angle of friction, ϕ, based on Eq. (9). In turn, the angle, ϕ, is correlated to the SPT value according to Eq. (10).

$$\nu = 0.1 + 0.3 \left(\frac{\phi - 25}{20}\right) \tag{9}$$

$$\phi = 27.1 + 0.3N - 0.00054N^2 \tag{10}$$

4. Case Studies

A total of six case studies are investigated to evaluate the influence of different soil-foundation modeling on the natural period, T, of framed structures. One of the case studies is examined without the modeling the soil mass beneath the foundation. Instead, fixed constraints are assigned to the concrete foundation. The rest of the case studies incorporate the soil-structure interaction through modeling the soil mass as indicated in Figs. 1 and 4. In the five case studies where the soil is modeled, different sandy soils characterized by different standard penetration test values have been studied. SPT values ranging from 10 to 50 are considered and the corresponding soil properties are obtained from the aforementioned relations depending on the SPT value. Table 1 presents the soil properties of the case studies.

Table 1. Sandy soil properties.

No.	SPT N Value	Angle of Friction, ϕ (degree)	Poisson Ratio, ν	σ_0 (kPa)	Shear Modulus, G (kPa)	Elastic Modulus, E (kPa)
1	10	30.1	0.176	100	78026	183479
2	20	32.9	0.219	100	98762	240683
3	30	35.7	0.260	100	113361	285612
4	40	38.3	0.300	100	125009	324898
5	50	40.9	0.338	100	134862	360824

Typical concrete raft foundation for all cases is defined in the finite element model with the geometric and material properties indicated in Table 2.

Table 2. Foundation properties.

Compressive strength, f_c' (MPa)	Width, B, (m)	Thickness, h, (m)	Moment of inertia, I (m^4)	Elastic Modulus, E (kPa)
28	3	0.60	0.0540	24870062

Table 3. Elements properties.

Frame	Compressive strength, f_c' (MPa)	Dimensions (m x m)
Beam	28	0.30x0.60
Column	28	0.30x0.60

To simulate the actual loads that act on each floor of the framed structure, a combination of both superimposed dead load and live load is applied on the beams of each floor as line loads. The values presented in Fig. 1 are applied in each of the cases studied. To better interpret the effect of different soil properties on the natural period of the system, the stiffness parameter, β, is introduced [35]:

$$\beta = \sqrt{\frac{B\,k}{4\,E_f\,I_f}} \tag{11}$$

where:

k is the soil subgrade modulus, in kN/m^3, and E_f and I_f, are the elastic modulus and moment of inertia of the footing, in kPa and m^4 respectively.

5. Results and Discussion

The results of the above case studies are presented in this section. The results of the natural vibration properties from the finite element model for the fixed-base model are presented in Table 4.

Table 4. Natural vibration properties of the case studies.

No.	SPT N Value	Natural Frequency, f_n (1/sec)	Natural Period, T_n (sec)	$T_n / T_{n\ \text{Fixed-base model}}$
1	Fixed base	0.664	1.506	----
2	10	0.812	1.232	0.818
3	20	0.814	1.229	0.816
4	30	0.816	1.225	0.814
5	40	0.816	1.225	0.814
6	50	0.817	1.224	0.813

The effect of different soil properties on the natural period of the framed structure is shown in Fig. 5. It can be seen, within the case studies where the soil mass is modeled, from Table 4 and Fig. 5 that the higher the SPT value of the soil, the higher the natural frequency of the building frame. Even though the influence is low, the results seem intuitive considering that higher SPT values imply stiffer soils that contribute to higher building frequencies. Although the comparison between the effect of soil types is reasonable, the effect of soil-foundation interaction and foundation constraint on the natural properties of the framed structure seems more interesting. While intuitive judgment may predict higher natural frequencies, and lower periods, for building frames with fixed-base models over models where soil mass is simulated, the results show the opposite. The natural periods of the structure in the five case studies where soil-structure interaction is considered represents about 81% of the model where fixed constraints are assigned to the foundation base. The reason behind this behavior is due to the effect of soil mass in the model. Despite that the fixity increases the stiffness of the system, the contribution of adding the soil mass to the system has a higher impact on the system resulting in a higher natural period and lower frequency.

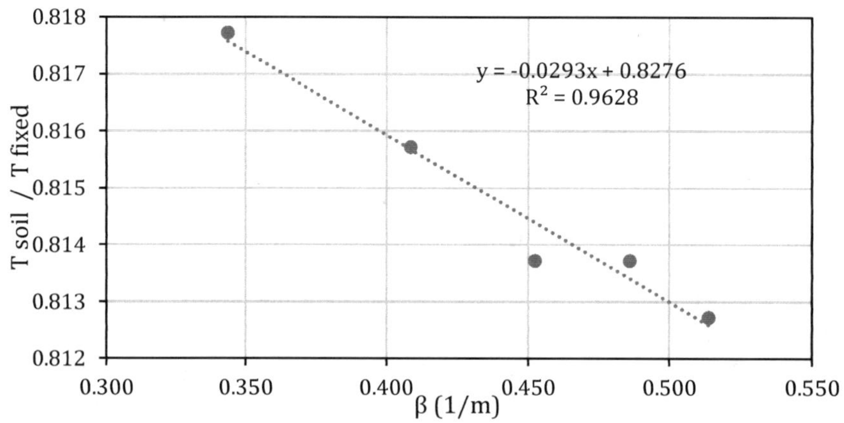

Figure 5. Relative natural period vs soil-foundation stiffness coefficient.

6. Conclusion

The effect of soil-structure interaction on the natural period of building frames is examined in this study utilizing sophisticated finite element models. A total of six case studies are investigated; one with a fixed-base foundation and the rest are modeled by simulating the soil mass beneath the foundation. A typical interior building frame has been imitated using the frame element while different sandy soils have been considered where soil-structure is modeled. Standard Penetration Test, SPT value, has been taken as an identification index for soil nature. All other soil properties have been estimated based on well-accepted correlations. Results show that the building frames modeled with a fixed-base foundation has approximately 22% higher natural period than where the soil mass is modeled. Variations of case studies that may include different span lengths, different column layout for floors, different floor levels, and a wider range of soil properties could be examined in future studies to investigate this ratio.

References

[1] R.W. Clough and J. Penzien, Dynamics of Structures., McGraw-Hill, 1975.
[2] UBC, Uniform Building Code, Conference of Building Officials, 1997.
[3] A. Williams, Seismic design of buildings and bridges, 3rd edition., Austin, Texas: Engineering Press, 2000.
[4] ASCE/SEI 7-16, Minimum Design Loads for Buildings and Other Structures, ASCE/SEI, 2016.
[5] V. Anand and S.S. Kumar, Seismic soil-structure interaction: a state-of-the-art review, Structures, Elsevier, 16, 317-326, 2018.
[6] A.S. Hokmabadi and B. Fatahi, Influence of foundation type on seismic performance of buildings considering soil-structure interaction, International Journal of Structural Stability and Dynamics, 16 (08), 1550043, 2016.
[7] V.T. Bhojegowda and K.G. Subramanya, Soil structure interaction of framed structure supported on different types of foundation, International Research journal of engineering and technology, 02 (05), 651-660, 2015.
[8] D. Guerdouh and S. Khalfallah, Effect of soil-structure interaction on seismic performance of frame structures (In Spanish), Revista de la construcciốn, 18 (02), 349-363, 2019.
[9] B. Belletti, A. Gasperi, A. Spagnoli and R. Valentino, Role of Soil-Structure Interaction on the Response of Precast RC Structures under Seismic Loading: Case Study, Practice periodical on structural design and construction, 22 (01), 4016014, 2016.

[10] M.B. Venkatesh and R.D. Deshpande, Analysis of RC building frame with raft foundation considering soil-structure interaction, International Research Journal of Engineering and Technology, 04 (05), 752-760, 2017.

[11] N. Sharma, K. Dasgupta, and A. Dey, Natural Period of RC Buildings Considering Seismic Soil-Structure Interaction Effects, in 16th Symposium on Earthquake Engineering (16SEE), Roorkee, India, 2018.

[12] A. Al Rumaithi, Characterization of dynamic structures using parametric and non-parametric system identification methods, Electronic Theses and Dissertations 1325, 2014.

[13] M. Mekki, S. M. Elachachi, D. Breysse, and M. Zoutat, Seismic Behavior of RC Structures Including Soil-Structure Interaction and Soil Variability Effects, Engineering Structures, 126, 15-26, 2016.

F. Behnamfar and H. Alibabaei, Classical and Non-Classical Time History and Spectrum Analysis of Soil-Structure Interaction Systems, Bulletin of Earthquake Engineering, 15 (03), 931-965, 2017.

[14] M. Barnaure and D. Manoli, Unfavourable Seismic Behaviour of Reinforced Concrete Structures Due to Soil-Structure Interaction, IOP Conference Series: Earth and Environmental Science, 362 (1), 012119, 2019.

[15] C.K. Wang, C.G. Salmon and J.A. Pincheira, Reinforced Concrete Design, 7th Edition., John Wiley & Sons, Inc., 2007.

[16] D. Darwin, C. W. Dolan and A.H. Nilson, Design of Concrete Structures, 15th Edition., McGraw Hill, 2016.

[17] O.C. Zienkiewicz and R.L. Taylor, The Finite Element Method, 5th Edition, Volume 1: The Basis, Butterworth Heinemann, 2000.

[18] L.H. Donnell, Beams, Plates, and Shells., McGraw-Hill, Inc., 1976.

[19] J.H. Oden, Mechanics of Elastic Structures., McGraw-Hill Book Company, 1967.

[20] S. Timoshenko and D. H. Young, Elements of Strength of Materials, 5th Edition., Van Nostrand Reinhold Company, 1968.

[21] T.R. Chandrupatla and A. D. Belegundu, Introduction to Finite Elements in Engineering, Prentice Hall of India, 1996.

[22] M. Shrikhande, Finite Element Method and Computational Structural Dynamics, PHI Learning Private Limited, 2014.

[23] A.K. Chopra, Dynamics of Structures, Theory and Applications to Earthquake Engineering, 5th Edition, Prentice Hall of India, 2017.

[24] R.D. Cook, D.S. Malkus, M. E. Plesha and R. J. Witt, Concepts and Applications of Finite Element Analysis, Fourth Edition, John Wiley & Sons, Inc., 2002.

[25] ACI318M, Building Code Requirements for Structural Concrete (ACI 318M-14) and Commentary (ACI 318RM-14), Farmington Hills: American Concrete Institute, 2019.

[26] H.C. Huang, Static and Dynamic Analyses of Plates and Shells: Theory, Software, and Applications., Springer-Verlag, 1989.

[27] K.J. Bathe, Finite Element Procedures., PHI Learning Private Limited, 1996.

[28] M.R. Gosz, Finite Element Method. Applications in Solids, Structures, and Heat Transfer., Taylor & Francis., 2006.

[29] R.D. Cook, Finite Element Modeling for Stress Analysis., John Wiley & Sons, Inc., 1995.

[30] S.R. Al Zaidee, H. A. Al-Baghdadi, and A. T. Fadhil, Effect of Columns Discretization on P-δ Analysis, Association of Arab Universities Journal of Engineering Sciences, 25(3), 126-136, 2018.

[31] S. Prakash, Soil Dynamics, McGraw-Hill Book Company, 1981.

[32] E.P. Popov, Mechanics of Materials, SI Version, 2nd Edition., Prentice-Hall International Inc., 1978.

[33] B.M. Das and Z. Luo, Principles of Soil Dynamics, Third edition., Boston, MA, USA, Cengage Learning, 2016.

[34] B.M. Das, Principles of Foundation Engineering, 7th Edition, CENGAGE Learning, 2010.

Erosion, Sediments Transport and Riverbank Stability: A Review

Asad H. Aldefae[1,a*], Rusul A. Al-Khafaji[1,b], Mohammed S. Shamkhi[1,c], and Haider Q. Kumar[1,d]

[1]Civil Engineering Department, Wasit University, Wasit, Iraq
[a]asadaldefae@uowasit.edu.iq, [b]msewan@uowasit.edu.iq, [c]rusul.amer892@gmail.com, [d]hqais@uowasit.edu.iq

Abstract. This paper is a review of the basic theoretical information needed to understand the riverbank stability, failure mechanism, and its causes. The first part describes the basic principles of river systems involving river hydraulics, sediment load, and channel morphology. This is pursued by a brief discussion concerning riverbank erosion that involves theoretical equations and numerical methods to quantify erosion. Finally, an in-depth discussion concerning riverbank stability that involves the types of failure mechanism, tension cracks, and riverbank settlement is conducted. Also, the paper reviews some previous studies on riverbank stability.

Keywords: Erosion, River, Riverbanks, Sediments, Hydraulic of flow.

1. Introduction

Rivers are the lifeline for hundreds of millions of peoples around the world. The sediment transport along the river section affecting the water quality and the storage capacity as well as the navigation of the river. Most of these tons of the sediments are coming from the collapsibility of the riverbanks particularly during the cyclic of flood-scarcity season. Physical and numerical modeling of rivers include the study of hydraulic system, sediment transport, and riverbank stability, however, these elements have not been studied in one model, where most of the models provide developed resolutions on hydraulic of flow and sediment transport, but they cannot analyze riverbank stability, so, more numerical studies are needed to evaluate and control the models [1]. This paper reviews the basic theoretical information needed to understand the riverbank stability, failure mechanism, and its causes. The first part describes the basic principles of river systems involving river hydraulics, sediment load, and channel morphology. This is pursued by a brief discussion concerning riverbank erosion that involves theoretical equations and numerical methods to quantify erosion. Finally, in-depth discussion concerning with riverbank stability that involves types of failure mechanism, tension cracks, and riverbank settlement. Also, the paper reviews some previous studies on riverbank stability. A numerical study is conducted to investigate the effect of seasonal variation of the river and the rise of groundwater level on the slope stability, the slope was designed over a period of 97 days with the most effective periods affecting the stability which is the decreasing of river level and the increase of groundwater level, the results showed that the factor of safety decreased by 13% to 17% at the end of the period [2].

A field study is conducted to describe the riverbank stability of a part of the Hilla river in Babylon province, the fieldwork included measuring the dimensions of the riverbank, identifying the erosional and sedimentation riverbank, and preparing a map in this regard [3]. The results of the study showed that the types of failures in the area of study are the alcove-shape failure, the slap-shape failure, and the toppling-shape failure, and found that the most important parameters which harm the stability of riverbank are the bank height, the bank slope, and the intensity of river meander. A study is carried out to investigate the effect of soil properties on the stability of the Shatt AL-Arab riverbank, where four study areas were selected [4]. After the laboratory tests were conducted to determine the soil properties of each area, cohesion, internal friction angle, and the weight density of soil, and for different depths, the Prokon software was used for slope stability analysis of the selected areas, the results of their study showed that all the slopes are stable but it is consistent in the degree of stability, and cohesion has a major role in conjunction with the bank height in determining the degree of stability, in addition to the effect of water flow level and groundwater level, from their study they obtained predictive equations that can be used to determine the slope stability in different conditions.

Principles of physical modeling are followed to investigate the stability of overhanging riverbanks using a transparent tank that has dimensions of 200 cm in length, 100 cm width, and 100 cm height [5]. The main purpose of their study was to compare the cantilever failure mechanism, where two types of material were used with three different densities. The results of their study showed that the mechanism of toppling failure is more likely to happen than the mechanism of shear failure. (A. Samadi et al. 2012) and performed numerical modeling using SIGMA/W software to simulate the stress-strain attitude in the bank of the model, the numerical results showed a good agreement between experimental and numerical data, however, experimental failures observed in a high-density cohesive soil were not repeated in simulations. An erosion measurement device is designed and built containing an acrylic flume with a length of 243 cm in length, 10 cm wide, and 5 cm height [6]. The purpose of the device was to measure the function of erosion (which referred to the relationship between the hydraulic shear stress and the rate of erosion) for a specific soil specimen, and water used from the red river in Winnipeg, Manitoba. Jianfar, 2014 [6] conducted a numerical modeling using Geo-Studio software to study the slope stability and seepage applying two scenarios one in the normal state of flow and the other in a state of flooding, based on previously available information on flow velocity and shear stress conducted by KGS group using Flow-3D software, Jianfar's results showed that the operation of floodgate has no effect on the stability of riverbank in the upstream as the factor of safety has approximated value in both of the two scenarios.

An experimental rectangular flume is made of Plexiglas wall that has dimensions of 10m x0.3cm x0.2m, to inspect the cantilever mechanism of the failure of cohesive riverbank's models and the fluvial erosion [7]. The slope was set to be 1/500, three types of soil were used which were classified according to the silt-clay content, their results showed that the initial stage of cantilever failure starts when the upper part of the river bank undermined due to the fluvial erosion at the lower part, followed by the appearance of cracks on the surface of the bank, then the cantilever failure occurs. Patsinghasanee et al. 2015 [7] applied a numerical model of their study consist of three grids, 1D grid for flow modeling in the lateral direction, a 1D grid for sediment transfer and representation of deformation in the lateral direction, and 2D grid for modeling the cantilever failure in the vertical and lateral direction, the experimental and simulated results showed a good coincidence. A study is conducted on the Tigris riverbank erosion in the city of Nu'amania using a two dimensional hydraulic mathematical RMA2 model, in addition to the use of static BSTEM software [8]. The purpose of their study was to find out the cause of riverbank erosion in the area to determine the best ways to address the problem of riverbank erosion that meets the hydraulic requirements at the lowest cost possible, where three types of treatments were compared which is the riprap, gabion, and vegetation. The result of their study showed that the treatment using riprap is the best solution due to the low cost, in addition to it reduces the flow velocity at the riverbank.

A horizontal flume has 1.7m length, 0.5m wide, and 0.6m deep is used, which was divided into two parts, one for the river, and the other for the bank, to investigate the tension cracks development due to change of water level and how these cracks effect on the riverbanks' stability. [9]. Their study investigated three elements, decline of river levels, primary water elevation, and the slope angle of the riverbank. Two experimental formulas were used to estimate the location of tension cracks and the angle of failure plane to verify the accuracy of the model used, where the model gave a good prediction in riverbank stability analysis.

An experimental test using 0.75m long, 0.25m wide and 0.25m deep slopping flume is used, with a rainfall simulator, and a strip laser device, to conduct an experimental erosion test, to investigate how the changes in the soil surface topography forms effect on the erosion practicability were estimated by the strip laser device [10]. Six slopes inclination of the flume were applied (5°, 10°, 15°, 20°, 25°, and 30°) and the rainfall simulator produced a 30-min with a rate of an equivalent kinetic energy 80 mm/hr. The variance between a digital elevation model of the different periods was used to gain the eroded soil volume and the conformable rating of the loss in the soil mass. The consequence was that the volume of eroded soil and sediment crops related to the slope and rainfall period. A flume made of acrylic material is manufactured, and consists of two segments, inflow segment of 50 cm long, and main channel segment of 150 cm long, 10 cm wide, and 15 cm high [11]. To study the factors affecting sediment connectivity of landslide materials, where they applied different angles for flow and channel gradient, and using different types of sediment with changing the water content for each one from (0 to 100, up 20%), the results showed that all landslides occurred when the water content was greater than or less than saturation and that the largest landslides were with the largest angle of flow and angle of the gradient.

2. River Systems

The primary explanation behind the complication of river engineering is that river flow has no fixed boundaries and geometry compared with, state, pipe flow, or open-channel flow. Also, the complication resulting from the change in roughness of the boundary due to the transfer of the sediment [12]. The main purpose that the rivers are serving is the transportation of water and sediment. The characteristics of flow and sediment load depend on two factors, climate, and geology of the region [13]. The discharge of the river depends on the climate in the region, where the composition of the river bed and banks depend on the geology of the region. River morphology is related to channel composition and geometry, and with the longitudinal profile, it is time-dependent and changes with discharge, sediment properties, and with the material of the bank [12]. River morphology is mainly based on climate and geology of the region, which in turn affect discharge and sediment load as shown in Fig 1, [14].

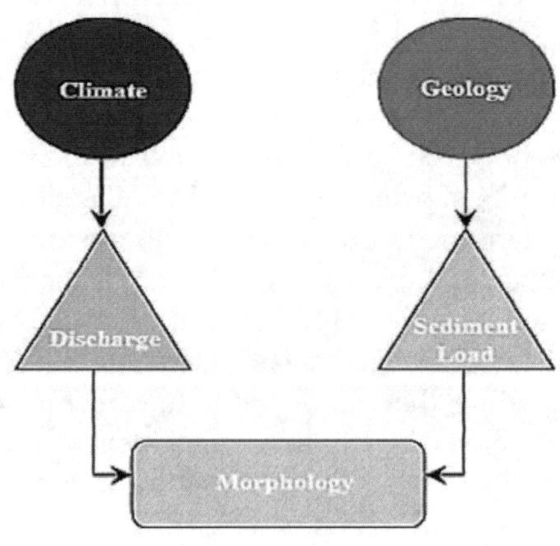

Figure 1. Variables affecting river morphology [13]

The operations that modify the morphology involve erosion or entrainment, conveyance of sediment load, and deposition [15].

2.1. Hydraulic of Flow

Corrosion of the bed and banks of a channel is the consequence of the shear stress that affects the cross-section of the channel. Shear stress depends on the fluid viscosity and velocity gradient as shown in Eq. 1 [16]. An increment in the velocity gradient leads to an increment in the shear stress.

IOP Conf. Series: Materials Science and Engineering **901** (2020) 012014 doi:10.1088/1757-899X/901/1/012014

$$\tau = \upsilon \frac{d_{\upsilon}}{d_{y}} \tag{1}$$

Where τ is the shear stress, v is the fluid viscosity, *dv/dy* is the velocity gradient. The velocity distribution in the open channel flow section is not uniformly due to the existence of the free surface and to the friction of the bed and banks of the channel. Other factors are also dominant, the velocity distribution such as the channel section shape, and the existence of the bends. The maximum velocity predominantly found at the free surface, where at the bed of the channel the velocity diminution to zero due to the frictional forces [17]. The utmost amount of erosion results from the highest shear stress at the bed and the toe bank region of the channel which causes the maximum velocity gradient. The effect of the inertial force and the fluid viscosity specify the attitude of open channel flow where the flow may be laminar or turbulent, in laminar flow the particles of water seem to move in definite streamlines and thin layers of fluid slip over neighboring layers, whereas in turbulent flow the particles of water stir in chaotic paths which causes great erosion [17]. Fig. 2 shows the velocity distribution in cases of laminar and turbulent flow.

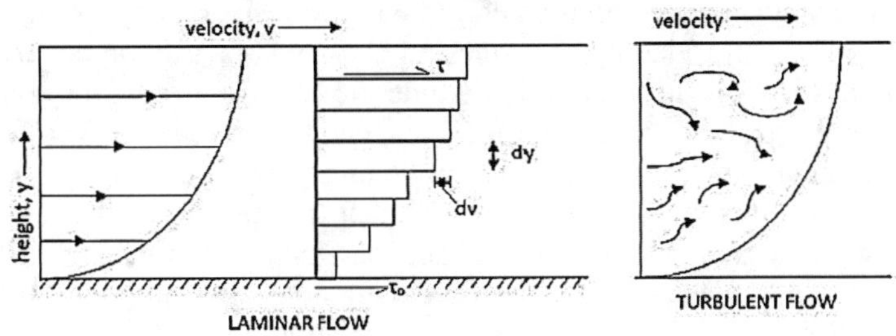

Figure 2. Velocity distribution in laminar and turbulent flow conditions [17]

To specify whether the flow is laminar or turbulent, the Reynold number should be calculated as expressed in Eq. 2, which represents the proportion of viscous and inertial forces [18]. The flow is laminar when the Reynold number is less than 500 and turbulent when it is greater than 2500.

$$\text{Re} = \frac{\rho VL}{\mu} \tag{2}$$

Where Re is the Reynold number, ρ is the density of water, V is the flow velocity, L is the characteristics length which represent by the hydraulic depth or the hydraulic radius in free-surface flows, μ is the dynamic viscosity. In laboratory flume, the disturbance that occurs at the entrance of the channel which is unavoidable causes arising in the water level in one side resulting in a spiral motion on this site only, but after a distance from the entrance, there will be a spiral motion on both sides of the channel, in this stage the spiral motion could be ignored for practical purposes in the straight channels because the shear stress on both sides will be in equilibrium in this state [17].

In the distribution of the mean velocity, the lines of equal velocity (isovels) are nearer at the boundaries as shown in Fig.3. In the center of the channel, the velocity is the highest due to the lack of friction as far from the boundaries [17]. The shear stress is greatest where the isovels are closer together, so in a wide shallow section, the shear stress is the highest at the bed, wherein narrow deep-section the shear stress is greatest at the banks [17].

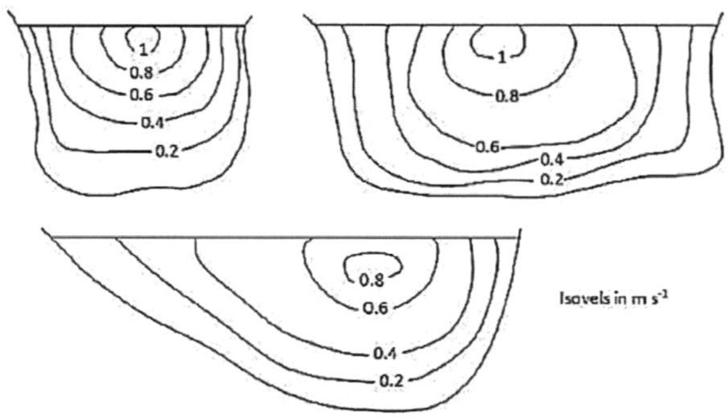

Figure 3. Distribution of lines of equal velocity in various channel cross-sections [17]

2.2. *Sediment Load*

Sediment transport typically due to a combination between gravity acting on the sediment, and/ or the movement of the fluid in which the sediment is entrained. The resisting force of the grains forming the boundary in alluvial streams controlled by the cohesion and angle of friction between the grains, when the driving force that caused by fluid motion, overcome the resisting force, the grains will be brought into motion. If the shear stress at a certain point is larger than the critical shear stress, the grains will be detached from its location, and the region is said to be scoured. There are two mechanisms of sediment transport, bedload and suspended load, in bed load sediment transportation, the ratio of relative tractive force τ/τ_{cr} at a point, is slightly greater than 1 [18]. In bed load transportation, the large grains of sediment (coarse sand and gravel) with size larger than 0.06 mm in diameter slide and roll on the bed, where could not be transported with flow due to its heavyweight [16]. Reciprocally, when the relative tractive force is much large, fine grains will be carried by the flow in suspension and move downstream by turbulence [18].

The suspended load which is composed of fine particles (silt and clay), requires little or no energy to be transported with the flow, which results in raising the efficiency of the river by decreasing the frictional losses of energy due to inner turbulence reduction [16]. The bedform affected by several parameters which are the flow velocity and the flow depth, slope of the bed, size of sediment particles, and the fall velocity of the particles, sediment particles will begin to move increasing the flow velocity [20]. When the mean velocity is greater than the critical velocity, the sediment particles will entrain and transport with fluid, and when the critical velocity is greater than the mean value, the sediment will drop out of suspension [16].

Hjulstrom curve expresses the relationship between the size of sediment particles and flow velocity for entrainment, transportation, and deposition as shown in Fig. 4, the upper line shows the erosional velocity or critical erosion velocity that needed to initiate sediment erosion, the lower line shows the fall or settling velocity, between the two lines there is the region of sediment transportation. Hjulstrom curve shows that noncohesive sediment such as sand needs low velocity to entrain from the bed, however, silt and clay need higher velocity to be entrained due to cohesive nature of clay particle that makes strong bonds among particles called electrochemical bonds, but once clay particles are entrained from the bed, they are easily maintained in suspension at a lower velocity than the velocity that needs to entrained the particles [15]. Hjulstrom curve also shows erosion which is the picking up of sediment and requires high velocity than transportation. In contrast, the deposition region of the diagram for coarse particles of sediment, where deposition happens if the velocity drawdown below the erosional velocity.

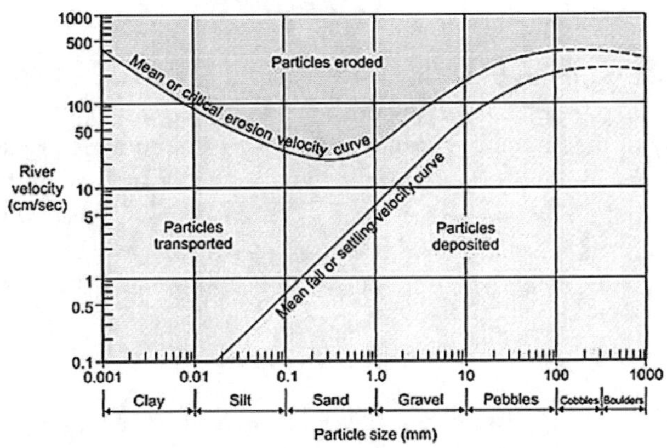

Figure 4. Hjulstrom curve for erosion, transportation, and deposition as a function of particle size and river velocity [19].

2.3. *Channel Morphology*
Channel morphology related to channel composition and geometry; it is time-dependent, changes with longitudinal profile, sediment amount and characteristics, discharge, and bank material properties [12]. Climate and geology of the region are secondary factors that influence the discharge and sediment type [13]. The precipitation amount and hence the discharge are specified by the hydrology, whereas the properties of the channel bed and banks and hence the sediment type are specified by the geology of the region. The channel form related to the processes that occur in the river which are erosion, sediment load transportation, and deposition [15]. For example, the flow velocity increases due to the discharge increase as a result of heavy rainfall, the channel restraint for this increase by extending the cross-sectional area to be able to carry the flow.

The cross-section degradation of the river depends on the response of different types of sediment to erosion [15]. Sediments that respond easily to erosion such as sand, produce wide and shallow channels, whereas clay and silt sediments produce steep and narrow channels [15]. When the discharge increases too quickly, the flow surpasses riverbank and the flooding occurs. Flow velocity decreases on the land surface and the deposit of the sediment, which are composed of the bed and banks of the flood plain. The coarse sediments deposited near the channel edge, while the finer and lighter sediments move away [21]. River flow starts from high regions towards lower and planer regions and then pours into the ocean [13]. The cross-section of the river and the sediment load differs from one region to another according to the difference of the topography of the earth. The amount and size of the sediment carried by flow have a significant impact on the river morphology when the concentration of sediments increases, deposition occurs on the bed due to the inability of discharge to carry the sediment load as a result of decreasing in river efficiency [13].

2.4. *Equilibrium*
Rivers are a system in dynamic equilibrium which means neither erosion nor deposition occurs, balancing of water flow and sediment transport. When river channels are altered under naturally dynamic hydrologic conditions, the river readjusts itself concerning morphology to reach its former balance or equilibrium. Some researchers have suggested the term of dynamic equilibrium or changing balance, where the river changes in some parts to reach an equilibrium state [15]. The state of the equilibrium of rivers can be achieved by the deposition of the soil particles and the erosion process. Many parameters affect directly of the riverbank's erosion such as the channel size, channel discharge, and the strength of the flow. Moreover, even of the river flow is stable, for a long period of flow, the sedimentation and the erosion of the riverbanks is an ongoing natural process [22].

3. Erosion

The removal process of soil particles from the bed and banks of the river, known as the erosion process [15]. Erosion concerns to many factors that could be categorized under three types, geometric factors that include the geometry of the channel, geomorphic factors that include the soil characteristics of the channel bed and banks, and hydraulic factors that include the flow characteristics. To determine the amount of eroded soil from the channel bed and banks, three variables should be known, which are the fluid shear stress, critical shear stress, and erosion rate. Previous researchers have presented several methods to calculate these variables; the following section will present some of these methods.

3.1. Fluid Shear Stress

Fluid shear stress is one of the most important factors desired to calculate the amount of erosion. It is known as the force per unit area in the flow direction [23]. Fig 5 illustrates the distribution of shear stress in the two-dimensional uniform flow in a channel, the direction of flow within x-axis, z-axis perpendicular to the flow. The stresses and forces implicate the horizontal component of the fluid weight W_x, shear stress τ on BC, and the hydrostatic forces on AB and CD. All forces implement in the x-direction. The hydrostatic forces are equal and opposite when the flow is uniform and W_x must be equiponderant by the shear force.

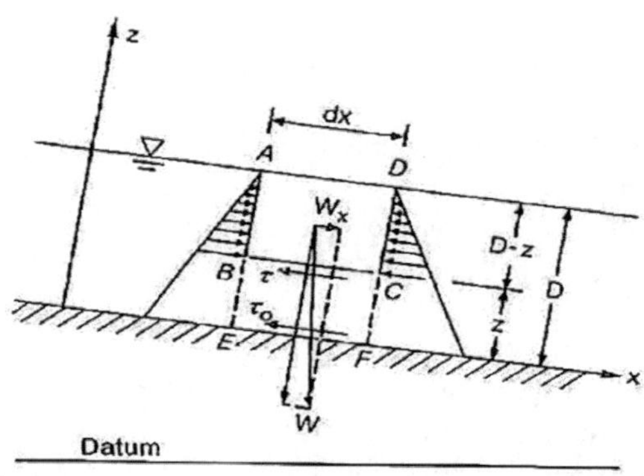

Figure 5. Schematic of forces acting on a unit volume of flow [21]

Fenton, 2005 presented the tractive force or drag force method, the tractive force resulting from the shear stress that is exerted by water flow on the channel bed and banks, occurs in straight channels with uniform flow. It is the most dependable method in channels that are exposed to erosion or scours in the boundary, where the erosion occurs as the forces that cause particle movement are larger than the resisting forces, this concept was reintroduced by [18]. The distribution of the tractive force or the shear stress on the channel bed and banks approximately considered to be equal to as shown in Eq. 3 and 4 respectively [24].

$$\tau_o = \gamma RS \quad (3)$$

$$\tau_o = 0.76\,\gamma RS \quad (4)$$

Where τ_o is the shear stress, γ is the specific weight of water, R is the hydraulic radius, S is the channel bed slope. Anderson, 1970 presented the same concept of the different of the shear stress in trapezoidal channels between the channel bed and banks, where the maximum shear stress occurs at the channel bed, and the minimum shear stress occurs at the top of the channel banks, as shown in Fig. 6 [25]. For this specific cross-section of side slope 2:1, the maximum shear stress on the bed and banks are 1.37γRS, and 1.08γRS, respectively.

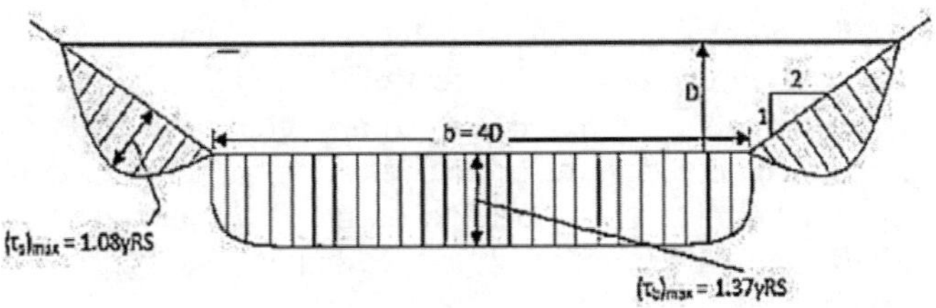

Figure 6. Shear stress distribution in the boundary of the trapezoidal channel [21]

The shear stress coefficient on the bed and banks of the channel was adjusted by [21], as a function of the ratio of the bed width to the water depth, Fig.7 (a and b) illustrates these coefficients for different side slopes.

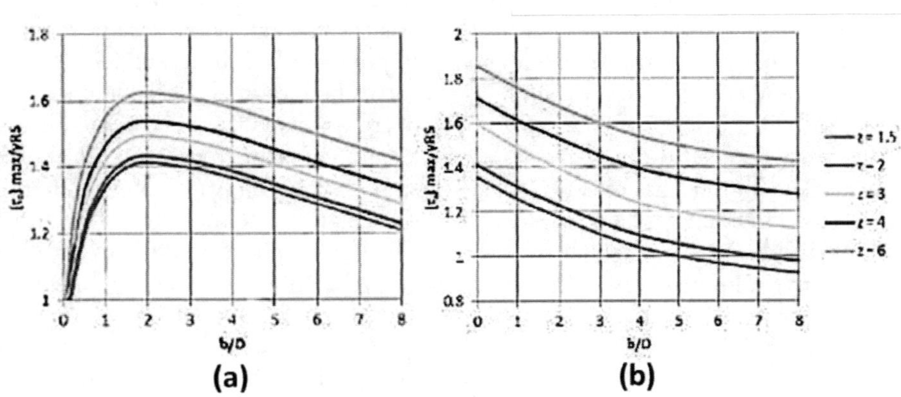

Figure 7. Maximum shear stress coefficient on the channel boundary with different side slope; (a) on the bed, and (b) on banks [21]

3.2. Critical Shear Stress

Soil particles begin to move when the force that causing movement is larger than the resisting forces. The resisting forces of soil particles depend on the soil properties, for example, in the non-cohesive soils like sand and gravel, the amount of particles resistance to erosion is the amount of gravity force to its submerged weight [15]. The amount of critical shear stress for non-cohesive soils can be determined from the Shields curve as shown in Fig. 8 which represents the relationship between soil particle diameters ranging from 0.1 mm to 10000 mm and the amount of critical shear stress ranging from 0.1 Pa to 10 Pa, this relationship based on several experiments conducted on a set of flat layers of sand. Shields curve has been developed by many researchers to include a larger range of particle diameters, but it nevertheless provides critical shear stress only on the channel bed [13].

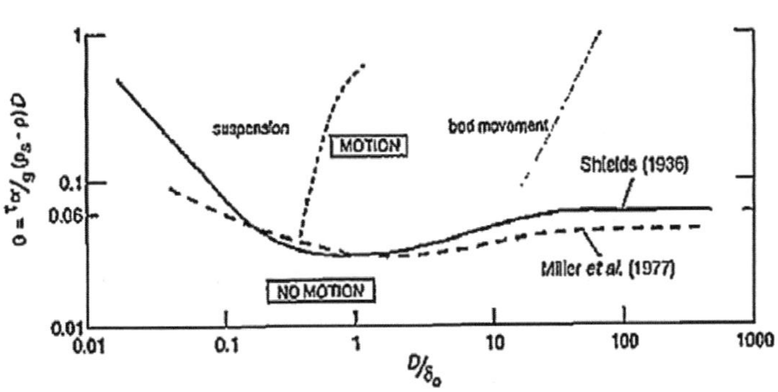

Figure 8. Shields curve [15]

IOP Conf. Series: Materials Science and Engineering **901** (2020) 012014 doi:10.1088/1757-899X/901/1/012014

The critical shear stress in the channel sides is different from that on the channel bed, where the particles in the channel sides slip down due to the effect of the weight component along the side slope, while the particles on the channel bed are exposed to the drag force in the direction of flow. To calculate the critical shear stress at the channel sides, the critical shear stress at the channel bed is multiplied by the factor *K*, which represent the drag force required to initiate movement on the channel sides relative to the tractive force required to initiate movement on the channel bed, Eq. 5 is used to determine the value of *K* factor, [26].

$$K = \sqrt{1 - \frac{sin^2\theta}{sin^2\emptyset}} \quad (5)$$

Where, θ is the side slope angle, and ∅ is the friction angle of soil. Critical shear stress for cohesive soils cannot be calculated from the Shields curve because the gravity force that causes the particles to slide down the side slope is much smaller than the cohesive force between these particles, so it is neglected. The resisting force for cohesive soil depends on the physical and chemical properties such as the chemical bonding between particles, interaction with pore, and eroding fluid [27].

A study was presented by [27] to develop a quantitative method to predict the critical shear stress and the rate of erosion for natural soil, 42 soil samples were collected from different stream-bank sites, half of the samples were taken from stable sites where no erosion was occurring, the other half were taken from sites where significant erosion was causing a problem with bank stability. Preliminary tests were conducted on all the 42 soil samples, and then based upon the preliminary test results, 30 samples were selected for detailed analysis which consists of flume erosion test, rotating cylinder erosion test, and some of the other preliminary tests. The results of the study showed that as the salt concentration of the eroding fluid decreased, the critical shear stress of the soil decreased and the amount of change of the erosion rate increased. From Fig. 9 that developed by [28] and revised by [9], the critical shear stress of soil can be calculated as a function of sodium adsorption ratio SAR, electrical conductivity, and soil salt concentration [27].

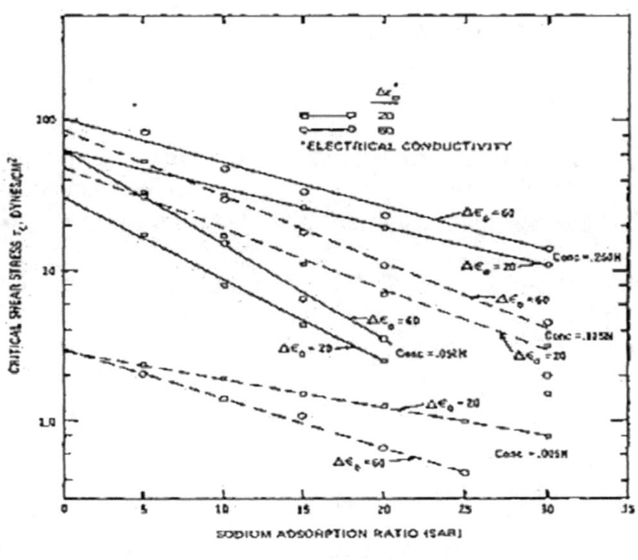

Figure 9. Critical shear stress as a function of SAR for different value of electrical conductivity and different soil salt concentration [27]

Sodium adsorption ratio is an indicator of soil salinity, which determines the ratio of sodium (Na) relative to calcium (Ca) and magnesium (Mg), as shown in Eq. 6 to determine the soil flocculation which improved water absorption. Calcium and magnesium are good soil flocculation while sodium is not. The term double-layer diffusion propagates on the surface of soil particles that are negatively charged, and which attract positively charged particles [30].

$$SAR = \frac{Na^+}{\sqrt{\frac{Ca^{++}+Mg^{++}}{2}}} \tag{6}$$

Another method for measuring critical shear stress was introduced by [31], using Erosion Functional Apparatus (EFA), where the soil sample is pumped using Shelby tube then flow is poured on it and at certain limits of velocity, after that, the time required to erode the soil is calculated and under the effect of a certain velocity, and the shear stress is determined using Eq. 7.

$$\tau = \frac{1}{8} f \rho V^2 \tag{7}$$

Where τ is the shear stress, f is the friction factor obtained from the Moody chart, ρ is the water density, v is the flow velocity in the flume. The results introduced by [31] showed that the critical shear stress for fine-grained soil ranging from 0 to 5 Pa. In flume studies, the critical shear stress is measured visually or graphically, visually, by measuring the shear stress at failure, as mentioned by [32] showed that the critical shear stress occurs when "pitting of the surface" occur, while [33] stated that the critical shear stress occur when "the water became cloudy", and when "general movement of the soil composing the channel bed was observed" as mentioned by [34]. To avoid inaccuracies in visual calculations of critical shear stress, the graphical representation can be used by plotting the shear stress versus the erosion rate with a best-fit line, the shear stress value that intersects with the point where no erosion occurs, represent the critical shear stress value [35]. Eq. 8 related between the critical shear stress for cohesive soil and the clay percent by weight, was presented by [34] through an experimental flume study conducted on 11 soil samples.

$$\tau_c = 0.493 \times 10^{0.0182 P_c} \tag{8}$$

Where, τ_c is the critical shear stress (Pa), P_c is the percent clay by weight.

3.3. Erosion Rate

Erosion rate is the amount of soil that lost over a while, once the applied shear stress on the soil overcomes the critical shear stress. The determination of the erosion rate can be useful in back-calculated to predict the bank retreat. Eq. 9 used to estimate the erosion rate depending on the excess shear stress, for fine-grained soil where the erosion occurs due to soil scour or overland flow [35], [36].

$$\varepsilon = K_d(\tau_a - \tau_c)^a \tag{9}$$

Where, ε is the erosion rate (m/s), τ_a is the applied shear stress (Pa), τ_c is the critical shear stress (Pa), a is a component commonly assumed equal to 1, K_d is the erodibility coefficient (m^3/(N.s)) calculated from Eq. 10 as a function of critical shear stress [39].

$$K_d = 0.2\, \tau_c^{-0.5} \tag{10}$$

The average applied shear stress can be calculated using Eq. 11.

$$\tau_a = \rho g d S \tag{11}$$

Where ρ is the density of water (kg/m^3), g is the gravity acceleration (m/s^2), d is the height of water above the midpoint of the section (m), S is the channel slope. Lane, 1955 Calculated the erosion rate using Eq. 12 which depending on the eroded sample height represented by (h), and the time desired to erode the soil (t), where the criterion height of the soil sample is 1 mm from the flume bottom [23].

$$z = \frac{h}{t} \tag{12}$$

A relationship between the excess shear stress and actual erosion rate was developed by [38], as illustrated in Eq.13 and 14. This relationship based on the test results of [29].

$$dB = \frac{223\times10^{-4}\tau_c e^{-0.13\tau_c}}{\gamma_s} \tag{13}$$

Where, dB is the initial erosion rate (m/min per unit area), τ_c is the critical shear stress (dynes/cm^2), γ_s is the unit weight of soil (kN/m^3). The actual rate of erosion was then determined by applying the excess shear stress, as shown in Eq. 14.

$$dw = dB\frac{(\tau-\tau_c)}{\tau} \tag{14}$$

Where, dw is the actual erosion rate (m/min), τ is the fluid shear stress (dynes/cm^2).

4. Conclusions

The paper presents the basic principles for the study of riverbank stability and failure, in terms of the failure mechanism and its causes. The conclusions are:

- Several factors contribute to the destabilization of the riverbank, which include hydraulic factors, geomorphic factors, and geometric factors.
- The climate and geology that differs from one region to another, play a major role in the degree of riverbank failure, where the climate controls the precipitation hence the discharge of the river, in addition to water level variation between summer and winter, also the effect of humidity, drought, ice, and melting. However, the geology controls the sediment load.
- The erosion rate of the riverbank is considered an important cause of riverbank failure, erosion rate depends on the fluid shear stress, and the critical shear stress. Previous researchers developed experimental and theoretical methods to calculate the critical shear stress which mainly depends on the material properties.
- The importance of the study of riverbank stability is to understand the reasons that led to failure, and thus to identify the necessary treatment that is compatible with the hydraulic requirements to prevent the recurrence of failure, also that to design stable slopes, one needs to know the changes that occur in the soil properties within the slope overtime under the effect of different factors that lead to failure.

Acknowledgment

The authors would like to thank the Wasit University, college of engineering, civil engineering department for supporting this research paper which is a part of an extensive experimental program of designing, and testing of flume apparatus.

References

[1] S.E. Darby, Modelling width adjustment in straight alluvial channels, John Wiley and Sons: Chichester (2000).
[2] J. M. Tutkaluk, The effect of seasonal variations in the Red River and upper carbonate aquifer on riverbank stability in Winnipeg, MSc Thesis, Department of Civil and Geological Engineering University of Manitoba, Manitoba (2000).
[3] J. H. Al-Jwthery, Stability of riverbank of part of Hilla River in Babylon governorate and its engineering effects. 26(1C) 37-47 (2008).
[4] S.Sh. Alhadi, B.N. Albadran, and N.Kh. Albahili, The effect of some soil properties on bank stability of Shatt Al- Arab river from Sindibad to Umm Al-Ressass islands. 24(1) 51-64 (2009).
[5] A. Samadi, E. Amiri-Tokaldany, M. Davoudi, and S.E. Darby, Experimental and numerical investigation of the stability of overhanging riverbanks, Geomorphology. 184(2013) 1-19.
[6] A. Jianfar, Evaluation of erosion rates and their impact on riverbank stability, MSc Thesis, Department of Civil Engineering University of Manitoba (2014).
[7] S. Patsinghasanee, I. Kimura, Y. Shimizu, and M. Nabi, Cantilever failure investigations for cohesive riverbanks, Proc. Inst. Civ. Eng. Water Manage. Proceedings of the Institution of Civil Engineers: Water Management. 170(2) 93-108(2017).
[8] H.A. AL-Thamir and T.M., AbdulAzeez, Two–Dimensional mathematical model to study erosion problem of Tigris River banks at Nu'maniyah, Journal of Engineering. 23(1) 112-35 (2017).

[9] C.H. Chen, T.Y. Hsieh and JC Yang, Investigating effect of water level variation and surface tension crack on riverbank stability. Journal of hydro-environment research. 15(1) 41-53 (2017).
[10] Y.-C. Wang and C.-C. J. W. Lai, Evaluating the erosion process from a single-stripe laser-scanned topography: A laboratory case study. 10 (7) 956 (2018).
[11] H. S. Kharismalatri, Y. Ishikawa, T. Gomi, R. C. Sidle, and K. J. W. Shiraki, Evaluating factors for controlling sediment connectivity of landslide materials: A flume experiment 11 (1) 17 (2019).
[12] P. Novák, A.I. Moffat, C. Nalluri, and R. Narayanan, Hydraulic structures. London; New York: Taylor & Francis, 2007.
[13] M. Morisawa, Streams: their dynamics and morphology. New York. McGraw-Hill Book Co., (1968).
[14] M. Morisawa. Rivers and valleys of Pennsylvania, revisited, Geomorphology. 2(1-3)1-22 (1989).
[15] K.S. Richards, Rivers form and process in alluvial channels. London; New York: Methuen (1985).
[16] D. Knighton, Fluvial forms and processes: a new perspective. London; Oxford University Press: New York, N.Y. Arnold (1998).
[17] V.T. Chow, Open-channel hydraulics, New York: McGraw Hill Book (1959).
[18] J. Fenton. Open channel hydraulics. Engineering Hydraulics and Hydrology. 54(6) (2005).
[19] H. Chanson. Hydraulics of open channel flow. Elsevier (2004).
[20] J.L. Florsheim, J. F. Mount, and A. J. B. Chin, Bank erosion as a desirable attribute of rivers, BioScience. 58(6) 519-529 (2008).
[21] H.H. Chang, Fluvial processes in river engineering, New York: John Wiley & Sons (1988).
[22] E.W. Lane, S. Leliavsky, Design of stable channels. New York, N.Y.: American Society of Civil Engineers, ASTM (1955).
[23] [24] O.J. Olsen, Q. L. Florey, Sedimentation studies in open channels: boundary shear and velocity distribution by the membrane analogy, analytic, and finite-difference methods. Denver, Colo., Bureau of Reclamation (1952).
[24] A.G. Anderson, A. S. Paintal, and J. T. Davenport, Tentative design procedure for riprap-lined channels, Washington: Highway Research Board, Division of Engineering, National Research Council (1970).
[25] K. Arulanandan, E. Gillogley, and R. Tully, Development of a quantitative method to predict critical shear stress and rate of erosion of natural undisturbed cohesive soils. Vicksburg, Miss.: U.S. Army engineer waterways experiment station, Geotechnical Laboratory (1980).
[26] A. Alizadeh, Amount and type of clay and pore fluid influences on the critical shear stress and swelling of cohesive soils, Ph. D. thesis. Davis. California, University of California–Davis (1974).
[27] R.T. Heinzen, Erodibility criteria for soils, Ph. D. thesis. Davis. California, University of California–Davis (1976).
[28] S. Barbour and D. Fredlund, Physico-chemical state variables for clay soils, in Proceedings, XII International Conference on Soil Mechanics and Foundation Engineering, Rio De Janeiro, Brazil. 1839-1843 (1989).
[29] J.L. Briaud, F.C. K. Ting, H.C. Chen, Y. Cao, S.W. Han, and K.W. Kwak, Erosion function apparatus for scour rate predictions, Journal of Geotechnical and Geoenvironmental Engineering. 127(2)105-113 (2001).
[30] J. W. Kamphuis and K. R. Hall, Cohesive Material Erosion by Unidirectional Current, Journal of Hydraulic Engineering. 109(1) 49-61 (1983).
[31] I.S. Dunn, Tractive resistance of cohesive channels. Journal of the Soil Mechanics and Foundations Division. 85(3):1-24(1959).
[32] E. T. Smerdon and R. P. J. A. E. Beasley, Critical tractive forces in cohesive soils, Agricultural Engineering Journal. 42(1) 26-29 (1961).
[33] G.J. Hanson and K.R. Cook, Development of excess shear stress parameters for circular jet testing, American Society of Agricultural Engineers. No. 972227(1997).
[34] E. Partheniades, Erosion and deposition of cohesive soils, Journal of the Hydraulics Division. 91(1) 105-39 (1965).
[35] G.J. Hanson and A. Simon, Erodibility of cohesive streambeds in the loess area of the midwestern USA, Hydrological Processes. 15(1) 23-38 (2001).

[36] A.M. Osman and C.R. Thorne, Riverbank stability analysis. I: Theory. Journal of Hydraulic Engineering. 114(2) 134-50 (1988)
[37] L. Fernando, The effect of flow induced erosion on riverbank stability along the Red River in Winnipeg, Department of Civil Engineering University of Manitoba Winnipeg, Manitoba (2007).

IOP Conf. Series: Materials Science and Engineering **901** (2020) 012015 doi:10.1088/1757-899X/901/1/012015

Assessment of the Three-Dimensional Model Produced by Photogrammetric and Geodetic Surveying Techniques

Yousif Hussein Khalaf[1,a*] and Nagham Amer Abdulateef[1,b]

[1]Department of Surveying Engineering, University of Baghdad, Baghdad, Iraq.
[a]yousif.hussain1976@gmail.com, [b]nagham.amer1984@gmail.com

Abstract. A combination of close-range photogrammetry techniques and geodetic surveying techniques are used to produce the three-dimensional model, which is one of the major tasks in the field of surveying at present because it is used in many engineering applications, industrial, medical and in cultural heritage objects which requires regular monitoring because of the high accuracy and reliability needed to conduct the measurements. The study area was in the university of Baghdad, specifically for the building of the university tower this research has been produced the three dimensional model of the tower depending on the overlapping ground images captured by a digital camera and the total station instrument used to measure the coordinates of ground points distributed on the university of Baghdad tower. The mathematical model used in this study is the direct linear transformation method to intensify ground points which are used to produce the three-dimensional model of the university tower. Also, a photogrammetric program is used to compute the object space coordinates from the point cloud three-dimensional modeling.

Keywords: Close range photogrammetry; geodetic surveying; DLT; MATLAB; Agisoft program-3D.

1. Introduction

Digital close-range photogrammetry is a branch of photogrammetry which has been widely used for non-topographic applications in industrial and engineering areas with the digital technology and corresponding algorithms [1]. The term of close-range photogrammetry is defined as the art, science, and technology of extracting information of an object's geometry and position from photographic images. this information may be used to recreate an accurate representation of an object in three-dimensional space, alternatively, it may be used to measure the change of an object position or shape [2]. This technique is used when an object to camera distance is not more than 300 meters as a maximum and a fraction of millimeters as a minimum. these limitations are assumed to distinguish between terrestrial, and close-range photogrammetry [3].

The close-range photogrammetry has a wide range of use and its applications are in many disciplines. such applications are different in terms of its importance some need high measurement accuracy and others don’t, the accuracy of these applications is a function of the elements of the photogrammetric system. these elements are the camera used, the accuracy achieved by any camera is a function of the principal distance, the format size, the error of image points, and the resolution of the image, the layout of the two camera stations, the layout of the two camera stations relative to the object is a function of the external orientation parameters. These parameters determine the distribution, transformation, and magnification of the image error to the corresponding object points [4]. The techniques of three-dimensional (3D) modeling have been advancing over the past few years. One of the main tasks of photogrammetry is to find the precise 3D coordinates of an object, these coordinates must be obtained from a stereo pair images. Features in the image can be observed and 3D coordinates are determined in a given coordinates system, under a mathematical model Direct Linear Transformation (DLT) [5].

2. Digital Close Range

Close range Photogrammetry is one of the most important subjects in geomantic engineering, according to the development in 3D modeling from a digital image, for engineering applications, which needs high resolution, many programmers have appeared which process the digital images [6].

3. Mathematical Models

The used mathematical model is the direct linear transformation (DLT) model which rely on means direct transformation from comparator coordinates of image points to object space coordinate [7]. This concept is particularly useful for imaging with non-metric cameras that have one in *x* image coordinate and the other in *y* image coordinate. The standard (DLT) equations are:

$$x = \frac{L1X+L2Y+L3Z+L4}{L9X+L10Y+L11Z+1} \quad (1)$$

$$y = \frac{L5X+L6Y+L7Z+L8}{L9X+L10Y+L11Z+1} \quad (2)$$

Where x and y are the measured coordinates of an image point. Fiducial marks in the focal length to define the axes of the photo coordinate system.[8]. The minimum number of points required to solve these equations using the least-squares method is six points. Each point provides two equations *X, Y, Z:* the object space coordinates system. L1, L2, …, L11: the eleven transformation coefficients ((DLT) parameters). Equations (1) and (2) can be reduced to the following form:

$$L1X + L2Y + L3Z + L4 - xL9X - xL10Y - xL11Z = x \quad (3)$$

$$L5X + L6Y + L7Z + L8 - yL9X - yL10Y - yL11Z = y \quad (4)$$

In matrix form

$$\begin{bmatrix} X1 & Y1 & Z1 & 1 & 0 & 0 & 0 & 0 & -x1X1 & -x1Y1 & -x1Z1 \\ 0 & 0 & 0 & 0 & X1 & Y1 & Z1 & 1 & -y1Y1 & -y1Y1 & -y1Z1 \\ X2 & Y2 & Z2 & 1 & 0 & 0 & 0 & 0 & -x2Y2 & -X2Y2 & -x2Z2 \end{bmatrix} \begin{bmatrix} L1 \\ L2 \\ \vdots \\ \vdots \\ L11 \end{bmatrix} = \begin{bmatrix} x1 \\ y1 \\ \vdots \end{bmatrix} \quad (5)$$

$$B.X = f \quad (6)$$

Where X is the vector of (DLT) parameters. (DLT) parameters can be obtained by using the principle of least squares method, we should consider solving the normal equation that can be derived from the observation, which can be expressed as [9]. The normal equation of the previous observation equation can be written as follows:

$$(BtB)X = (Btf) \quad (7)$$

By multiplying both sides of Eq. (7) by by $(B^tB)^{-1}$ and reducing we obtain:

$$X = (BtB) - 1\ Btf \quad (8)$$

After computing the DLT coefficients for all images, the ground coordinate of any object point is computed. The formulas for computing object space coordinate from (DLT) coefficients is as follows [10]:

$$(L1 - xL9)X + (L2 - xL10)Y + (L3 - xL11)Z = x - L4 \quad (9)$$

$$(L5 - yL9)X + (L6 - yL10)Y + (L7 - yL11)Z = y - L8 \quad (10)$$

If the solution is done by using Matrix (which mean measuring the ground coordinates for one point in two photos) the equations will be as the following: -

IOP Conf. Series: Materials Science and Engineering **901** (2020) 012015 doi:10.1088/1757-899X/901/1/012015

$$\begin{bmatrix} L_1^{(1)} - xL_9^{(1)} & L_2^{(1)} - xL_{10}^{(1)} & L_3^{(1)} - xL_{11}^{(1)} \\ L_5^{(1)} - yL_9^{(1)} & L_6^{(1)} - yL_{10}^{(1)} & L_7^{(1)} - yL_{11}^{(1)} \\ L_1^{(2)} - xL_9^{(2)} & L_2^{(2)} - xL_{10}^{(2)} & L_3^{(2)} - xL_{11}^{(2)} \\ L_5^{(2)} - yL_9^{(2)} & L_6^{(2)} - yL_{10}^{(2)} & L_7^{(2)} - yL_{11}^{(2)} \end{bmatrix} \begin{bmatrix} X \\ Y \\ Z \end{bmatrix} = \begin{bmatrix} x^{(1)} - L_4^{(1)} \\ y^{(1)} - L_8^{(1)} \\ x^{(2)} - L_4^{(2)} \\ y^{(2)} - L_8^{(2)} \end{bmatrix} \quad (11)$$

4. The Agisoft Program

Agisoft PhotoScan is a stand-alone software product that performs photogrammetric processing of digital images and generates 3D spatial data [6]. After opening the digital photos in (Agisoft) control points are selected and their ground coordinate are interred, then the processes of creating the 3D model is extracted, the processing of aligning photos are done to calculate exposure stations.

5. Experimental work

This research aims to compute the 3D coordinates the are used to produce three models according to the following steps:

1) Selecting a baseline with a length of 10 m.
2) Measuring the local 3D coordinates of 178 control points and checkpoints on the University of Baghdad tower, total station instruments are used to fix the coordinates of the control points and baselines with millimetric accuracy in a predesigned network [11].
3) Using Nikon D5200 camera to take a stereo pair for the tower with an overlap equal to 100%.
4) Measuring the digital photo coordinates to compute DLT parameters.
5) Using DLT program (MATLAB language) to intensify points, 250 points have been intensified points that are inaccessible.
6) The property of symmetric in the shape of Baghdad university tower has been used to compute the coordinates of the other side and then get a 3D drawing of the tower.
7) The 3D design has been produced by using AutoCAD, QT Modeler programs, and Agisoft program.
8) The control points were precisely measured by using Topcon total station ES 105. A baseline of ten meters has been chosen with two end stations A and B. The 3D coordinates of station A assumed locally to be (5000, 5000, 40) meter, then the total station has been mounted on A and backsight on B, the 3D coordinates of the control points have been measured directly by the total station. The control points and checkpoints have been distributed on the tower same of these points are shown in Fig. 1.

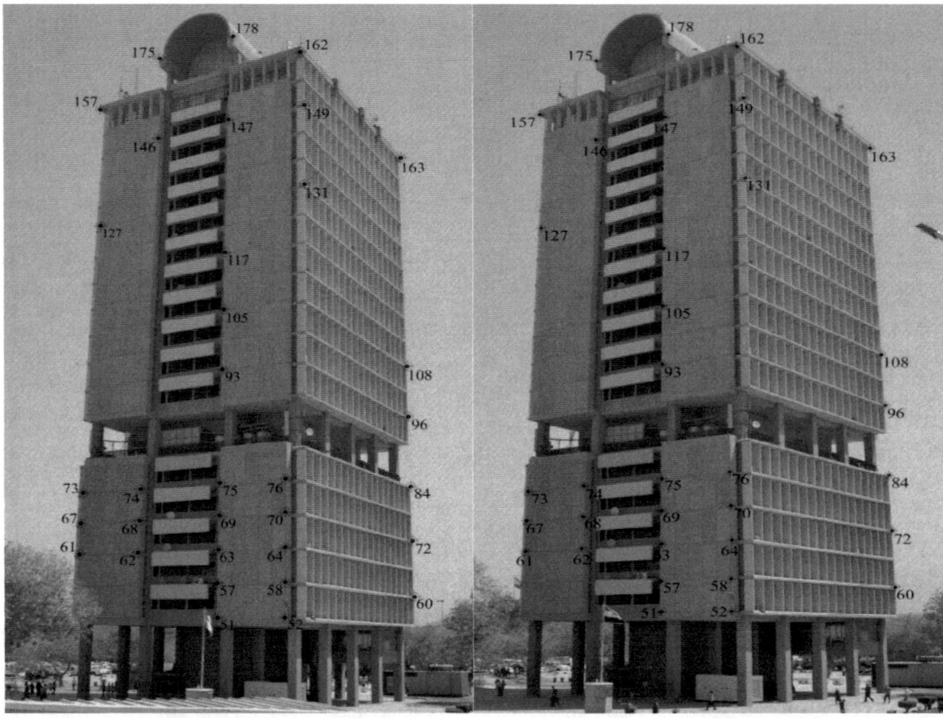

Figure 1. Stereopair of photos for the Baghdad university tower.

To establish the DLT, at least six control points are required. If more than six points are used, so that and improved solution can be arrived at by using least. square. In this research 23 points are used to compute DLT parameter the measured ground coordinates of these points are listed in Table 1.

Table 1. The measured object coordinates of the control points

Point	Northing	Easting	Elevation	Point	Northing	Easting	Elevation
Occ	500.0000	500.0000	40.0000	127	513.6983	598.8347	35.1839
B.S	508.6860	499.9978	39.9550	131	512.9127	598.5543	35.1790
63	538.4229	595.4597	35.1676	60	512.5939	599.2462	35.1811
67	537.6394	595.1849	35.1799	72	511.5658	604.4550	35.1982
68	537.3381	595.9473	35.1938	84	510.7608	604.1059	35.1823
69	532.2048	593.0862	35.1875	96	510.4714	604.9048	35.1902
73	531.4138	592.7802	35.1900	108	509.4389	610.0368	35.1944
74	531.1309	593.6083	35.2160	163	508.6574	609.7711	35.1878
75	524.2462	590.0477	35.1813	63	508.3522	610.5832	35.1718
51	523.4825	589.7672	35.1773	67	538.3408	595.3009	43.0456
52	523.1548	590.5008	35.1870	68	537.6541	595.2289	43.0604
57	517.9772	587.6124	35.1812	69	537.3600	596.0064	43.0612
58	517.1893	587.3590	35.1902	73	532.1634	593.0882	43.0527
147	516.8882	58.1155	35.1881	74	531.4112	592.8137	43.0620
175	515.8454	593.2299	35.1866	75	531.1327	593.5400	43.0467
93	515.0512	592.9373	35.1752	51	524.2568	590.1744	43.0721
117	514.7514	593.7130	35.1773				

23 control points distributed on the University of Baghdad tower are chosen for computing the (11 DLT) parameters. The photo coordinates were measured in the stereopair which is illustrated in figure (1). The results of image measurements have been listed in Table (2)

Table 2. The measured photo coordinates of the control points.

Point No.	Left image		Right image	
	x_1 (mm)	y_1 (mm)	x_2 (mm)	y_2 (mm)
63	166	321.9	167.3	328.7
67	112.6	310.5	118.2	318.7
68	135.5	309.2	139.1	316.8
69	166.3	307.3	167.4	314.2
73	113.4	297.1	118.7	305.6
74	136.1	295.4	139.4	303.1
75	166.6	292.8	167.6	299.8
51	165.3	352.3	167.1	358.8
52	193.1	352.3	193.1	358.2
57	165.6	337	167.2	343.5
58	193.2	336.3	192.9	342.3
147	170.3	131.7	168.7	140.9
175	143.6	104.4	143.8	115.8
93	167.6	242.2	167.7	249.7
117	168.9	190.4	168.2	198.8
127	120.4	178.9	123.6	189.9
131	200.7	160.2	197.8	167.2
60	242.9	342.8	251.5	347.6
72	242.3	318.1	250.3	322.7
84	241.7	294	249.1	298.5
96	241	263.3	247.7	267.5
108	240.3	240.8	246.5	244.9
163	238.3	150.2	242.7	154.4

The DLT Eqs. (1) and (2) were applied to compute the 11 DLT parameters and this process has been done by preparing a program in Matlab language as in Fig. 2. Matlab can be used for math computations, modeling and simulations, data analysis and processing, visualization and graphics, and algorithms development [12]. The results of the Matlab program are listed in Table 3.

Table 3. DLT parameters of the left and right photos.

Par.	Left photo	Right photo	Par.	Left photo	Right photo
L_1	-0.40018	-0.43797	L_7	0.55700	0.51761
L_2	0.65625	0.58083	L_8	328.66370	341.26267
L_3	-0.09993	-0.08759	L_9	-0.00168	-0.0015
L_4	-129.80528	-67.63666	L_{10}	-0.00026	-0.00039
L_5	-0.61599	-0.57830	L_{11}	-0.00050	-0.00049
L_6	-0.08596	-0.14603			

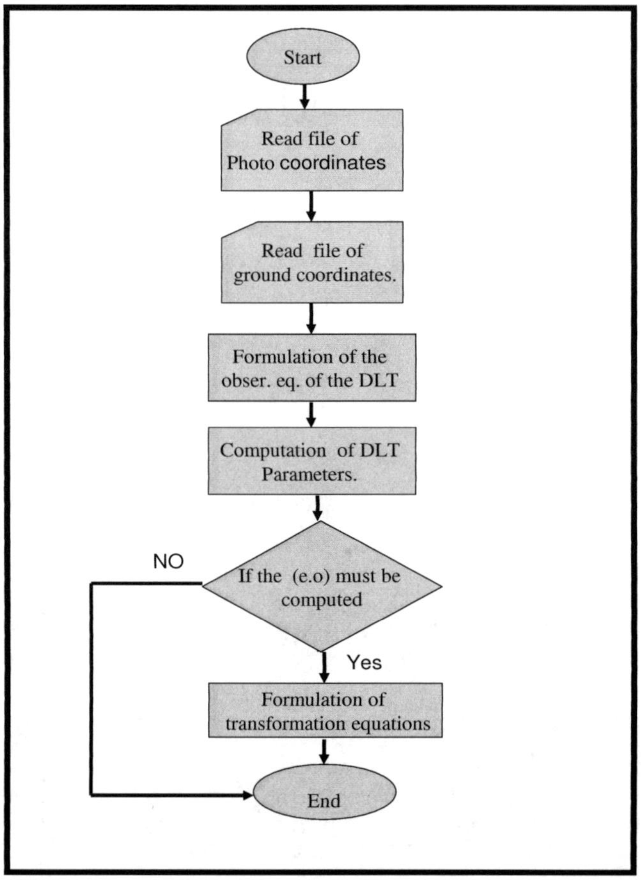

Figure 2. Flow chart of the computation of DLT parameter.

The three-dimensional coordinates of any point appear in overlapped photos can be computed by using equation (11) after computing the 11 DLT parameters of and the photo coordinates in a stereopair. 23 checkpoints were selected to validate the resulted accuracy of the 3D coordinates by determining the Root Mean Square Error (RMSE) which was found to be about RMSE = 2.2 cm as shown in Table 4.

Table 4. The coordinates of checkpoints and their RMSE.

Point No.	X (m)	Y (m)	Z (m)	ΔX (m)	ΔY (m)	ΔZ (m)
63	588.620	524.357	50.328	0.091	-0.027	0.025
67	594.238	538.601	53.646	-0.955	-0.050	-0.012
68	591.872	532.284	53.625	-0.119	0.032	0.009
69	588.683	524.374	53.585	0.034	-0.046	0.050
73	594.076	538.551	56.897	0.076	-0.010	0.025
74	591.701	532.551	56.883	0.055	0.070	0.034
75	588.955	524.431	56.934	-0.271	-0.104	-0.011
51	588.685	524.374	43.787	0.017	-0.040	-0.057
52	586.383	518.041	43.801	-0.075	0.040	-0.054
57	588.773	524.398	47.068	-0.068	0.055	-0.017
58	586.252	518.017	47.079	0.055	0.069	-0.021
147	588.786	524.365	98.723	-0.031	-0.037	-0.073
175	591.989	532.573	110.016	-0.075	0.082	0.026
93	588.712	524.414	68.945	0.011	-0.066	0.007
117	588.721	524.378	82.232	0.016	-0.038	-0.068

127	594.206	538.624	88.837	-0.039	-0.101	-0.074
131	586.688	516.367	88.870	-0.056	-0.022	-0.108
60	610.747	507.304	47.109	-0.163	-0.098	-0.062
72	610.634	507.259	53.611	-0.054	-0.044	0.0005
84	610.458	507.219	60.145	0.101	-0.001	0.050
96	610.464	507.143	68.858	0.090	0.055	0.071
108	610.505	507.138	75.5002	0.053	0.055	0.038
163	610.410	506.855	105.077	-0.035	0.033	-0.064
RMSE	±0.025		±0.024		±0.019	
	Total RMSE			±0.022		

After applying the DLT equations, the 3D coordinates of the intensified points are computed and then these coordinates could be used to produce the 3D model of the Baghdad University tower this model is produced in this research by using two programs Auto CAD and QT modeler this drawing could be illustrated in Fig. 3.

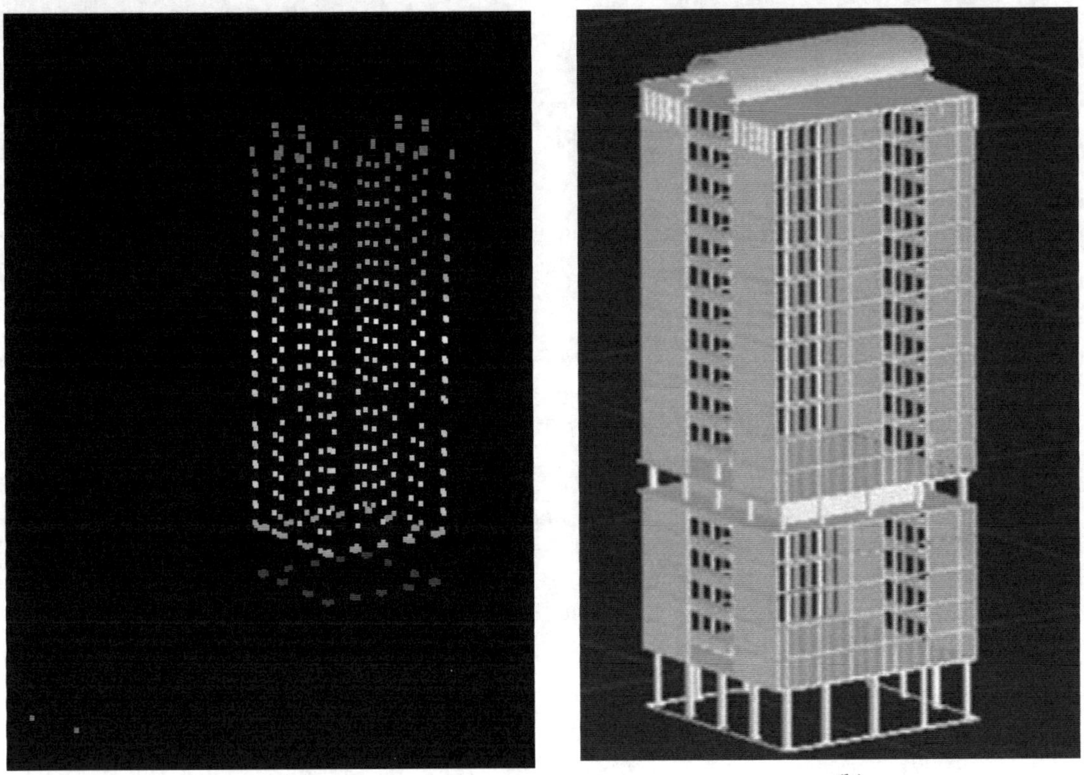

(a) (b)

Figure 3. (a) 3D model for the University tower by using (QT Modeler) (b) 3D Model by using (Auto CAD).

After aligning the camera photos point cloud is built to calculate 3D coordinates of the model university of Baghdad tower, then points cloud mesh will create the 3D structure of the model and filtering is required to remove all unneeded noise for university of Baghdad tower, finally, the texture is made as shown in Figs. 4a and 4b.

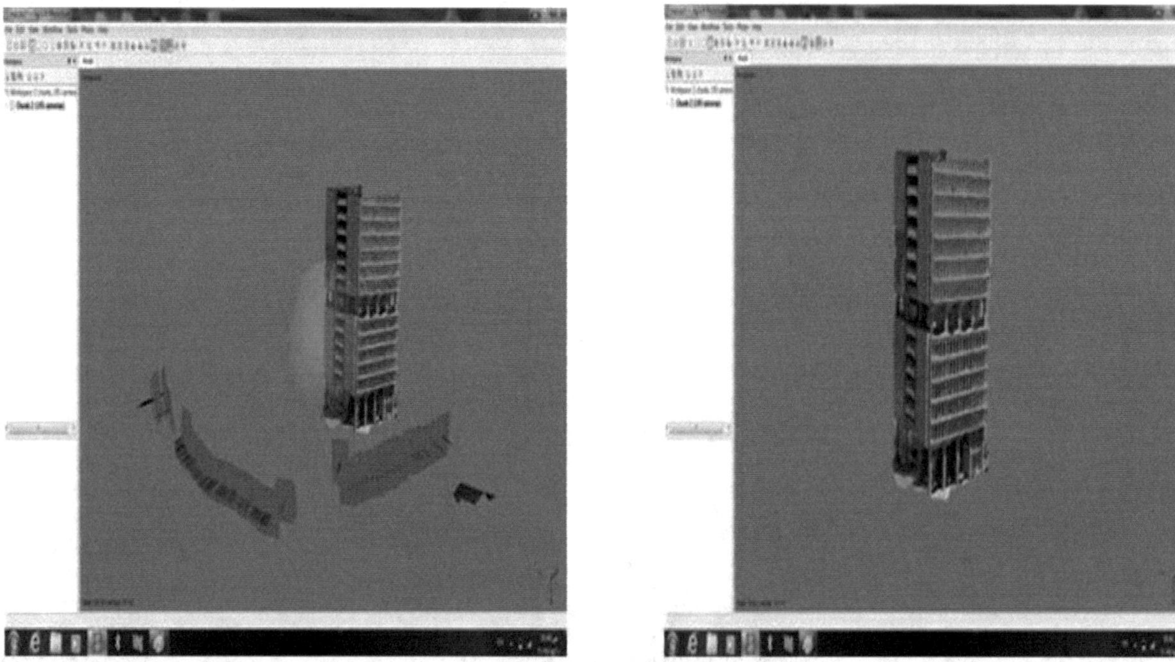

(a) **(b)**

Figure 4. (a) Filtering and building mesh for the University of Bagdad Tower, (b) 3D Model texture University of Baghdad Tower.

6. Conclusion

From the results obtained in this research, we concluded that the presented technique of DLT gives good accuracy which is suitable for many engineering applications, like architectural engineering and industrial engineering. The DLT could be used to recreate the 3D model of an object which is one of the important applications in close-range photogrammetry especially when the objects are inaccessible. The presented technique in this research can save time and effort because of the use of DLT equation which was easier than the other mathematical model in programming, the use of laser total station (without reflector), and digital camera. The other adopted techniques are represented by computing the 3D coordinates obtained from the Agisoft program is gave a centimetric accuracy of the point cloud in the 3D modeling of the tower but this photogrammetric technique is needed a high specification computer because there are millions of points of the point clued in the 3D modeling.

References

[1] Li, X., (1999). Photogrammetric Investigation into Low- Resolution Digital Camera Systems, Doctoral Thesis, University of New Brunswick, Canada, 1999.
[2] Mikhail, E., M., Bethel, J. S, Glone, J. C, Wiley, J., and Inc, S. (2000). Introduction to Modern Photogrammetry.
[3] Parmehr E., G., and Azizi A., (2004). A Comparative Evaluation of the Potential of Close Range Photogrammetric Technique for the 3D Measurements of the Body of a Nissan Patrol Car, ISPRS-Helsinki.
[4] Remondino, F., (2006). Image-Based Modeling or Object and Human Reconstruction, Doctoral Thesis, Aristoteles University, Thessaloniki, Greece, Zurich.
[5] Wolf, R., and Dewitt A., (2000). Elements of Photogrammetry with Application in GIS, 3rd edition.
[6] Wong, X., and Clarke T.A, (1998). Separate Adjustment of Close Rang Photogrammetric Measurements, city University, London.

[7] Parmehr E., G., and Azizi A., (2004). A Comparative Evaluation of the Potential of Close Range Photogrammetric Technique for the 3D Measurements of the Body of a Nissan Patrol Car, ISPRSHelsinki.

[8] Satchet, M., S., (2004). Positioning by Analytical Photogrammetry with Unknown Camera Parameters, MSc. Thesis, College of Engineering, Baghdad University.

[9] Sharqi, M., M., (2002). Accuracy Assessment of Digital & Analytical Close Range Photogrammetric Techniques, MSc. Thesis, College of Engineering, Baghdad University.

[10] Abulateef, N., A., (2009). Accuracy Evaluation of Digital Close Range Photogrammetry by Free Adjustment Method, MSc. Thesis, College of Engineering, Baghdad University.

[11] Bashar S., Nagham A., Yousif H.,(2019) Out of Plumb Assessment for Cylindrical-Like Minaret Structures Using Geometric Primitives Fitting. ISPRS Int. J. Geo-Inf. 8, 64.

[12] Yousif H., Nagham A.,(2018). Orthophoto Production from Aerial Photograph by using Matlab and GIS. International Journal of Civil Engineering and Technology (IJCIET) Volume 9, Issue 9.

IOP Conf. Series: Materials Science and Engineering **901** (2020) 012016 doi:10.1088/1757-899X/901/1/012016

Analysis of Hyperspectral Remote Sensing Images for Extraction Geological Rock Types Maps by Geospatial Techniques

Hussein Sabah Jaber[1,a*], **Zainab Fakhri Merzah**[2,b]

[1]Department of Surveying, University of Baghdad, Baghdad, Iraq
[2]College of Science for Women, University of Baghdad, Baghdad, Iraq
[a]husseinsabah00@gmail.com, [b]zanaib2013@gmail.com

Abstract. Hyperspectral images involve a wealth of data, but interpreting them needs an understanding of exactly what properties of ground materials that it is trying to measure and how to relate to the measurements made by the hyperspectral sensor. The purpose of the study is to use hyperspectral remote sensing data for finding of geological locations and the creation of geological rock types map. In this research, the Advanced Spaceborne Thermal Emission and Reflection Radiometer (ASTER) image has been used and analyzed using geospatial techniques such as ENVI software. These techniques include: enhancement, de-correlation stretch, band ratio and classification. All these techniques have been performed to produce geological rock kinds map. Also, 3D surface for ASTER data can be displayed using ENVI software.

Keywords: Hyperspectral, ASTER, band ratio, de-correlation stretch, classification technique.

1. Introduction

Hyperspectral data sets usually contain about 100 to 200 spectral bands of narrow bandwidths (5-10 nm), while, multispectral data usually contain about 5 to 10 bands of large bandwidths (70-400 nm) [1]. Figure 1 represents the assessment of hyperspectral and multispectral data of vegetation. In this Fig, it is indicated the more information of reflectance in hyperspectral data [2].

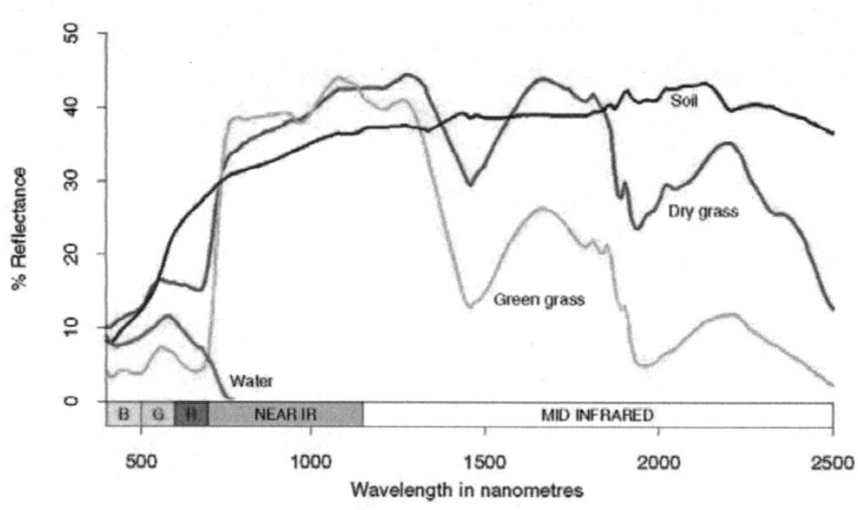

Figure 1. Evaluation of hyperspectral and multispectral data of a vegetation. [2].

Hyperspectral is usually denoted as a data cube with spatial data in the X-Y axis, and spectral data denoted in the Z-axis. Spectroscopy is a part of physics focused on the transmission, measurement and interpretation of electromagnetic spectra [3]. There are various uses which can take benefit of hyperspectral data as following:

1) Atmosphere: such as water vapour and cloud.
2) Ecology: such as leaf water.
3) Geology: such as rock types and soil types.

The objective of this research is to derive base-level exploration of hyperspectral remote sensing data that can be imported into digital field data capture environment to serve as base maps for geologic mapping to produce of geological rock types map using geospatial techniques.

2. Data source of Hyperspectral Imagery

In this research, ASTER data is used. ASTER is an imaging instrument onboard Terra. These data are used to create full maps of reflectance and elevation [4]. ASTER, in contrast to Landsat ETM, has no channel in the blue wavelength region, so the proposed band combination best mimics a true colour composite image displaying the area in its natural colours. ASTER sensor data as shown in Table 1.

Table 1. ASTER sensor information.

Satellite	Sensor	Bands	Range of spectral	Size of scene	Res. of pixel
	VNIR	1-3	0.52 - 0.86 μm		15 m
ASTER	SWIR	4-9	1.6 - 2.43 μm	120 X 150 km	30 m
	TIR	10-14	8.12 - 11.65 μm		90 m

The data set provided consist of a subset of an ASTER scene for the Lake Victoria Gold district in Tanzania where the data contains 15 bands. Displaying these data by using ENVI software showed the first 14 bands contain the VNIR_SWIR_TIR bands of ASTER, whereas band 15 in this data stack contains the DEM generated from the ASTER stereo pair. Also, colour composites generate a true colour composite image by displaying band 2 in red and band 1 in green and blue as shown in the Fig.2.

Figure 2. Displaying ASTER image data by using ENVI software.

IOP Conf. Series: Materials Science and Engineering **901** (2020) 012016 doi:10.1088/1757-899X/901/1/012016

3. Methods of Analysis

There are some processes of this research can be explained as follows:

3.1. Enhancement of ASTER image

The purpose of enhancement is to offer better contribution for additional computerized image processing [5, 6]. So, there are several kinds of enhancement methods in ENVI software. Three enhancement methods used to find clear visualization of the image are (linear, linear 2%, and equalization) as shown in the figure (3a, b and c). After using these methods, it can be noticed that the equalization method is a better visualization of the image.

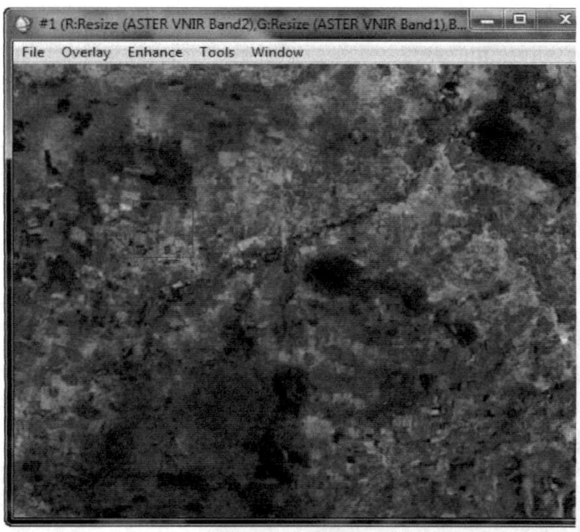

a. Linear algorithm

b. Linear (2%.) algorithm

c. Equalization algorithm

Figure 3. Enhancement of ASTER image.

3.2. De-correlation Stretch technique

The ASTER SWIR bands are rather narrow in terms of wavelength covered which depended on the area imaged often results in high correlation data redundancy between the bands [7]. It can be seen that this image is dominated by grey tones as shown in the left side of the Fig.4. De-correlation stretching technique is required to overcome the high connection in data sets and create a more colourful composite image [8, 9]. De-correlation stretching technique takes three bands, calculates the corresponding three principal components, then stretches the DN values maximally along the three principal axes and finally rotates the data back to the original reflectance/radiance data space as shown in the right side of the Fig. 4.

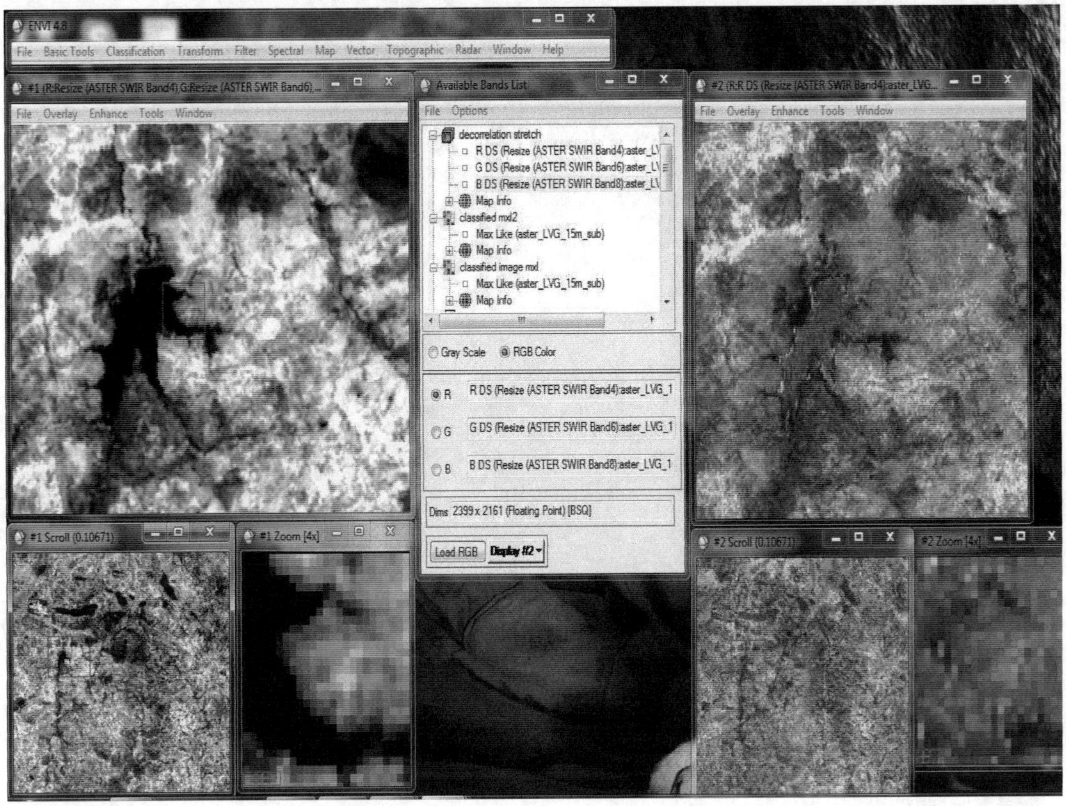

Figure 4. Image before (grey tones) and after applying de-correlation stretch (colourful composite).

3.3. Band Ratio method

Band ratios aim to improve the spectral differences between bands and to decrease the effects of topography [10, 11]. Dividing one spectral band by another produces an image that offers relative band intensities [12]. To compute band ratios, it is required a numerator band and a denominator band. The band ratio is the numerator divided by the denominator [8, 13]. The combination is three as follows:

a) Image band 3/band 1.
b) Image band 4/band 6.
c) Image band 8/band 9.

3.4. Image spectral classification technique

Image classification is a vital stage of digital image analysis. Two core classification methods are supervised and unsupervised classification [14, 15]. In this research, Maximum Likelihood Classification (MLC) as one of the algorithms for supervised classification has been used because there is an intersecting of spectral signs, so a statistical choice criterion to help constraining pixels to be assigned to the class of highest probability. Figure 5 (a and b) shows the training of the classifier to recognize the particular type of Region of Interest (ROI). Seven classes of Region of Interest (ROI) were selected on the study area image based on training samples that indicted seven classes in this area as follows: Lower Nyanzian, Congioramate, Chert Tuff, Hornfels, Apllte, Upper nyanzian, and Granite.

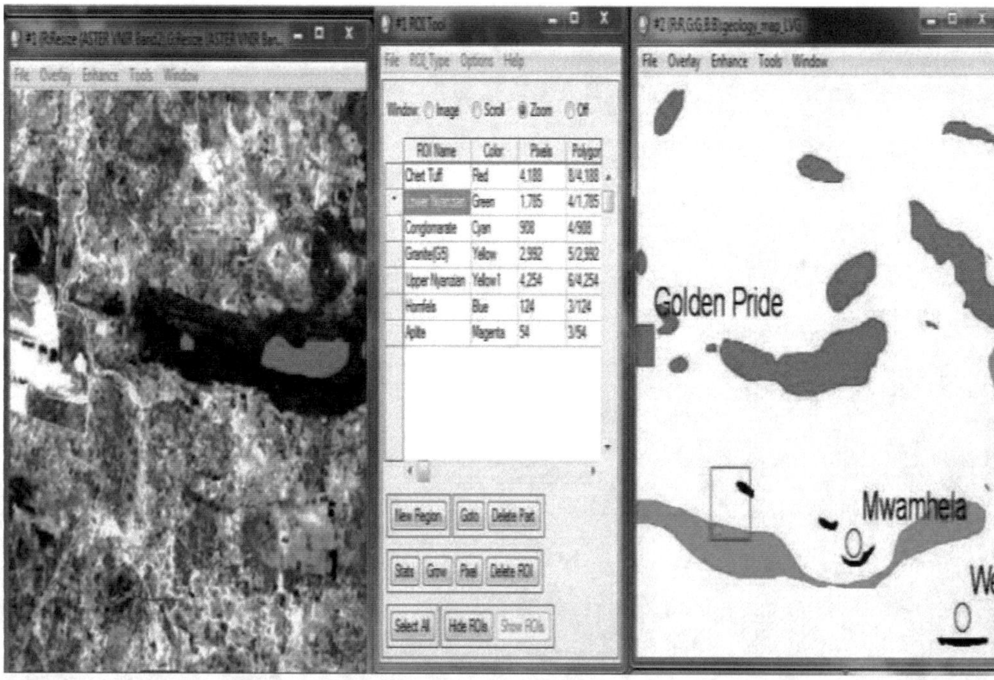

Figure 5a. Classes of (ROI).

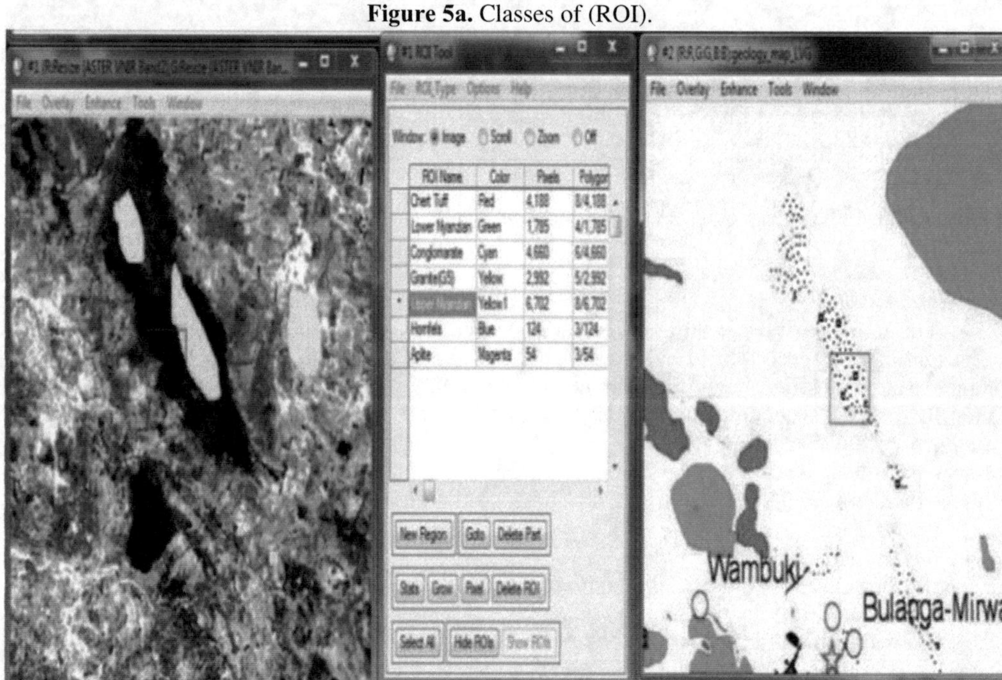

Figure 5b. Training samples classes of region of interest (ROI).

IOP Conf. Series: Materials Science and Engineering **901** (2020) 012016 doi:10.1088/1757-899X/901/1/012016

4. Results and discussion

The result of three ratios for iron oxides, clay minerals, and carbonate/sulphate minerals are shown sequentially in the Fig. 6 (a and b) of grey and RGB scale from 1 to 3. So, using this technique provides information about the range of concentration of mineral. After that, based on training samples, the image classified using maximum likelihood classification to create geological rock types map as shown in Fig. 7. After that, existing DEM for ASTER data can be displayed as a wireframe with greyscale or colour image draped over it using 3D surface view tool in ENVI software as shown in Figure 8. This 3D model can be used to extract further information such as slope and aspect for the study area in the future. The results were validated by using overall accuracy and kappa coefficient where overall accuracy equals (83%) and kappa coefficient equals (0.82). These results showed that accepted because of the kappa coefficient was near to 1 [14,15].

a. Grey scale

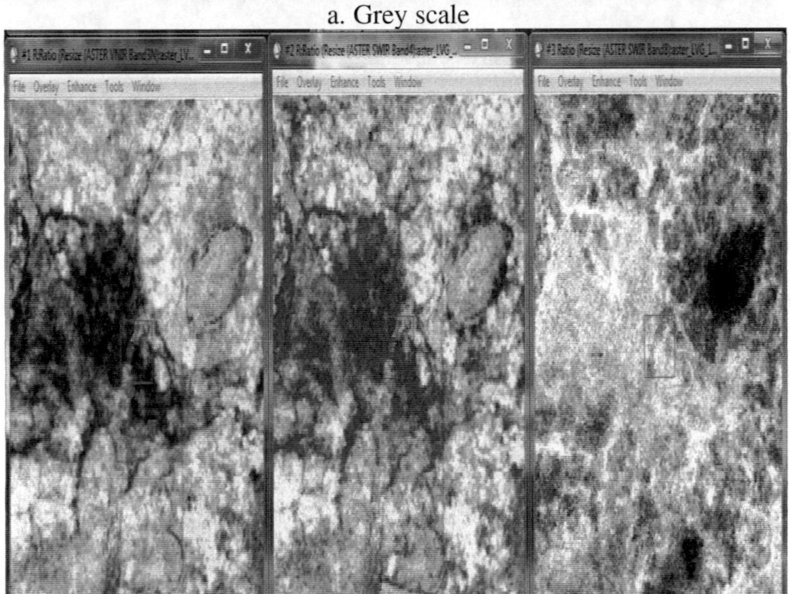

b. RGB scale

Figure 6. Band ratio image.

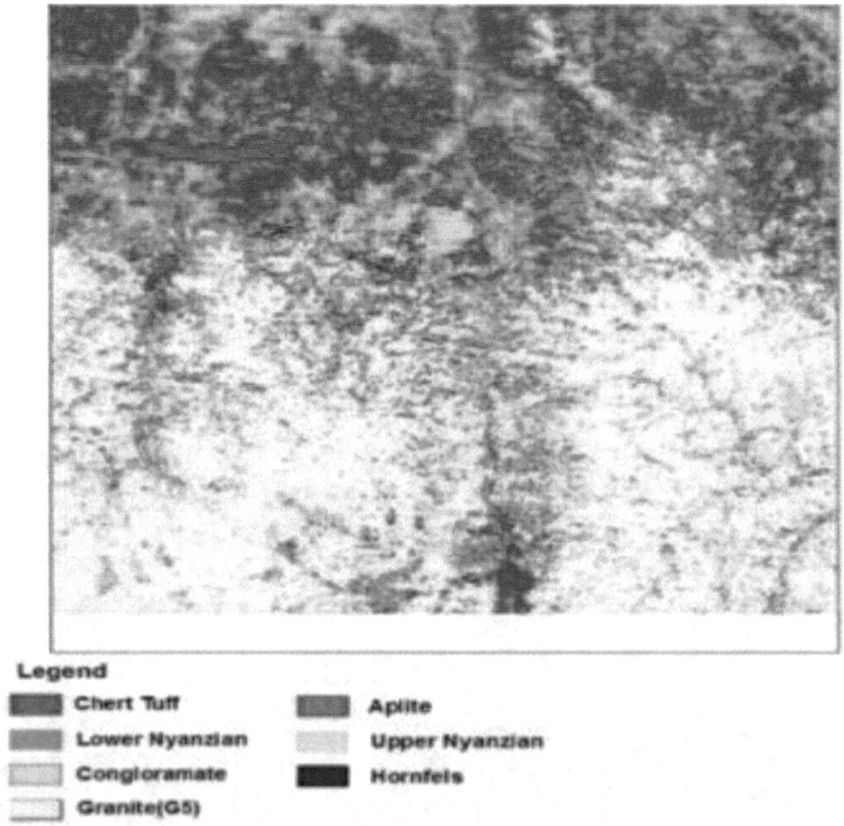

Figure 7. Geological rock types map.

Figure 8. 3D surface view of ASTER data.

5. Conclusions

Hyperspectral remote sensing is an efficient system that provides large data sets, which has up to 200 spectral bands. These numbers of bands give us a wide range to extract the information of the spectrum. Hyperspectral can be applied in various applications. In this research, ASTER data has been used and image processing by using ENVI software has been applied to this data. Furthermore, performance classification is one of more interesting technique to get the classified image for detection and production geological rock types map. The accuracy of classification of ASTER data can be more accurate by using topographic information band. Also, 3D surface for ASTER data can be displayed using ENVI software.

References

[1] P. Hardin and A. Hardin, Hyperspectral remote sensing of urban areas. Geography Compass, 7(1), 2013, pp. 7-21.
[2] A. Michael and R. Bhaskar, ASTER Users Handbook, Version 2, (2015).
[3] S.R. Delwiche and M.S. Kim, Hyperspectral imaging for detection of scab in wheat. Proceedings of SPIE 4203, Biological Quality and Precision Agriculture II, 2000, pp. 13–20.
[4] C.H. Bock, G.H. Poole, P.E. Parker, and T.R. Gottwaldov, Plant disease severity estimated visually, by digital photography and image analysis, and by hyperspectral imaging, Critical Reviews in Plant Sciences, 29(2), 2010, pp. 59–107.
[5] R.E. Kennedy, P.A. Townsend, J.E. Gross, W.B. Cohen, and P. Bolstad, Remote sensing change detection tools for natural resource managers: Understanding concepts and trade-offs in the design of landscape monitoring projects, Remote sensing of environment, 113, 2009, pp. 1382–1396.
[6] R.A. Naidu, E.M. Perry, F.J. Pierce, and T. Mekuria, The potential of spectral reflectance technique for the detection of Grapevine leaf roll-associated virus-3in two red-berried wine grape cultivars, Computers and Electronics in Agriculture, 66, 2009, pp. 38–45.
[7] R.L. Graham, An efficient algorithm for determining the convex hull of a finite planar set, Information Processing Letters, 1, 2000, pp. 132–133.
[8] T.M. Liliesand, and R.W. Kiefer, Remote sensing and image interpretation, 3rd Ed, John Wiley and Sons, Inc. Toronto (1994).
[9] J.M. Latin, J. Douglas, and P.E. Green, Analysing multivariate data. China: Machine Press, 2003, pp. 38–40.
[10] D. Lu, P. Mausel, E. Brondizio, and E. Moran, Change detection techniques. International journal of remote sensing, 25, 2004, pp. 2365–2401.
[11] F.W Nutter, M.L. Gleason, J.H. Jenco, N.C. Christians, Assessing the accuracy, interpreter repeatability, and inter-rate reliability of disease assessment systems. Phytopathology, 83, 2003, pp. 806–812.
[12] X. Fang, X. Zhu, Z. Wang, G. Zhao, Y. Jiang, and Y.A. Wang, Hyperspectral characteristics of apple leaves based on different disease stress, Remote Sensing Science, 2(3), 2014.
[13] R.T. Mahlein, P. Welke, H.W. Dehne, L. Plumer, U. Steiner, and E. Oreke, Development of spectral indices for detecting and identifying plant diseases, Remote Sensing of Environment, 128 (21), 2013, pp. 21–30.
[14] S.R. Delwiche, Classification of scab and other mold-damaged wheat kernels by near infrared reflectance spectroscopy, Transactions of the ASAE, 46(3), 2003, pp. 731–738.
[15] H.S. Jaber, Classification of hyperspectral remote sensing for production minerals mapping using geological map and geomatics techniques, International Journal of engineering and technology (UAE), 7 (4.20), 2018, pp. 480-484.

Geoecological Maps of Surface Water Pollution Akmola Region of Kazakhstan Based on Hydrochemical Analyzes

Beisenova Raikhan[1,a], Saspugayeva Gulnur[1,b], Zhamangara Aizhan[1,c*], Ramazanova Makhabbat[2,d], Tulebekova Assel[1,e]

[1]L.N. Gumilyov Eurasian National University, Kazakhstan
[2]Universidade Portucalense, Portugal
[a]raihan_b_r@mail.ru, [b]gulnur_erzhanovna@mail.ru, [c]kashagankizi@mail.ru,
[d]ramazanova@upt.pt, [e]krasavka5@mail.ru

Abstract. Ecological and geographical map that characterizes the introduction and storage in the natural or built environment of physical, chemical, and biological substances in concentrations exceeding the natural (background) values and adversely affecting human health, fauna, flora, and other components of the environment. Maps may reflect contamination of the earth, atmosphere, surface water and groundwater, soils, vegetation, wildlife, and others. They show the sources of pollution (solid waste, air emissions of gases and aerosols, drains and diversion of utilities and others), pollution factors (impact of power plants, transportation, industrial production, animal husbandry, and so on), distribution of natural and man-made pollutants (mechanical, physical, chemical, biological), and others. The levels of pollution often characterize the performance of exceeding the maximum permissible concentration (MPC), the maximum permissible emissions (MPE), and other regulations.

Keywords: maximum permissible concentration (MPC), the maximum permissible emissions (MPE), phosphates, nitrates, nitrites, ammonium nitrogen.

1. Introduction

Water pollution, as well as air pollution, is a complex and highly dynamic process. The concentrations of the various pollutants present in the water, characterized by a complex and time depend on the dynamics:

- Intensity of the ponds;
- Rate of self-purification and precipitation;
- The volume of water mass, the nature and speed of its movement;

Each of these factors is relatively independent of contamination from the other and has its dynamics. These contaminants have been deposited into water bodies with wastewater from industrial and agricultural enterprises, municipal areas with surface runoff due to runoff from contaminated areas during the deposition from the atmosphere, from the secondary chemical transformation processes of pollutants from natural sources. Information sources and methods of mapping surface water contamination are different for different card sizes. For the creating a survey of small-scale maps are enough published in yearbooks information about the long-term average levels of contamination at gauging stations, as well as the volume and structure of discharges by city. In this case, linear marks (rivers) and ranges (for lakes and reservoirs, expressed on a scale map) classes are characterized by the quality of water for stretches of major rivers and lakes, the levels and composition of pollution, the volume and composition of the discharges, technogenic load on water objects.

In the study of diffuse pollution sources in rural areas, the mapping area was divided into watersheds particular order depending on the scope of the research. Mapping of surface water pollution based on field measurements expeditionary research is less common due to the high mobility of the water environment and, accordingly, the rapid variability in pollution. Pollution of various components of the natural environment is one of the most obvious manifestations of the impact of society on nature. In its structure and dynamics, pollution has global trends, but it is manifested uniquely for each section of the territory; occurs under the influence of both technogenic and natural factors in almost all components of the geosystem. Therefore, it is extremely urgent to develop methods for cartographic support of a scientifically based assessment of the state of territories by identifying the pollution structure of various components of the environment. Cartographic approaches are most effective at the local level. This scale of the study allows us to combine the detail of the displayed information with the clarity of the transfer of the general laws of development of pollution processes, and when using modern geoinformatics, it significantly increases the efficiency and effectiveness of the analysis of environmental and geographical information.

The purpose of our study is the practical implementation of integrated mapping of pollution of the territory of Lake Burabay and Ulken Shabakty for its environmental and geographical assessment using modern geographic information technologies. Following the purpose of the work, the following tasks were set:

- give a general description of geographic information systems;
- identify the level of pollution of the Lake Burabay and Ulken Shabakty;
- develop a map-scheme of pollution of Lake Burabay and Ulken Shabakty using the ArcView 9 geographic information system.

2. Materials and Methods

Ecological mapping is the science of methods for collecting, analyzing, and mapping the information on the state of the human environment and other biological species on the environmental situation [1]. Accurate and reliable data play a crucial role in monitoring the environmental situation. The source of information for environmental mapping is remote sensing data, qualitative and quantitative data, expeditionary and stationary studies of the state of the components of the environment, and the state of bio-indicators. The relevance of the use and the distinctive role of modern cartography in the study of environmental and geographical problems is that it allows using maps based on the principles of spatial-temporal modeling to study the properties of natural complexes, their changes in time, communication and spatial relationships [2, 3].

Today, surface water pollution caused by the intensive development of industrial production and population growth has become a global problem. Industrial and agricultural enterprises, as well as public utilities, consume large amounts of water, which after use re-enters the environment, contaminated with chemical, mechanical, biological substances. Depending on the source of pollution, such waters are called industrial wastewater or domestic wastewater. To display surface water pollution on an environmental map, it is advisable to use the method of localized diagrams. Chart icons in the form of multi-colored and different-sized columns convey the chemical composition and amount of pollutants present at the sampling site [2-6]. With the expansion and deepening of environmental measures, one of the main areas of GIS application is monitoring the consequences of actions taken at the local and regional levels. Sources of updated information may be the results of ground-based surveys or remote observations from air transport and space. The use of GIS is also effective for monitoring the living conditions of local and introduced species, identifying cause-effect chains and relationships, assessing the favorable and adverse effects of environmental protection measures on the ecosystem as a whole and its components, making operational decisions on their adjustment depending on changing external conditions [3,4].

GIS greatly simplifies the publication of any kind of cartographic products. Using the built-in software language (for example, ARC / INFO ARC Macro Language (AML)), one can write programs to automatically create any type of printed card, graph, chart, and table. Also, simple software products (such as ArcView GIS) allow anyone, even inexperienced users, to view and directly operate on the data contained in the GIS database. Using such simple and easily accessible programs, any user can read and print maps (recorded, for example, on CD-ROM in GIS format ARC / INFO) [5-6]. GIS has certain characteristics, which rightly allow us to consider this technology as fundamental for processing and managing information. GIS tools far exceed the capabilities of conventional mapping systems, although naturally, they include all the basic functions of obtaining high-quality maps and plans. The GIS concept itself contains comprehensive capabilities for collecting, integrating, and analyzing any data distributed in space or tied to a specific place. If it were necessary to visualize the available information in the form of a map, graph or diagram, supplement or modify the database, integrate it with other databases - the only right way would be to access the GIS. In the traditional view, the possible limits of integrating heterogeneous data are artificially limited [7-9].

GIS allows researchers to go much further. Demographic information, information on land ownership, welfare and income of the population, volumes of capital investments and investments, zoning of the territory, condition of the bread market, etc. can be added to the above data set. As a result, it becomes possible to determine the effectiveness of planned or ongoing conservation measures, their impact on people's lives and agricultural economies. Researchers can go even further and add data on the spread of diseases and epidemics, establish whether there is a relationship between the rate of degradation of nature and human health, and determine the possibility of the emergence and spread of new diseases. In the end, it is possible to accurately assess all the socio-economic aspects of any process, for example, reducing the area of forestland or soil degradation [11-12].

GIS is successfully used to create maps of key environmental parameters. In the future, upon receipt of new data, these maps will be used to identify the extent and rate of degradation of flora and fauna. When entering data from remote, in particular satellite, and conventional field observations, it can be used to monitor local and large-scale anthropogenic impacts. Data on anthropogenic pressures should be superimposed on zoning maps of the territory with highlighted areas of particular interest from an environmental point of view, for example, parks, reserves. Assessment of the state and rate of degradation of the environment can also be carried out on the test plots allocated on all layers of the map. [14].

The functional integrated capabilities of GIS in the most explicit form are manifested and conducive to the successful conduct of joint interdisciplinary research. They provide the unification and overlapping of each other of any data types, if only they are displayed on the map. Such studies include analysis of the relationship between population health and a variety of (natural, demographic, economic) factors; quantitative assessment of the influence of environmental parameters on the state of local and regional ecosystems and their components; determination of landowners' income depending on the prevailing soil types, climatic conditions, remoteness from cities; identification of the number and density of distribution areas of rare and endangered plant species, depending on the height of the area [13-15].

3. Results

Figure 1 is a map of pollution in the lake Ulken Shabakty. On the map are given in terms of sampling, the results of hydrochemical and hydrobiological analysis identified a class of water quality of the reservoir. It was found that in the majority of samples Lake Ulken Shabakty refers to moderately polluted, as indicated in the sampling sites purple circle. According to hydrobiological analyzes found in the lake Burabay flowering reservoir for all sampling points, reflected on the map in green.

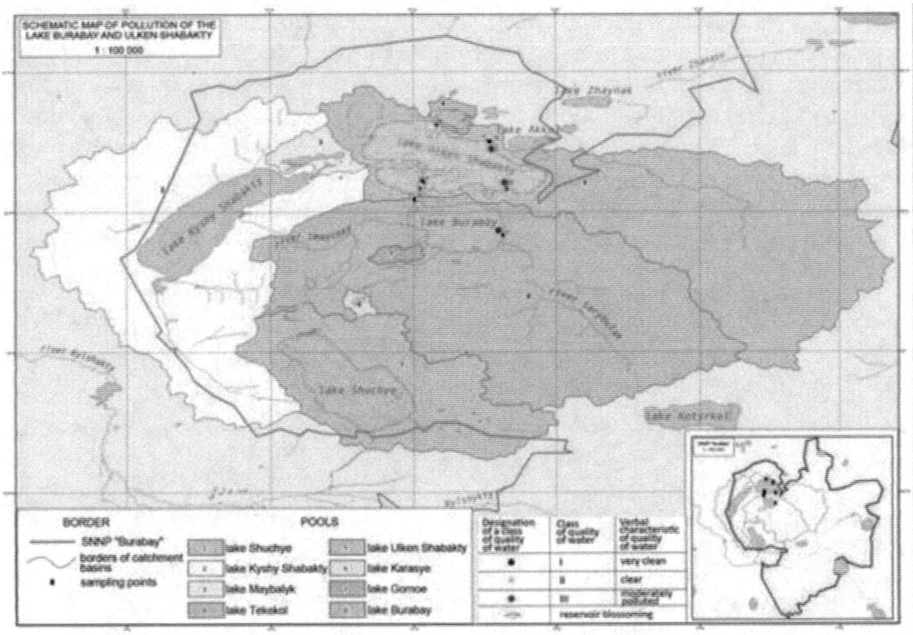

Figure 1. Map of pollution in the lake Ulken Shabakty for Q3 2013.

The following types of environmental maps were constructed based on long-term hydrochemical analyzes in the sampling points and lakes Burabay, Ulken Shabakty. Fig. 2 shows a schematic map of the average concentrations of phosphate per year in mg/dm2. The analysis shows that the concentration of phosphates in the lake Burabay was sharply exceeded in 2011 and the lake Ulken Shabakty phosphate since 2009 gradually increases from 0.004 to 0.01.

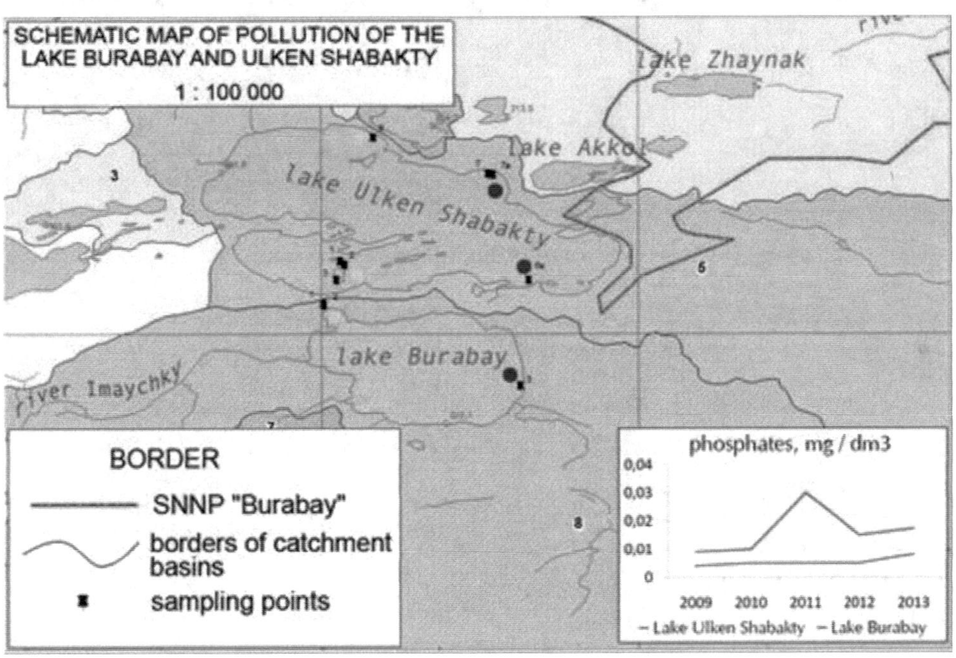

Figure 2. Schematic map of the indicators of phosphates in the lake Burabay and Ulken Shabakty.

In the next Fig. 3 was given changes in the nitrate concentration in mg/dm^2 for 2009-2013. In the sampling points, which shows a sharp increase in nitrite in 2011, and low levels of nitrite in the lake Ulken Shabakty. Nitrate concentration as major ions containing nitrogen, is of great importance in the assessment of rivers and lakes is given in the following Fig. 4. On the map is a visible upward trend in nitrate levels in both lakes. With this trend in 2011 was a sharp increase in the concentration of nitrates in Lake Burabay and the lake Ulken Shabakty. The concentrations of nitrates had elevated in nitrate levels in lakes in 2012, then again in 2013. The concentration of ammonia was increased from 2009 to 2011 and then in 2012 she went again and the lake Burabay and Lake Ulken Shabakty (Fig. 5).

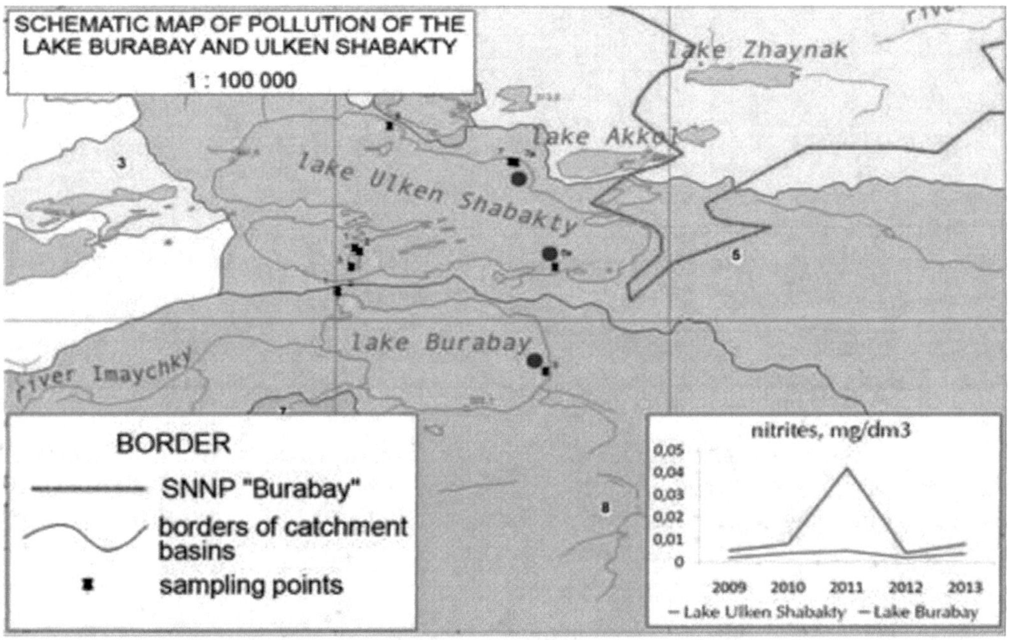

Figure 3. Schematic map of the indicators of nitrite in the lake Borovoe and Ulken Shabakty.

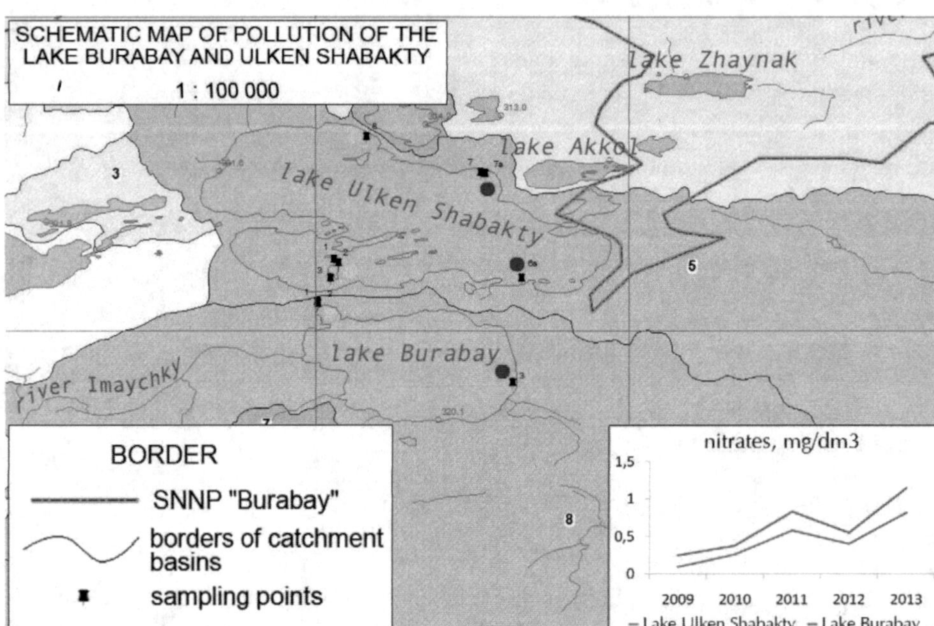

Figure 4. Schematic map of the indicators of nitrite in the lake Burabay and Ulken Shabakty.

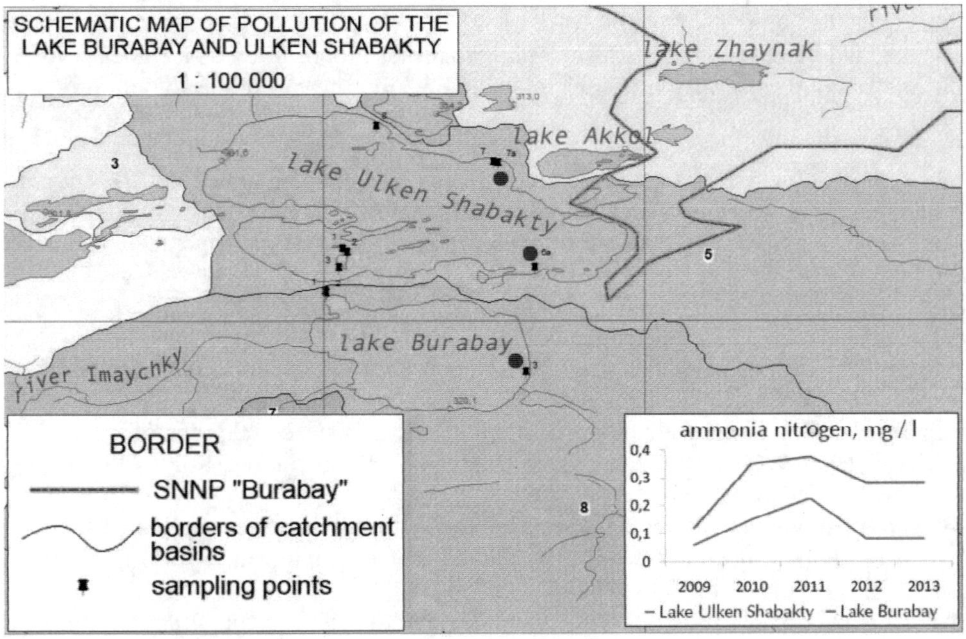

Figure 5. Schematic map of the indicators of ammonium nitrogen in the lake Burabay and Ulken Shabakty.

4. Discussion

Thus, started to develop pollution maps allows us to reflect data obtained from field studies in the form of complex maps for better integration of diverse environmental information. Integrated card fully and accurately represents the ecological state of the territory, make it possible to take fully into account changes in the nature and their impact on human, allow displaying a set of interrelated objects and develop recommendations for the organization and monitoring "flowering" of water.

- Development of maps of water pollution, drinking, and fishing industry.
- Creation of integrated environmental maps to reflect the status of water bodies and to identify the localization of environmentally favorable and unfavorable regions.

When monitoring surface water quality, observe the following principles: comprehensive and systematic observation of the terms of their consistency with the characteristic hydrological situations determination water-quality uniform methods. Mass development of some algae in reservoirs, lakes, and ponds leads to a malicious "blooming" of water. Harmfulness of mass development of algae, especially cyanobacteria (blue-green algae), is the production of a large number of dangerous to human health and animal potent toxins, reducing water quality, aesthetic appearance of a violation of the reservoir, and others. The products of cyanobacteria were hardly utilized by representatives of the upper trophic level. Algal blooms, ultimately, leads to an overall degradation of aquatic ecosystems. This phenomenon in recent years in the world is becoming one of themselves negative processes affecting the functioning of all living things. "Blooms" entails a chain of negative consequences in the form of a fish kill, mass mortality of benthos, plankton and neuston animals and waterfowl and mammals. Also, another problem for a person associated with a change in flowering water it is smell and taste in the water security.

Despite numerous studies conducted by foreign scientists, the study of the genesis of toxic and odor-producing substances and controls on toxic organisms remains an urgent problem and has a national character in many countries. In the future, it might turn into a global problem. Analysis of hydrochemical state of water lakes Ulken Shabakty and Lake Burabay (control), enabled to establish changes in the environment, or anticipating the coming of the "blooming". In the first two reservoirs during field studies, we noted the "blooming" of water, but in such a lake Burabay phenomenon was not observed, we find that the water content before flowering nitrite nitrogen is reduced and the ratio of nitrogen to phosphate is reduced in the range of 30-65. However, this phenomenon should be studied more detailed.

It is noted also that the "bloom" in natural waters is always observed high water temperatures from 23-25°C up to 28°C and increasing the pH from 8 to 8.5. At the "bloom" there is a tendency to reduce the content of dissolved oxygen, which increases immediately after the end of the "blooming".

5. Conclusion

According to the results of research, the following conclusions can be drawn out: Development of an electronic map of contamination of water bodies will be a significant contribution to improving the quality of the monitoring, modeling, and forecasting of ecological status of potable water, fishery and recreational facilities.

References

[1] I.A. Sutorikhin, Mathematical-cartographic support for the creation of information-analytical system "Water and ecology of Siberia. IA Sutorikhin, OV Lovtskaya, IN Rotanova//Computational technologies. - 12 T - Special Issue 3, (2007) 66-72.
[2] Geoinformatics: Trainee for schools., Edited by V.S. Tikunova. - M.: Publishing Center "Academy". 2005, pp. 475-480
[3] C.W. Fetter, Applied Hydrogeology, Prentice-Hall, 4th Edition. 2001, pp. 125-132
[4] J.P. Wilson, Fotheringham, A.S. The Handbook of Geographic Information Science, Malden, MA: Blackwell Publishing., (2007) 3-6
[5] M.Y. Dubinin, Open Desktop GIS: an overview of the current situation /M.Yu.Dubinin/ Newsletter GIS Association. №5 (72), 2009, pp.20-27.
[6] A.S. Tulebekova, Control equipment for pile test according to American and Kazakhstan standards, Modern Applied Science, 2015, pp. 192-198.
[7] A.S. Turashev, A.S. Tulebekova, T. Muzdybayeva, and E.T. Uzak, Analysis of the interaction between construction and soil ground under influence of the static load. 14th Asian Regional Conference on Soil Mechanics and Geotechnical Engineering, 2011.
[8] M.F. Goodchild, Twenty years of progress: GIScience in 2010, Journal of Spatial Information Science, 2010 DOI:10.5311/JOSIS.2010.1.2.
[9] V.V. Shcherbakov, Information systems. Structure GIS techniquesю Creation and use. Ekaterinburg, (2002) 9-11
[10] S.L. Dingman, Physical Hydrology, Prentice-Hall, 2nd Edition, 2002, pp.235-245
[11] Y. Korolev, Overall Geoinformatics - M: "The Academy". 2001, pp. 450-458
[12] S. Harbaugh, Current and future geographic information system projects within the eastern Gulf of Mexico, and the potential for one universal application. Int. Oil Spill Conference, British Columbia, Canada, 2003.
[13] K.S. Meiramkulova, M. Zhumagulov, G.Y. Saspugayeva, Zh. Zhakupova, M. Mussimkhan, Treatment of poultry slaughterhouse wastewater with combined system. Potravinarstvo Slovak Journal of Food Sciences, 13(1), 2019, pp.706-712
[14] N.Y. Ramazanova, Z.G. Berdenov, S.K. Ramazanov, N.B. Kazangapova, S.M. Romanova, S.T. Toksanbaeva, J. Wendt, Landscape-geochemical analysis of steppe zone basin Zhaiyk, News of the National Academy of Sciences of the Republic of Kazakhstan, Series of Geology and Technical Sciences, 2019, № 4(436) pp. 33-41
[15] G.S. Aidarkhanova, G.E. Saspugayeva, K.B. Massenov, E.T. Abseitov, K.M. Satova, Ecological assessment of the quality of crop production supplied to the markets of major cities in Central Kazakhstan. The study of contamination of agricultural plants radionuclides and heavy metals, Oxidation communication. №2, 38, Bulgaria, 2015 pp.900-904

IOP Conf. Series: Materials Science and Engineering **901** (2020) 012018 doi:10.1088/1757-899X/901/1/012018

Methods of Spatial Interpolation in The Representation of Soil Properties/Nineveh Governorate Soil Case Study

Suhaib H. Khadir Taha[1,a] * and Osama M. Najim[1,b]

[1]Geography Department, Human Education College, University of Mosul, Mosul, Iraq.
[a]Sahlah80@gmail.com, [b]smslmn408@gmail.com

Abstract: Spatial interpolation is a process used to guess cell values at points or locations where real models are not available. The principle of interpolation is based on Spatial autocorrelation and Spatial Dependence, which measures the degree of relationship or dependence between points or objects near and far and by using statistical techniques to address the effects of data values and the relationship between the various interrelated. The study relied on more than one method of spatial interpolation (Kriging - Inverse Distance Weighted-Radial Basis Functions-Global Polynomial-Local Polynomial). To represent data for one of the most important elements of the natural environment and the soil and within the spatial space of the governorate of Nineveh, In the present study, some of the physical properties of the soil are expressed (soil texture, pH, bulk density, and chemical components such as (organic matter, total Bases, cation exchange capacity, exchangeable sodium percentage, calcium carbonate, available water capacity, and electrical conductivity based on (92) points distributed according to the results of the spatial interpolation test based on the mean bias error values and root mean square error, which was used as a criterion for the differentiation between the methods of completion used in the study. The study also showed that the Kriging method is suitable for the parts of spatial completion of soil properties. It was possible to reach (27) maps to represent some of the physical and chemical properties of the soil of the study area.

Keywords: Geostatistical analysis, GIS, soil Characterization, spatial Interpolation, maps.

1. Introduction

Spatial statistics technologies have proven to be highly effective in studying many elements of the natural environment with a wide area span that makes it difficult to survey and cover as well as "the material cost in the case of its study, but Interpolation Methods have provided solutions to such cases and have become used In many studies, especially after the results matched with the results of the prediction of traditional methods, Spatial Interpolation is a smart guessing process in which the researcher or user attempts to make a reasonable estimate of a value in a continuous field and in natural places where the value has not been measured, It is an important way to reach the aim of estimate the values of unmolded interfaces by applying point views. Where statistical techniques are used to address the effects of data values and the relationship of the different incoming among them, and the self-correlation analyzes provide quantitative estimates of the spatial relationship between each of the two data taking into account the distances between them, so the effect on the converged sites is more than the sites that are far from each other The other one.

The research aims to use spatial prediction methods to study some of the physical and chemical soil properties so that it is possible to estimate the field data not studied and express the spatial change in the values along the surface of the study. Nineveh governorate was chosen as a study area to know the spatial distribution of (soil tissue, organic matter (OM), (pH), bulk density (BD), total exchangeable bases (TEB), cation exchange capacity (CEC), exchangeable sodium percentage (ESP), calcium carbonate (CaCO3), gypsum, available water capacity (AWC), and electrical conductivity (EC) based on (92) randomly distributed points. While the importance of the research is summarized in finding the best representation for the distribution of some physical and chemical soil properties in the study area, and by using spatial interpolation methods (spatial statistics) in the geographic information systems environment where they save time, effort and reduce costs. The study problem lies in the ability to benefit from the accuracy of methods Spatial completion and its effectiveness in representing some physical and chemical properties of the soil, which are available within the programs of geographic information systems, and conducting various statistical Solutions to choose a method of interpolation that is appropriate to distribute the values of soil properties in Nineveh Governorate. To achieve the aim of the study, a quantitative analytical research approach was followed by providing the required data on the phenomenon studied with a view to interpolation it spatially, as well as making use of the geographic information systems technology, programs, and spatial analytical tools.

2. Data Sources and Programs Used in The Study

The sources of data used in the current study are:

1) Land Sat 8 ETM+
2) Digital Elevation Modal
3) HWSD (Harmonized World Soil Database (version 1.2.2009) (FAO) Food and Agriculture Organization of the United Nations, (IIASA) International Institute for Applied Systems Analysis, ISRIC-World Soil Information, (ISSCAS) Institute of Soil Science–Chinese Academy of Sciences, (JRC) Joint Research Centre of the European Commission) .
4) Soil data for some areas of Nineveh Governorate from (Ministry of Water Resources -National Center for Water Resources Management - Department of Environmental Studies.

The programs used in the study are:

2.1.Global Mapper v.18.

It is considered a GIS application in data processing and is characterized by its ease of use, with access to an unparalleled variety of spatial data, and was used to integrate and process Images, various maps and the digital height model

2.2.Arc GIS v10.4

It is an integrated set of applications of geographic information systems, and provides tools for spatial analysis and is used to build information layers, conduct spatial analysis and producing maps. The area of Nineveh Governorate is about (37863) km^2. The study area represented by Nineveh Governorate is located in the northern and northwestern part of Iraq, bordered to the north by Dohuk Governorate and Lake Mosul Dam, while it is bordered to the east by the governorates of Erbil and Kirkuk, and south by the provinces of Salahuddin and Anbar and to the west of the Syrian Arab Republic, and accordingly is limited to The study area between the two latitudes (34 55) - (30 37) north and between longitudes (41 25) - (44 25). Figure 1 shows the study area and locations of boreholes.

3. Natural Characteristics of The Study Area

3.1. Geology.

As the type of soil prevailing in a region is determined by the type of parent rock (the original material), which depends to a large extent on the prevailing rock and its characteristics and development over time [1]. There is a group of rock formations and sediments in Nineveh Governorate that reflect different sedimentary environments, extends the chronological age from the Cretaceous period from the second geological time to the Holocene era from the fourth geological time, and the characteristics of these formations and sediments of different rocks between sandy, mud, alluvial, limestone, gypsum and halite [2]. The properties of these rocks with other environmental variables in the study area were reflected in determining many of their physical and chemical properties.

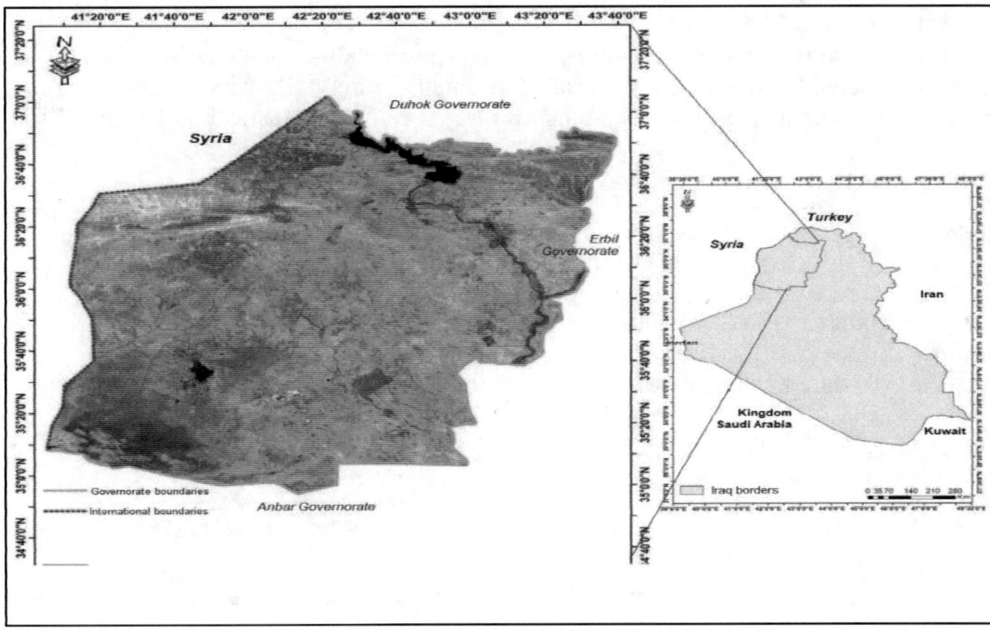

Figure 1. Location of study area.

3.2. Surface

In terms of terrain and slope, the study area is divided into each of the flat sector and its lands have few and medium slopes (0-7.9%) according to the (Zink) classification and constitute an appropriate environment for soil development and stability and of great thickness and is the most widespread form of sloping surfaces in Nineveh Governorate with an area (35.875450) km^2, 94.75%, Compared to the other sector (the folds) whose lands have high slopes (8-30)%, as these surfaces do not allow the soil to develop as their moisture content is low, so we find soil with shallow rocky or naked soils in steep slopes [3].

3.3. Climate

The climate is the active factor among the factors of soil formation and is essential in all biochemical reactions in which it occurs. The erosion factors depend entirely on the elements of rain, temperature, and wind in weathering and transporting materials [4]. The study area is characterized by its transitional location between three types of climatic regions according to the classification (Koppen) Mediterranean climate (Cs) and prevails in the northern and northeastern regions and the Sinjar Heights, and the semi-arid climate (Bs) and dry (Bw) in most parts of the study area [5], In general, the climate of Nineveh Governorate is hot dry summer, and rainy in winter The temperature rates are varied, due to the effects of location, height, and different atmospheric pressure, and accordingly, the roles of each heat factor in mechanical and chemical weathering (rock debris and rocky crumbs) that the various soil formation factors in the region depend on, and the role of falling rain Some elements of the melt this rock fragment mixed with water to carbon dioxide, such as calcium carbonate, sodium, magnesium, and calcium sulfate. To keep gravel, sand, clay, clay, iron oxides, and lime.

3.4. Natural Vegetation Cover

The vegetation cover its importance lies in that it works to stabilize the soil and protect it from erosion, not to mention the direct contribution to increasing the concentration of organic matter inside the soil after its death and its decomposition, and that the natural plant in the study area consists mainly of herbaceous plants, where it covers the largest part of the conservative lands, while The trees form a limited part, in addition to other plant species [6]. The contribution of the vegetation to the composition of the soil of the study area varied according to its density and durability.

4. Spatial Interpolation Methods

To compare the spatial interpolation methods, the five representation methods (Kriging, IDW, Radial Basis Functions, Global Polynomial, and Local Polynomial) in this study were compared to choose the best method for all variable data and present the results according to statistical parameters [7].

1) **Mean Bias Error (ME)**: Equation of errors in the results is:

$$\text{Mean Bias} = \frac{1}{N}\sum_{1}^{N}(M - O) \quad (1)$$

Where
N represents the number of data or points
M The estimated value of the studied property
O represents the measured value of the studied property.

2) **Root Mean Square Error (RMSE)**: It is used to evaluate the model's susceptibility to graphical representation [17].

$$\text{RMSE} = \sqrt{\frac{1}{N}}\sum_{i=1}^{N}(x_p - x_m)^2 \quad (2)$$

Where
N represents the number of data or points
X_P is the estimated value represents studied models
Xm is the measured value represents the studied models.

To find out the efficiency of spatial interpolation methods, compare the efficiency of the method in integrating data through error results (mean error and the square root of the error square resulting from using the method) [10].

a) When the value of the mean error or the bias error is equal to zero or close to zero, then this means that the method fits the data, that is, the larger the value it means less representation and greater bias, incompatibility [11].
b) The small value of the square root of the error square indicates that the estimate of the results is accurate and appropriate.

4.1. Data Distribution Patterns

The process of determining the quality and distribution of data helps in the appropriate selection of statistical methods and methods used with geographic information systems based on the results. Most data follow the symmetric natural distribution, in which all measures of central tendency apply to each other and their values are equal. The natural distribution is not achieved in most data for the natural properties of soils that may contain anomalous or extreme values or large values within a small set of values or vice versa when the graph or distribution is twisted Iterative towards outliers. In this case, it should not be sufficient to describe the distribution in measures of central tendency, but rather it needs another measure that expresses the extent of the distribution in the distribution from the natural or the degree of (Skewness) and (Kurtosis) [12]. The frequency distribution is crooked, positive if the extreme data are large values, the long end of the curve extends to the right and the measures of central tendency become mean then the mode, and negative, if the extreme data are small values, then the long end of the curve extends to the left and the central tendency measures become the mode and then the median Then the mean. Figure 2 below shows the shapes of the three distributions curves for the positions of the arithmetic mean for the mean, median, and mode [13,14].

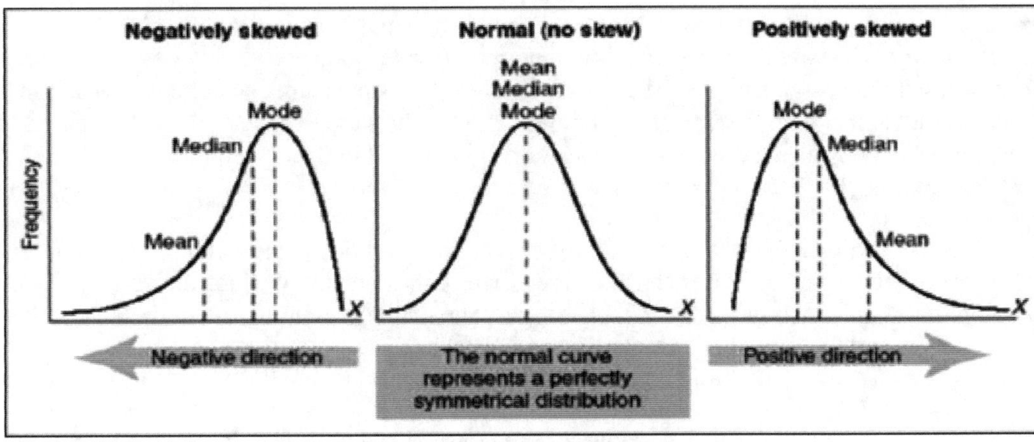

Figure 2. Samples distribution curves.

5. Data Analysis

Geographic information systems meet with geostatistic at the common point of interest in conducting analytical operations on quantitative data, calculating statistical measures, and knowing the movement of contrasting models within one region. The next stage is choosing the best programs to produce the results and display. The study of the region begins by collecting data on its elements followed by arrangement and coordination to facilitate its statistical analysis to obtain the information. The statistical approach uses two types of statistical applications for data [8].

1) **Descriptive analysis** and is concerned with the digital description of a particular society (study models) so do not withdraw its results to another environment (the rest of the study area).
2) **The inferential analysis** includes the sampling process any small sample (study models) representing the large community (study area) provided that the final results are approximate and within the limits of a statistically calculated error.

Sample consistency the problem of extracting reliable information from the sample is concentrated in determining its conformity with the study community (study area) and although it is impossible to confirm conclusively from that, we can come up with appropriate estimates for calculating the standard error. The process of calculating the standard error includes identifying the amount of difference or fluctuation in the measurements that the sample provides us with the averages for the different samples (central tendency measures) and calculating the standard deviation mathematically from the averages and determining the pattern of data distribution for each variable, whether natural or abnormal (dispersion measures) and then Determine the value and direction of deviations (data distributions) [14].

5.1. Steps and Modus Operandi

The steps of the work in the study were to set up the basic data approved in the study, which is the soil data of the study area which was obtained from the Harmonized World Soil Database (HWSA), and the soil data obtained from the Ministry of Water Resources-National Center for Water Resources Management - Department Environmental studies (paper data) for some regions of the study. And then exporting that data in the format (XLS.) To the Arc Map v10.4, and projecting it in the form of points bearing each of them (XY) values and drawing a shapefile for each point projected with the entry of data for each point separately carrying each point the physical and chemical properties, to select the ideal method for the spatial interpolation of the data, statistical values (central tendency measures and dispersion measures) have been found for each of the soil properties to know how to distribute data and make the necessary transformations for them, and choose an appropriate method for spatial completion based on the values of the mean error and the square root of the mean error and apply that to methods Spatial interpolation available in the (Geostatistical analysis) window within the Geographic Information Systems program, and choosing an appropriate data method with high accuracy and less .

The spatial analyzer ArcGIS v10.4 provides several tools or spatial interpolation methods for creating a surface cell network of raster data. There are two interpolation techniques:

- **Deterministic** to design surfaces based on measured points or mathematical equations, a method based on the extension of cell symmetry or similarity, such as the inverse of the weighted distance.
- **Geostatistics** it depends on Statistical processors and is used in the design of advanced models of surfaces, and is characterized by including accuracy calculations for points estimated as in the kriging method.

It is evident from the above that the spatial interpolation methods present in the program's spatial analyst follow different methods to reach the desired map of the data, all of which are correct when they meet the conditions on which they were established and the map representing the study area is closer to reality. It contains only a very small percentage of Mistakes. The result that can be reached from all methods gives us maps that vary from one to the other and may share small details or move away from them a lot To choose the ideal way to represent the soil data for the study area and represent it more closely to the truth, rely on the statistical values of the physical and chemical properties data for the soil of the study area as shown in Table (1) through which we decide how to distribute the data and then how to deal with it to choose the best appropriate method for mapping the completion maps For each of the characteristics of the studied soil, the extraction of statistical parameters for the comparison between spatial integration methods, which are the mean error and the square root of the mean error square as shown in Table 3 shows the mean error values and the square root of the error square, which were used to compare spatial interpolation methods.

Table 1. Values of the statistical measures of the soil of the study area.

Properties (Depth)(m)	Statistical measures							
	Count	Min	Max	Mean	Median	Std. Dev.	Skewness	Kurtosis
Top Sand% (0-30 cm)	92	16	43	32.6	35	7.878	-1.258	3.251
Sub Sand (%) (30-100 cm)	92	0	40	30.6	35.5	9.662	-1.547	4.466
Top Silt (%) (0-30 cm)	92	29	47	39.39	40	5.994	-0.5252	1.985
Sub Silt% (30-100 cm)	92	0	45	35.16	37	8.0688	-2.7218	12.407
Top Clay% (0-30 cm)	92	18	55	28.141	23	12.629	1.391	3.238
Sub Clay% (30-100 cm)	92	0	57	30.65	27	13.148	0.708	3.488
AWC (mm/m) (0-100 cm)	92	15	150	110.38	125	45.563	-0.563	1.610
Top Bulk Density (0-30 cm)	92	1.25	1.65	1.427	1.39	0.1177	0.550	2.1495
Sub Bulk Density (30-100 cm)	92	0	1.76	1.394	1.41	0.2927	-3.321	17.47
Top OM (0-30 cm)	92	0.37	0.96	0.5405	0.47	0.1582	1.226	3.730
Sub OM (30-100 cm)	92	0	0.53	0.298	0.22	0.122	0.1014	2.0874
Top pH (0-30 cm)	92	7.6	8.3	7.9022	7.9	0.1176	0.5277	5.003
Sub_ pH (30-100 cm)	92	0	8.2	7.7054	8	1.428	-5.195	28.239
Top CEC_{Soil} (0-30 cm)	92	10	44	19.72	16.5	11.424	1.284	3.249
Sub CEC_{Soil} (30-100 cm)	92	0	43	17.77	15	11.842	1.11	3.257
Top TEB_{Soil} (0-30 cm)	92	10.4	50.8	25.182	24.65	12.245	0.239	1.913
Sub TEB_{Soil}(30-100 cm)	92	0	58.3	30.649	31.45	17.672	-0.0859	1.323
Top ESP (0-30 cm)	92	1	41	4.641	3	7.355	3.788	16.65
Sub ESP (30-100 cm)	92	0	43	15.837	17	12.228	0.3017	1.623
Top $CaCO_3$ (0-30 cm)	92	0.2	31.6	14.09	9	12.645	0.501	1.449
Sub $CaCO_3$ (30-100 cm)	92	0	35	12.183	9	10.05	0.641	2.187
Top $CaSO_4$ (0-30 cm)	92	0	15.1	4.859	1.1	6.652	0.890	1.816
Sub $CaSO_4$ (30-100 cm)	92	0	23	7.87	2.5	9.915	0.836	1.791
Top ECe (0-30 cm)	92	0.1	22.3	2.44	1	4.335	3.98	18.26
Sub ECe (30-100cm)	92	0	14.8	3.504	1.5	3.66	1.29	4.33
Top USDA Tex (0-30cm)	92	1	3	1.79	2	0.433	-1.034	3.210
Sub_USDA Tex (30-100cm)	92	0	3	2.33	3	0.952	-1.022	2.583

IOP Conf. Series: Materials Science and Engineering **901** (2020) 012018 doi:10.1088/1757-899X/901/1/012018

6. Discussion of Results

The Kriging method was chosen to produce Spatial Interpolation maps as appropriate to the natural properties of the soil according to the results of the Spatial Interpolation test as in Table (2) as well as its ability to deal with data whether it is of natural or abnormal distribution (logarithmic distribution or BOX distribution - COX) as Kriking method is classified as a spatial and statistical method and depends on the principle of correlation between the variables in the sample values and the spatial direction of the distribution of these points, and takes into account both the distance, degree, and direction of the change between the known points, It also has several mathematical models (Spherical, Circular, Ordinary, Universal, Exponential,...) that give it the ability and flexibility to deal with the data, which gave the best results and all the studied soil variables and proved their high effectiveness according to the results in their application to soil properties data of study area.

Table 2. The mean error values and Root Mean Square Error.

Interpolation Methods	Soil properties	ME	RMSE
Krigking/$CO_{kriging}$	Top Sand, %	0.12	0.44
	Sub Sand, %	0.09	0.670
	Top Silt, %	0.05	0.41
	Sub Silt, %	0.01	0.45
	Top Clay, %	0.006	0.40
	Sub Clay, %	1.1	0.30
	AWC, mm/m	0.688	-0.033
	Top Bulk Density	-0.003	0.11
	Sub Bulk Density	-0.0009	0.2
	Top OM, %	-0.0002	0.15
	Sub OM, %	-0.0007	0.16
	Top pH	-0.003	0.4
	Sub pH	0.07	1.120
	TOP_CEC_Soil	-0.003	0.38
	SUB_CEC_Soil	-0.010	0.39
	TOP-TEB_Soil	-0013	0.48
	SUB-TEB_SOil	-0.03	0.44
	TOP-ESP	0.007	0.32
	SUB-ESP	0.033	0.41
	TOP_CaCo3	0.02	0.5
	SUB-CaCo3	-0.002	0.38
	TOP-CaSo4	0.001	0.26
	SUB-CaSo4	-0.001	0.43
	TOP-ECe	-0.01	0.29
	SUB-ECe	0.0009	0.3
	T_USDA_TEX	0.01	0.41
	S_USDA_TEX	0.01	0.84

As a result, according to the final on the previous stages were obtained maps of the spatial interpolation shown in Fig. 3 which represents some physical and chemical properties in the study area and on two levels (surface (0-30 cm) and subsurface (30-100 cm)).

IOP Conf. Series: Materials Science and Engineering **901** (2020) 012018 doi:10.1088/1757-899X/901/1/012018

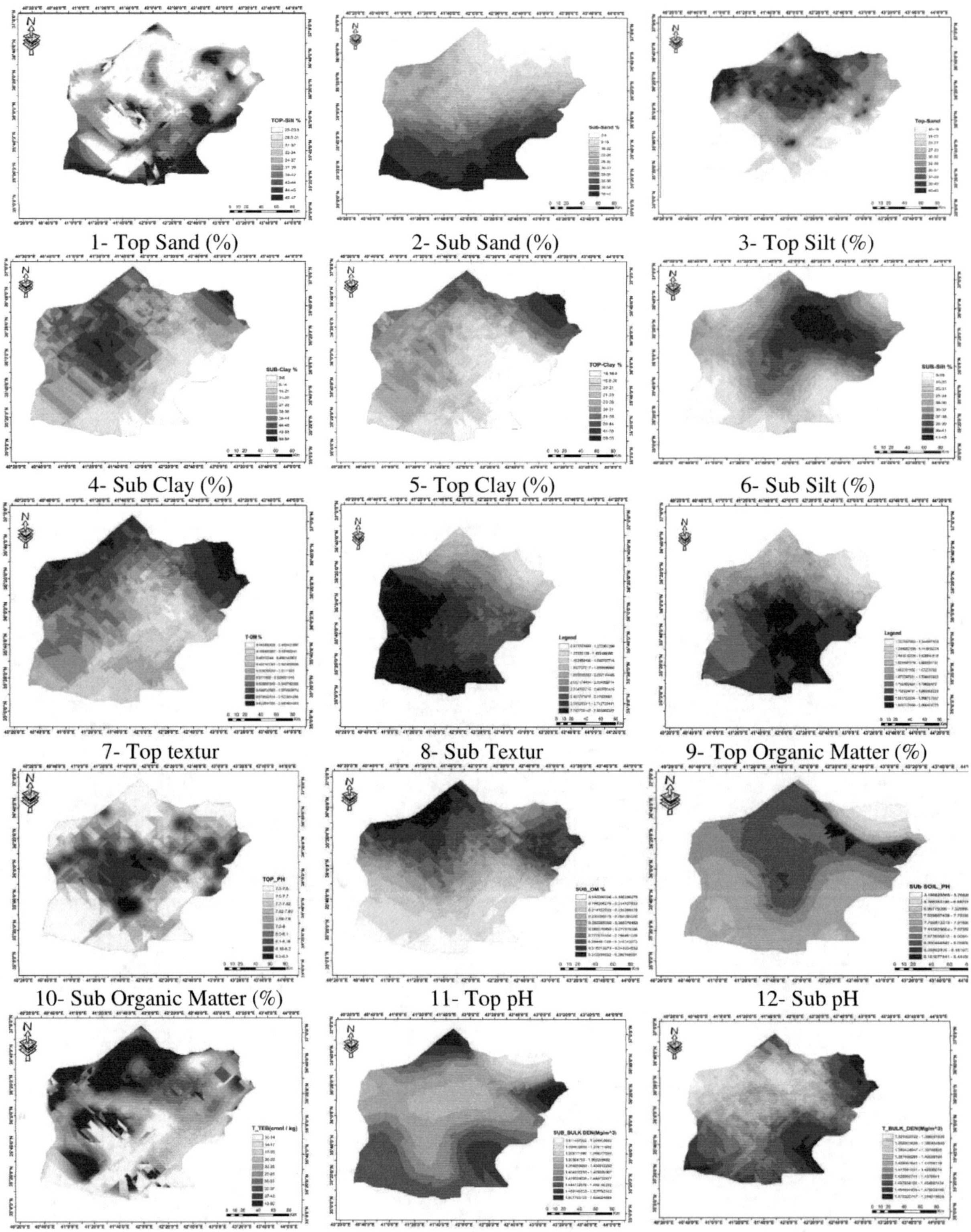

1- Top Sand (%) 2- Sub Sand (%) 3- Top Silt (%)
4- Sub Clay (%) 5- Top Clay (%) 6- Sub Silt (%)
7- Top textur 8- Sub Textur 9- Top Organic Matter (%)
10- Sub Organic Matter (%) 11- Top pH 12- Sub pH
13- Top Bulk Density 14- Sub Bulk Density 15- Top Total Exchangeable Bases

Figure 3. Some physical and chemical properties of soil in the study area.

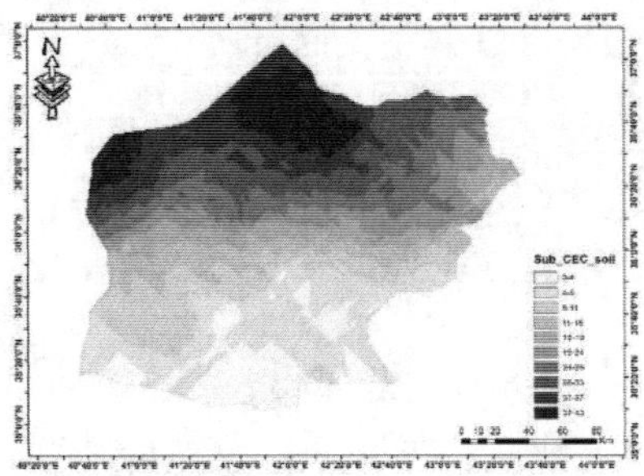
16- Sub Total Exchangeable Bases

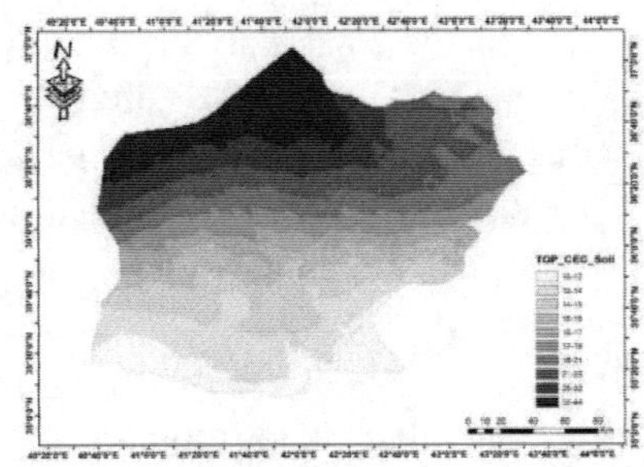
17- Top Cec

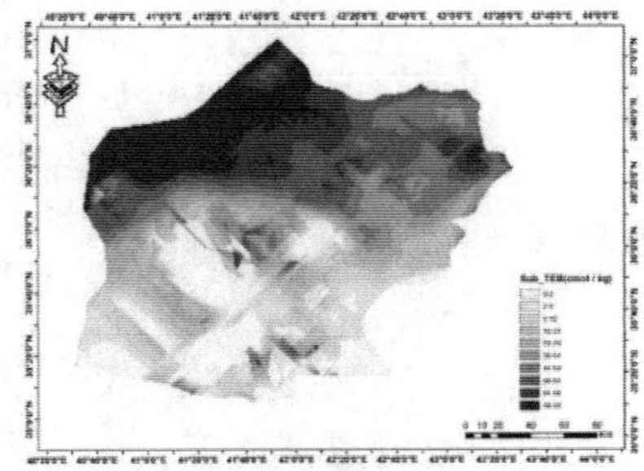
18- Sub Cec

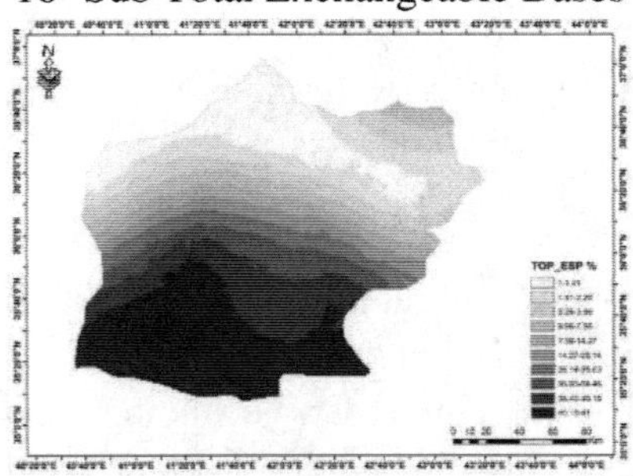
19- Top Esp

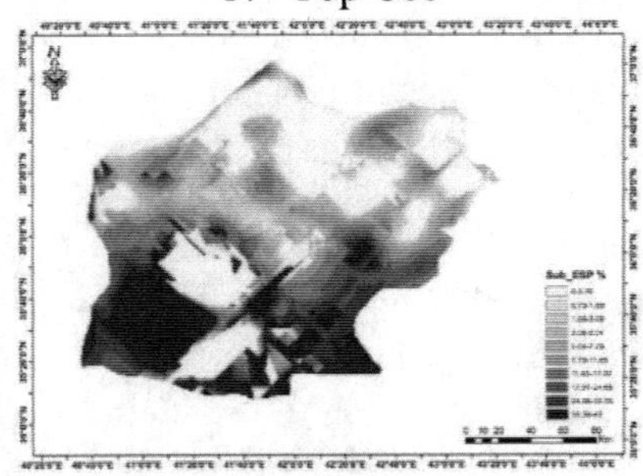
20- Sub Esp

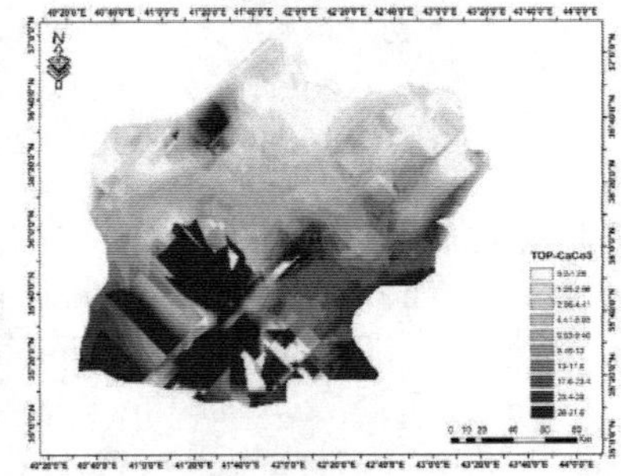
21- Top_Caco3

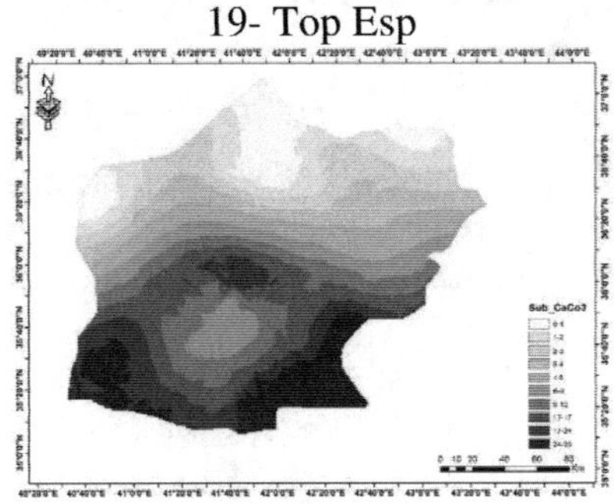
22- Sub $CaCO_3$

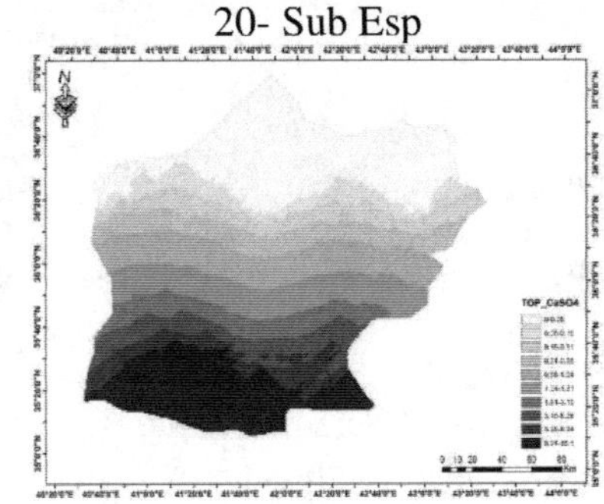
23- Top $CaSO_4$

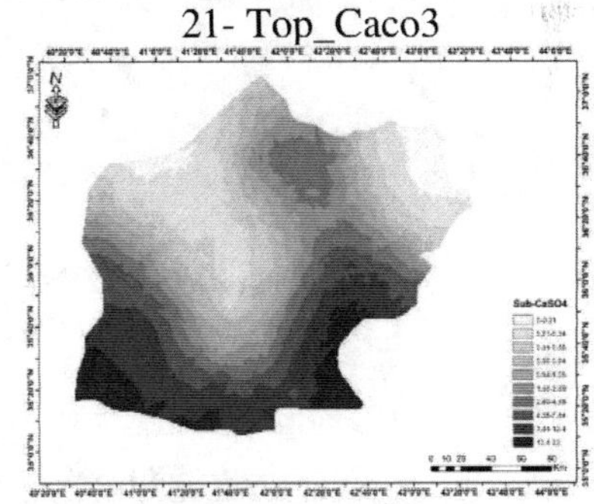
24- Sub $CaSO_4$

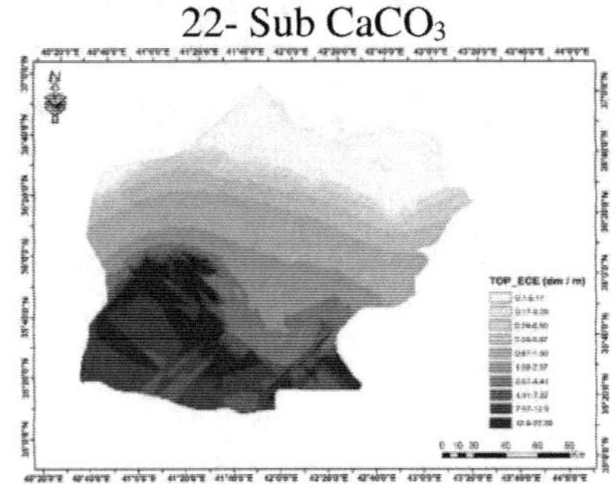
25- Top Ece

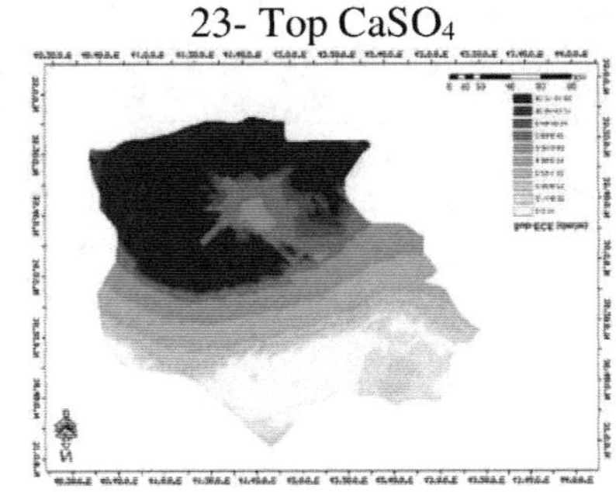
26- Sub Ece

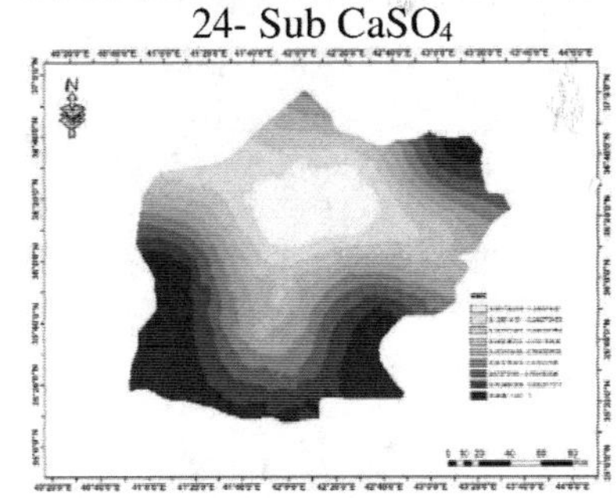
27- Awc

Figure 3. Continued.

7. Conclusions and Proposals

- The distribution of the data studied in the study area was not a natural distribution of most of the physical and chemical soil properties except for the (pH), (silt) and (Bulk density) properties, so logarithmic distribution was used in the data distribution process, to make it appropriate in Spatial Interpolation processes.
- The possibility of preparing Spatial Interpolation maps to represent soil properties using geological statistics and according to the results of the mean error and the square root of the mean error of the data, and the process took place within the GIS software with high accuracy.

- The kriging method was used in the Spatial Interpolation process, as it uses more than one mathematical and statistical formula to reach the desired result.
- The study showed, through the values of the statistical measures and iterative scales, as well as in the results of the Spatial Interpolation values and the final maps represented, the variation of the characteristics of the studied properties of the soil between the surface and subsurface, due to the different influencing conditions.
- The study showed that there is a discrepancy between the parts of the study area in its physical and chemical properties of the soils in the maps representing the study area and they are identical to the prevailing natural data in the study area between the diversity of geology, geomorphological, climatic and natural layers of vegetation.
- The study also showed that the values of most of the studied soil properties in the study area are acceptable and within certain specifications except for the gypsum characteristic.

Recommendations

- The study recommends applying this type of study at larger levels and more models so that its results are closer to simulating reality and comparing the results obtained in the field.
- The study recommends increasing attention to this field of scientific and modern technology and its introduction in various fields and other environmental studies and linking it to other branches of science as it creates a database that can be used by decision-makers.

References

[1] M.M.S. Al-Mesmari, Wadi Al-Slayeb Basin - Study in Geography of Soil, Libyan International Journal, College of Education Al-Marj, University of Benghazi, (2016).
[2] T. Buday, The Regional of Iraq Stratigraphy and Palecography, Dar Al- Kutib Publishing house, University of Mosul, Iraq (1980).
[3] S. Marin, GIS solution in, Natural Resource Management, Ed. Renewable Natural Resource Foundation and, Natural Academy of Sciences Natural Resource, Council, Washington, (1999).
[4] A.H. Al-Shalash, Geography of Soil, Baghdad University Press, (1985).
[5] S.B. Al-Najmawi, Nineveh Governorate Climate, Master Thesis (unpublished), Department of Geography, College of Education, University of Mosul, (2001) p. 45
[6] K.S. Al-Juburi, The effect of climate change on the development of vegetation in Nineveh Governorate, Ph.D. thesis, Department of Geography, College of Education, University of Mosul, (2015).
[7] A. Hani and S. A.H. Abari, Determination of cd, Zn, K, pH, TNV, Organic material and Electrical Conductivity (CE), Distribution in Agricultural Soils using Geostatistics and GIS (case study: western of Natanz –Iran). World Academy of Science, Engineering and Technology, No.60 22-24.
[8] R. Webster and M.A. Oliver, Geostatistics for environmental scientists. John Wiley & Sons, 2007.
[9] E. Verfaillie, V. Van Lancker, and M. Van Meirvenne, M., Multivariate geostatistics for the predictive modelling of the surficial sand distribution in shelf seas, Continental Shelf Research, 26(19) 2006 2454-2468.
[10] NWillmott, C. J. and K.Matsuura (2005). Advantages of the mean absolute error (MAE) OVER the root mean square error (RMSE) in assessing average model performance. climate Research, vol.30, pp.79-82.
[11] S.A. Moasheri, A. S. Goshki and A. Parsaie, "SAR" Qualities parameter persistence by a compound method of geostatic and artificial neural network (Case study of Jiroft plain), Int. J. Agri. Crop Sci., 6(3) 2013 157-166
[12] K. Johnston, J.M. Ver Hoef, K. Krivoruchko, and N. Lucas, Using ArcGIS geostatistical analyst. Vol. 380. Redlands: Esri, 2001.

[13] Kroulik, M., A Slejska, D, Kokoskova, and V.Venclova(2005).SoilElectrical Conductivity as one Possible Toole for Predicting of Cirsium Arvense Infestation Occurrence. Herbicides, Theory and Applications WWW.intechopen.com.

[14] L. Brodský, V. Vaněk, L. Borůvka, and J. Száková, Consistency of spatial dependence of soil chemical properties in two fields: a geostatistical study." Plant Soil Environ, 50(11) 2004 507-512.

Application of Nanotechnology in Iraqi Construction Projects

Maha Walhan Jumaah[1,a*] **and Meervat Altaie** [1, b]

[1]Department of Civil Engineering, University of Baghdad, Iraq
[a]maha.walhan@yahoo.com, [b]meervat3@yahoo.com

Abstract. The great potential of nanotechnology in the construction sector, through many applications, such as innovative building materials and intelligent building systems. The paper aims to determine the level of knowledge for Nanotechnology in Iraqi Construction Projects. These knowledge levels are essential to develop construction project management, then identification obstacles to application Nanotechnology in Iraqi projects. The closed questionnaire using items of question measured by five-points Likert scale, including structured–questionnaires (n = 68), interviews (n = 12). Lacks the set of knowledge currently required to be qualified and effective in practice for use in the construction industry.

Keywords: Construction Nanotechnology, sustainability building, civil engineering.

1. Introduction

Project management problems take on particular–importance in-nanotechnology because of the innovative character of the projects-to be implemented [1]. Understanding and controlling materials at the level of individual atoms and molecules in nanotechnology within the range of 1-100 nm (10^{-9} m) and the creation of materials, systems, and devices with unique properties and new functions [2]. Nanotechnology signifies the main opportunity for the construction industry to develop new products [3], high performance, Promote quality, energy efficiency, economic sustainability by balancing environmental requirements and reduce maintenance costs [4]. Nano modifications increase the strength of current or innovative materials when applied in the construction industry .Improved physical properties (strength and durability) [5], Reduced construction costs due to the cheapness of raw materials [6], More Sustainable Construction [7, 8].

1.1. Nanotechnology for concrete

The compressive strength for concrete increases via added nano-SiO_2 at an early age [9]. Enhanced pore volume distribution can occur by filling the pores between cement particles with nano-scale. Besides; can close tiny cracks on the concrete surface by Using fiber sheet containing nano-silica particles and hardeners permeate [10]. Adding a small amount of carbon nanotubes to the Portland cement developed the mechanical features. Moreover, increase the compressive and flexural strength to the reference samples deprived of the reinforcement [11,12]. Admixture fly ash (industrial waste products) with the cement led to reduce the requirement for cement and funding to sustainability, also improves concrete durability and strength [13].

1.2. Nanotechnology for steel

Copper nanoparticles added to steel reduce surface unevenness and increase safety thus avoiding fatigue cracking [14]. Weld toughness increased when added Nanoparticles of magnesium and calcium [15]. MMFX2 is produced by MFX Steel Corp, nanostructure modified steel, USA. MMFX2 steel has a lower cost than stainless steel and corrosion resistance similar. To date, the MMFX Steel has got a certificate for general construction use throughout the United States [16]. Steel cables are used in the construction of bridges and pre-cast of concrete, and stronger cable materials will reduce costs and construction time [5].

1.3. Nanotechnology for Coatings

Paint containing nanoparticles or nanolayers was developed for: an anti-corrosion or caring coatings for components; thermal control, self-cleaning, antibacterial coatings for work surfaces; anti-reflection coatings for glass/windows; and more durable paints and anti-graffiti coating for buildings such as concrete walls [13] and structures 10]. Epoxy nanocomposite coatings with SiO_2, Fe_2O_3, TiO2, ZnO, and clay nanotubes were successfully produced on a steel substrate at a concentration of the total weight of epoxy resin [17].

1.4. Sensors Nanotechnology for

Micro electrical and Nanomechanical systems sensors(MEMS) have been advanced in production to observe and control the materials/structure performance and the environment condition [18]. Incorporated MEMS sensors can be consumed to observe a range of concrete properties, including temperature, humidity, humidity, chloride ions, water-related degradation, and carbon dioxide [13].

1.5. Nanotechnology for glass

TiO2 coating captures and destroys organic and inorganic air pollutants and bacterial membranes through a photocatalyst process. Self-cleaning assets for Nano TiO2 useful in windows. Also, NanoTiO_2 coverings can be applied to building exteriors to inhibit pollutants, and thus reduce a facility's keep costs [15].

1.6. Nanotechnology for wood

Nanotubes or nanofibers in composite wood, make elements (wood tissue) twice as strong as steel. Harvesting these nanofibers will clue to a different paradigm in sustainable construction because production and use will be part of a renewable cycle [6].

1.7. Nanotechnology for Insulation

Efficient insulation methods and materials in buildings is an important aspect of sustainability through efficient use of energy. Buildings consume more energy produced, and households account for a quarter of carbon emissions, which come from meeting space heating needs [14]. Air gel which has a treble density of air but has great strength and insulation capabilities. Air gel, currently used in solar collectors, is solid, but so transparent that it looks like a hologram [19].

2. Methodology

2.1.Information and Sample

In this study the questionnaire consists of two main parts, the first part is the level of knowledge of nanotechnology in Iraqi construction industry (8 factor). Part two especially obstacles of implementation nanotechnology in Iraqi construction projects (13 factor). The five-dimensional Likert scale was used in the questionnaire, the rating from small scale (1) to high scale (5) including –questionnaires (n = 68), interviews (n = 12). The target group are engineers who work in construction design, supervision, implementation, and maintenance engineers (architecture, civil engineers, mechanical engineers, chemical engineers) and any related engineering specialty.

2.2.Measurements

The first step measures the validity and reliability of the questionnaire by Cornbrash's Coefficient Alpha method and Half Split method. After that, finding the descriptive Statistics (The mean, Standard deviation) then finding Relative Important Index (RII), 0ne sample t-taste. The second step includes the test of hypotheses by the inferential statistics test, ANOVA test (One-way Analysis of Variance) [20].

2.3.Statistical Tools
The Statistical-Package for Social-Sciences (SPSS version 23) is used in the analysis of data.

3. Results and Discussion

3.1. The validity and reliability of the questionnaire
As shown in Table 1, the result of validity and reliability for the questionnaire by Cornbrash's Coefficient-Alpha method and-Half-Split technique [21]. The coefficient alpha ($C\alpha$) and Guttman split-half Coefficient was calculated for two parts, the result ensures the reliability of the questionnaire, because the reliability outcomes are higher than 0.6.

Table 1: The validity and reliability of the questionnaire.

Part	Cronbach's Coefficient Alpha	Guttman split-half Coefficient	Number of factors
The level of knowledge of nanotechnology	0.806	0.91	8
Obstacles to the application of nanotechnology	0.853	0.873	13

3.2. The Descriptive Statistics
The knowledge level of nanotechnology containing eight questions to evaluate the level of knowledge by specialists in Iraq. And obstacles to the implementation of nanotechnology covering 13 questions. The five-dimensional Likert scale was used in the questionnaire, the rating from small scale (1) to high scale (5) including –questionnaires (n = 68). This data was presented to the participants' views. As shown in Table 2 the descriptive statistics of the questionnaire results.

3.3. Relative Importance Index (RII) of nanotechnology in Iraq
Offered to the participants and take their views about these factors then analyzed. The analysis by SPSS 23, includes a one-sample t-test (2-tail), P-value, Relative Importance Index RII, and rank factors that are offered in Table 3. Citing of values that ranking based on the highest RII, mean and the lowest SD, if some items have similar RII, Means, as in the case of (B6, B1), (B8, B2), and (B12, B3).

It was important to determine the neutral value of RII and compares the total RII with the neutral value of RII. Consequently, the neutral value for RII is $(3/5)\times100 = 60\%$. Based on all of that, and as shown, the level of knowledge for nanotechnology in Iraq as Low level. Regarding results for all items of the part of Nanotechnology obstacles, they show that the Mean for all those items equals 4.11 and the total RII equals 82%, which is greater than 60% Also, the value of t (tabulated t), at a degree of freedom 67and significance level = 0.05, equals 1.996, as shown the value of t-test greater than the value of t (1.996), P-value of all items a smaller than the significance level 0.05.

3.4. Hypothesis related to respondents' profiles
By taking into account the percentage knowledge of nanotechnology, One-way Analysis of Variance (ANOVA) provides a parametric statistical test of whether the Means of several groups (more than two) are equal or not (by using the F-ratio). As shown in table 4.

The critical value of F at degree of freedom = [(K-1), (N-K)] at the significance (probability) level (α) = 0.05, df = [(4-1), (68-4)] = [3, 64] and critical value of F at significance (Probability) equals 2.75. We used one-way ANOVA to test the differences among the opinions of the respondents taking into account the Percentage knowledge of nanotechnology. Regarding F-test, the significance values for A-part P-value < 0.05. Also, the values of the F-test for A-part are greater than the critical value of F 2.75.

Table 2: Statistical information for the questionnaire

No.	Factor	Mean	Sta. Dev.
	The level of knowledge of nanotechnology		
A1	I have read some research and studies about nanotechnology.	2.15	0.919
A2	Some of my college courses at University talked about nanotechnology.	1.68	0.969
A3	I have High-rate information about nanotechnology.	2.16	0.940
A4	I have an idea on how nanotechnology is applied in the engineering construction industry.	2.34	1.016
A5	I know that the application of nanotechnology in the engineering construction industry has a great impact.	2.72	0.844
A6	I use nanotechnology in my engineering field.	1.93	1.238
A7	I believe that nanotechnology is an important building industry and builds construction in Iraq.	3.51	0.970
A8	I think that Nanotechnology has a positive impact on the sustainable environment.	3.88	0.873
	Obstacles to the application of nanotechnology		
B1	Lack of knowledge of nanotechnology and its importance in the engineering field by stakeholders.	4.19	0.966
B2	Lack of knowledge of how nanotechnology is applied in the field of design and construction projects by engineers.	4.15	0.868
B3	High costs for processing nano construction materials and equipment for examination.	4.09	0.893
B4	There are no government systems that fully support the application of nanotechnology in Iraq.	4.29	0.947
B5	Engineers prefer to work with traditional materials.	3.74	1.217
B6	The absence of scientific training courses The engineers know the benefits and importance of nanotechnology in engineering projects.	4.19	1.055
B7	The unwillingness of engineers to learn new techniques at work.	3.43	0.759
B8	The weak level of government financial funding directed to promote the purchase of nanomaterial.	4.16	1.101
B9	The lack of interest of civil society members in scientific research within its research institutions.	4.07	0.919
B10	Absence of the role of companies and the private sector in financing scientific research on nanotechnology and not to benefit from the research outputs of scientists.	4.40	0.849
B11	There is no support for nanotechnology research and studies. The private sector from industrial and investment companies to establishing private laboratories or providing financial support to universities to carry out joint development research projects. There are two benefits for the company providing support, benefiting from research and training facilities for its employees within universities. Which becomes a community partnership.	4.25	0.887
B12	The absence of a teamwork culture and cooperation between the engineering specialties of the research bodies.	4.12	0.970
B13	Lack of a culture of public awareness and dissemination of nanotechnology in audio, visual, and visual ways.	4.35	0.748

Table 3. Relative Importance Index (RII), t (2-tail), P-value, and rank factors.

Factor	RII	t-value	P-value	Rank
The level of knowledge of nanotechnology				
A1	0.43	7.657	0.000	6
A2	0.34	11.266	0.000	8
A3	0.43	7.352	0.000	5
A4	0.47	5.369	0.000	4
A5	0.54	2.731	0.008	3
A6	0.39	7.153	0.000	7
A7	0.70	4.378	0.000	2
A8	0.78	8.335	0.000	1
All A- item	0.51	5.337	0.000	-
Obstacles to the application of nanotechnology				
B1	0.84	10.169	0.000	5
B2	0.83	10.892	0.000	8
B3	0.82	10.045	0.000	10
B4	0.86	11.266	0.000	3
B5	0.75	4.982	0.000	12
B6	0.84	9.314	0.000	6
B7	0.69	4.631	0.000	13
B8	0.83	8.701	0.000	7
B9	0.81	9.629	0.000	11
B10	0.88	13.571	0.000	1
B11	0.85	11.617	0.000	4
B12	0.82	9.500	0.000	9
B13	0.87	14.907	0.000	2
All B-items	0.82	16.086	0.000	-

Table 4. ANOVA -test results regarding the current the Percentage knowledge of nanotechnology.

Part	*F*- test (ANOVA)	*P*-value (Sig.)	Mean			
			Less than 25% N=32	25% to less than 50% N=25	50% to less than75% N=9	More than75% N=2
All A-item	18.108	0.000	2.148	2.760	3.222	3.875
All B-items	0.408	0.748	4.108	4.166	4.042	3.731

4. Conclusions and Recommendations

Based on the results of the study, the level of awareness of nanotechnology is low by engineers in the Iraqi construction. Also, there is a statistically significant dissimilarity between academic professionals and engineers working in the fields of design, implementation, and supervision. And we found the obstacles of nanotechnology are substantially affecting the adoption of nanotechnology in Iraqi construction. Academic institutions and universities should play a key role in introducing engineers to nanotechnology concepts, functions, and benefits. This can be done by conducting various workshops and providing technical training courses in the application of nanotechnology in construction properly.

References

[1] V.P. Meshalkin, O.V. Stoyanova, and M.I. Dli, Project management in the nanotechnology industry: Specifics and possibilities of taking them into account, Theor. Found. Chem. Eng. 46 (2012) 50–54.

[2] Luís Manuel Coelho Guerreiro, Nanotechnology In Construction : Towards Structural Applications, (2015).

[3] F.H. Halicioglu, The Potential Benefits of Nanotechnology for Innovative Solutions in the Construction Sector, Nanotechnol. Constr. 3(2009) 209–214.

[4] O. Figovsky and D. Beilin, Green nanotechnology, Green Nanotechnol., 8104(2017) 1–538.

[5] A. Abdelnaieem, Towards sustainable architecture with nanotechnology, Hous. Build. Natl. Res. Cent. HBNRC.(2010), pp. 1–17.

[6] I. Iavicoli, V. Leso, W. Ricciardi, L. L. Hodson, and M. D. Hoover, Opportunities and challenges of nanotechnology in the green economy, Environ. Heal. A Glob. Access Sci. Source, 13(2014),1–11.

[7] B. Bhuvaneshwari, S. Sasmal, and N. R. Iyer, Nanoscience to Nanotechnology for Civil Engineering – Proof of Concepts, Recent Res. Geogr. Geol. Energy, Environ. Biomed. Nanosci., July(2011), pp. 230–235.

[8] J. Teizer, M. Venugopal, W. Teizer, and J. Felkl, Nanotechnology and Its Impact on Construction: Bridging the Gap between Researchers and Industry Professionals, J. Constr. Eng. Manag. 138(2012) 594–604.

[9] P. Zhang, J. Wan, K. Wang, and Q. Li, Influence of nano-SiO_2 on properties of fresh and hardened high performance concrete: A state-of-the-art review, Constr. Build. Mater., 148(2017) 648–658.

[10] S. Abulmagd and Z. A. Etman, Nanotechnology in Repair and Protection of Structures State-of-the-Art, J. Civ. Environ. Eng. 8(2018).

[11] W. D. A. Jalil, A. J. Haider, and A. M. Hasan, Self-compacting concrete in digital fabrication and construction: A review, no. January (2018) p. 020005.

[12] V. K. Ganesh, Nanotechnology in Civil Engineering [interaktyvus], Engineeringcivil.Com, 8(2013) 12.

[13] A. Sev and M. Ezel, Nanotechnology Innovations for the Sustainable Buildings of the Future, Int. J. Civil, Archit. Struct. Constr. Eng. 8(2014) 846–856.

[14] C.Vigneshkumar, Study on Nanomaterials and Application of Nanotechnology and Its Impacts in Construction, Discovery. 23(2014) 8–12.

[15] N. Rajput, Nanotechnology In Civil Engineering and Construction: A Review," Int. J. Res. Eng. Appl. Sci. (IMPACT FACTOR – 5.981). 5(2015) 208–214.

[16] M. Devendra and K. Manjunathachari, Study on Nanotechnology in Civil Engineering Structures, Int. Adv. Res. J. Sci. Eng. Technol.. 2(2017) 1–5.

[17] T. A. Nguyen, T. H. Nguyen, T. V. Nguyen, H. Thai, and X. Shi, Effect of Nanoparticles on the Thermal and Mechanical Properties of Epoxy Coatings, J. Nanosci. Nanotechnol. 16(2016) 9874–9881.

[18] A. K. Rana, S. B. Rana, A. Kumari, and V. Kiran, Significance of Nanotechnology in Construction Engineering, Int. J. Recent Trends Eng.. 1(2009) 6–8.

[19] D. A. Altun and B. Örgülü, Towards a Different Architecture in Cooperation with Nanotechnology and Genetic Science : New Approaches for the Present and the Future, Archit. Res. 2014, 4(1B) 1-12. 4(2014) 1–12.

[20] N. J. Salkind, Encyclopedia of research design (hardcover). London, the UK: SAGE Publications, Inc. Sign in (2010).

[21] S. B. Cecilie Katrine Utheim Grønvik, Atle Ødegård, Factor Analytical Examination of the Evidence-Based Practice Beliefs Scale: Indications of a Two-Factor Structure, Open J. Nursing. 6 (2016) 3–4.

Evaluation of the Productivity of Ready Mixed Concrete Batch Plant Using Artificial Intelligence Techniques

Hussein T. Almusawi[1, a *] **and Abbas M. Burhan**[1,b]

[1]Civil Engineering Department, University of Baghdad, Baghdad, Iraq.
[a]Husseinengineer9@gmail.com, [b]abbasm.burhan@coeng.uobaghdad.edu.iq

Abstract. An accurate prediction of the productivity of the batch plant is considered an essential element for control and planning the construction project. The actual productivity values of construction equipment in the site not consistent with the nominal ones. Therefore, it is necessary to make a comprehensive evaluation of the nominal productivity of equipment concerning the effected factors and then re-evaluate them according to the actual values. This research involved investigation of the ready mixed concrete batch plant to evaluate the productivity with the actual values. Artificial intelligence techniques that are represented by Artificial Neural Network (ANN) and Support Vector Machine (SVM), in addition to the statistical technique, represented by multi-linear regression (MLR), were used as tools for modeling the actual productivity and produce the predicted model for the productivity. Three of accuracy measurements, correlation coefficient (R), mean absolute error (MAE) and root mean square error (RMSE), were used to develop the model and to make a comparison between the actual and predicted productivity. The researcher developed three mathematical models of MLR, ANN, and SVM used to predict the productivity of ready mixed concrete batch plant and also the results showed that the artificial intelligence techniques were more precise than those calculated by the conventional techniques, and SVM model was the best generalization than ANN model in both the training and the validation data.

Keywords: Construction; batch plant productivity; multilinear regression; artificial intelligence; support vector machine.

1. Introduction

Productivity is considered as an important indicator that effecting the selection of any equipment. Construction equipment is one of the resources needed for the completion and success of construction projects. Recently, Contractors undertaking many kinds of construction activities that need various types and sizes of equipment, and usually, they invest the biggest value for costs of the project, particularly for equipment-heavy projects. The productivity of equipment plays a major role in time and cost estimation of any construction project. The utilization of equipment includes increasing Production rate, low total cost, carry out activities that cannot be carried out manually, maintain the planned production rate when there is a shortage of labor, maintain high-quality standards, and so on. Thus, the correct choice and use of equipment contribute to the economy, safety, quality, speed, and timely completion of the project [1]. Otherwise, delays in project implementation may occur due to incorrect selection of equipment, unavailability of equipment on time, poor technology, and wrong mechanization.

Many researchers have defined the productivity in different ways according to the nature of the work, all agreed that the productivity is the rate of outputs to inputs [1]. Productivity can be defined as a measure of transforming resources of an individual, firm, industry, or overall economy into goods, services, and generates income. The success of construction projects is closely related to the production of machinery [3,4]. Machine manufacturers generally provide optimal hourly production from their machines to users. This ideal hourly production is called nominal hourly production, which is distinctly different from actual hourly production in construction projects. Actual production depends mainly on the conditions of the sites of the project. Estimation of actual production and hence the discrepancies between the nominal and actual production rate is an essential element in assessing the time and cost required to finish construction. To estimate production, it is very important to know how different the project site conditions affect the machinery production [2].

2. Methodology

One of the most important methodologies used in solving engineering problems is modeling. Many variables can be analyzed and modeled with similar data, and the results can be used in several applications [6]. In this paper, two types of forecasting systems were employed to establish the predicted model of productivity to estimate wet ready mixed concrete (WRMC) plant production, and dry ready mixed concrete (DRMC) plant production. The first one is a statistical method which is represented by multi-linear regression (MLR). The second one is an Artificial Intelligence (AI) which is represented by Artificial Neural Network (ANN) and support vector machines (SVM).

2.1.Model Database

The factors that involved in this context are based on the actual data of batch plant which conducted during observation and documentation processes of the concrete production. Generally, the data are classified into input or independent variables and output or dependent variables. The independent variables of DRMC are; times of cement filling (N_{cement}), times of aggregate filling ($N_{aggregate}$), the width of the conveyor belt of the aggregate (W_{belt}), power of motor for conveyor belt (PWbelt), the distance between the truck mixer location and the project site (D), speed of the truck mixer (Struck), traffic situation (T) and the number of truck mixers (No. truck mixers). The independent variables of WRMC are; width of the conveyor belt of the aggregate (W_{belt}), power of the motor for the conveyor belt (PWbelt), times of the loading of the mixing unit (Nmixer), distance between the batch plant location and the project site (D), speed of truck mixer (Struck), traffic situation (T), number of the truck mixers (No. truck mixers). While the output or dependent variables are the actual productivity for both (DRMC) and (WRMC).

2.2.Multiple linear regression model (MLR)

Multiple linear regression model (MLR), can be defined as a statistical process to estimate relationships between the variables and it has been employed to develop a statistical model that used for production the forecasting model of the productivity. Eq. 1 represents the general form of Multiple Linear Regression equations and the figure (1) explains the linear regression modeling:

$$y = w_{0+} \; w_1 x_{1+} w_2 x_{2+} w_p x_{ip+} \varepsilon_i \qquad (1)$$

Where: x_1, x_2... x_p: are independent variables (input factors), w_0, w_1... w_p are the coefficients in the linear relationship, and ε_1, ε_2... ε_i are the errors that create scattering around the linear relationship at each of the i = 1 to n observations

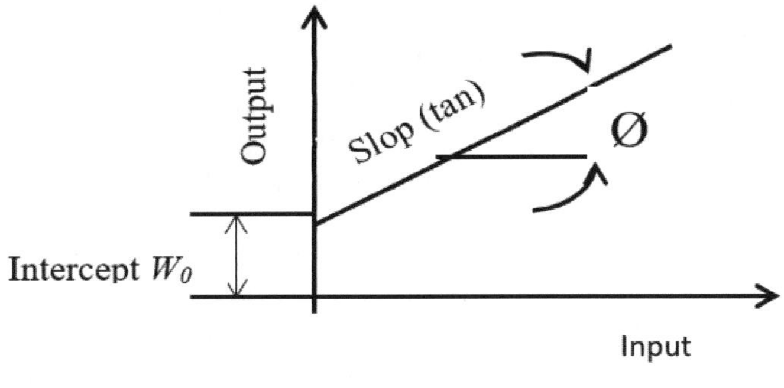

Figure 1. Linear regression modeling.

2.3.Artificial neural network (ANN)

Artificial neural networks (ANNs) or conduction systems are computational systems that inspired by the biological neural networks that form human brains. It is a form of artificial intelligence that tries to mimic the function of the nervous system and the human brain [7]. the concept of ANNs was introduced in 1943 by the neurophysiologist Warren McCulloch and the mathematician Walter Pitts, researches into applications of ANNs have. However, a neural network is an algorithm used in cognitive tasks, such as improvement and learning, which are based on concepts derived from research into brain nature. The artificial neural network has been evolved and employed as an alternative to regression analysis since the propagation algorithm proposal [8]

It can be learned from several input patterns and associated output patterns. through the process named training, the network becomes able to provide solutions to new problems even if incomplete Generally, two kinds of ANNs tutorials, feed-forward and feedback of ANN. The feedback network itself feeds and is well appropriate for solving problems optimization, while the feed-forward network transmits the information in a forward direction only.

Fig. 2 shows an example of the neural network, consisting of an input layer, a hidden layer, and an output layer, all linked with each other.

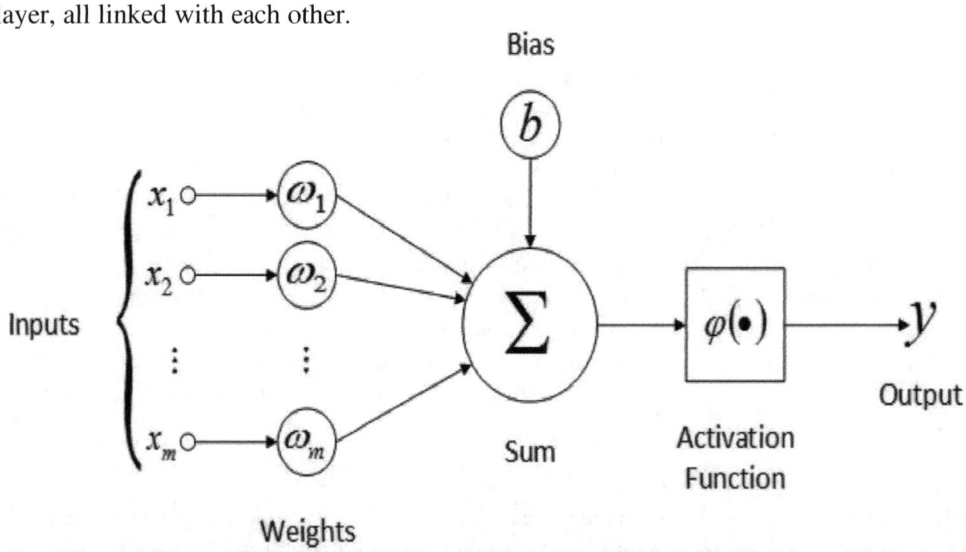

Figure 2. Basic structure of an ANN model.

The weighted amount of input is transported to the neurons hidden, in which they are converted using the activation function. The outputs of the neurons hidden, in turn, act as inputs to the resulting neurons as they undergo another transformation. The output of the anterior neural network for feeding is given by one hidden layer and one neural network for output by the Eq. 2.

$$Yt = f_o\left[\sum_{j=1}^{HN} WO_j * f_h\left(\sum_{i=1}^{m} WH_{ij} . X_{it} + b_j\right) + b_o\right] \quad (2)$$

Where Yt is the output, f_h, f_o are the hidden and the output neuron activation functions, HN is number of hidden neurons, WO_j is the weight of the correlation between jth neurons hidden and outputs neuron, m is the number of input neurons, WH_ (ij) is the weight of the relationship between ith inputs and jth hidden neurons, X_{it} is the input, b_j is the bias of the hidden jth neurons, and b_o is The resulting neuron bias. Support vector machine (SVM). SVM employs alternative ε insensitive loss of function which adjusted to involve a measure of distance [9]. However, present the SVM, a brief overview would be Introduced on how to create SVM for the problem of regression [10]. Assume y an observer variable with a real value, f(x, α) so α∈A is a set of real functions including the regression function f(x, α0). Now, if set data of training $\{(x_1, y_1), \ldots\ldots, (x_i, y_j)\ x \in R^N, y \in r\}$ is prepared, which x independent variable (inputs), y dependent variable (outputs), Rn space of N-dimensional vector and r is the space of one-dimension vector and has a linear function (3);

$$f(x, \alpha) = (w.x) + b \tag{3}$$

where w is the modifiable vector of weight and b is the numerical threshold. An optimal regression function is gained by reducing experimental risk as in Eq. 4;

$$R_{emp} = \frac{1}{l}\sum_{i=1}^{l} |y_i - f(x_i, \alpha)|_\varepsilon \tag{4}$$

The loss, error, function L_ε with an ε insensitive zone defined as in Eq. 5

$$L_\varepsilon = \begin{cases} 0, & \text{if } |y_i - f(x_i, \alpha)| \le \varepsilon \\ |y_i - f(x_i, \alpha)| - \varepsilon, & \text{otherwise} \end{cases} \tag{5}$$

Where ε represents the zone that the regression function lies within. When the expected value is within the region, the loss becomes zero and if the expected value is outside the area, the loss will be equal to the absolute value of the deviation ε [11]. Equivalence to minimize the performance which reflects the objective of SVM to discover the function of f(x,α) which has the greatest deviation of ε from the actual output, for all data of training, [12]. Figure 3 explains ξ* and ξ which can be defined as the parameters represent the slack variables that determine the amount the errors on the samples or They are minimal and upper limits on system output [13]. The constant c is a predefined value, C is greater than 0, that determines trade-off among the flatness f and amount which the deviations more than ε are allowable [14].

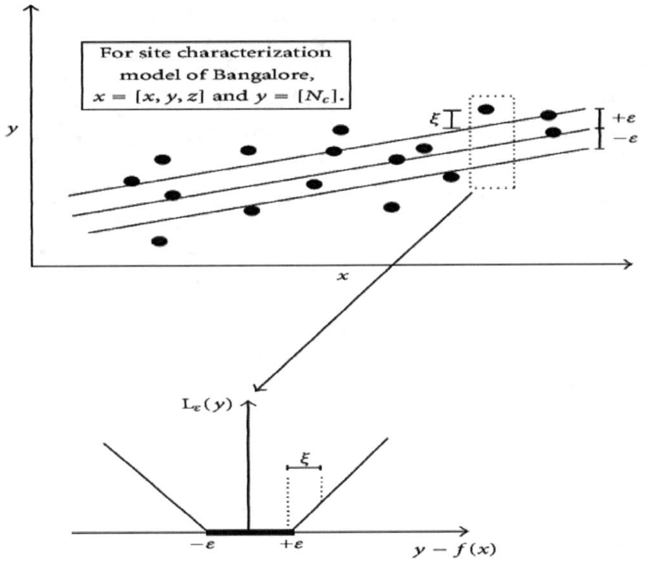

Figure 3. The accuracy ε and Slack Variable ξ in SVM.

In respect to the proposed predictive model of the delay in this study, the linear regression doesn't appropriate. And therefore, kernel (K) for nonlinear mapping is employed to remap the data in the space with a higher-dimensional feature to perform a linear regression [15] as in Fig. 4. [16] and formula (6).

$$f(x) = (w_0.x) + b_0 \tag{6}$$

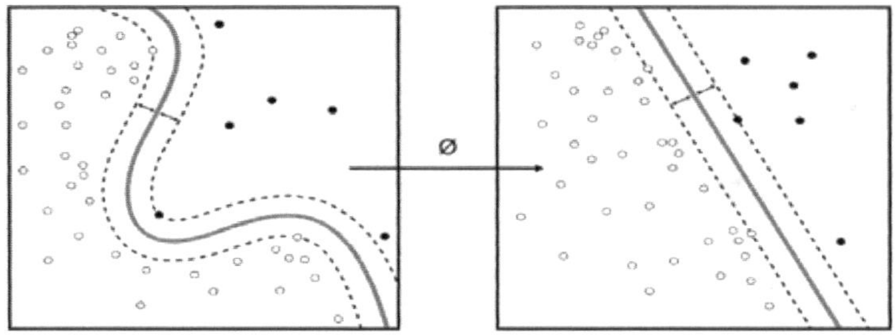

Figure 4. Non-linear SVM conception.

Many kernels utilized for non-linear regression such as Gaussian function and multilayer perceptron, as shown in Table 1.

Table 1. Popular kernel function

Type of Classifier	Kernel Functions
Linear, dot product, kernel	$K(x, x_i) = (x^T x_i)$
Complete polynomial of degree *d*	$K(x, x_i) = [(x^T x_i)]^d$
Gaussian RBF	$K(x, x_i) = e^{\frac{1}{2}[(x-x_i)^T \Sigma^{-1}(x-x_i)]}$
Multilayer perceptron	$K(x, x_i) = \tanh[(x^T x_i) + b]$
Inverse multi quadric function	$K(x, x_i) = \frac{1}{\sqrt{\lVert x - x_i \rVert^2 + B}}$

3. Results and Discussion

As mentioned previously, the gathering data of dry and wet ready mixed concrete (RMC) were employed to establish the three mathematical models. The first model was depended on the statistical technique which represented by the regression using the statistical product and solutions services (SPSS V. 25) software. The second model was based on (ANN) using Neuframe 4 software. While the third one was depended on support vector machine (SVM) using (WEKA 3.9.1©2016) software.

3.1. MLR Model

A multiple linear regression (MLR) can be used to establish the statistical model and produce the predicted equation for both dry ready mixed concrete (DRMC) and wet ready mixed concrete (WRMC). The final MLR model was established using statistical product and solutions services (SPSS). Table 2. summarizes the factors that were used in developing the MLR equation, the weights and the standard error

Table 2. The summary of DRMC model using MLR

Factor	Unstandardized Coefficients	
	Beta (β)	Std. Error
Constant	-3.667	6.388
W_{belt}	82.695	18.703
P_{Wbelt}	0.084	0.689
N_{cement}	-0.231	0.415
$N_{aggregate}$	-0.891	1.657
D	-0.457	0.311
S_{truck}	0.147	0.064
T	-1.475	1.815
No. $_{truck\ mixers}$	0.447	0.621

IOP Conf. Series: Materials Science and Engineering **901** (2020) 012020 doi:10.1088/1757-899X/901/1/012020

Eq. 7 represents the final model used to estimate DRMC productivity.

$$\text{DRMC Productivity} = 82.69W_{belt} + 0.084\ PW_{belt} - 0.231N_{cement} - 0.891\ N_{aggregate} - 0.457\ D + 0.147\ S_{truck} - 1.475\ + 0.447\ N_{truck\ mixers} - 3.667 \quad (7)$$

Where:
DRMC Productivity is the dry ready mixed concrete productivity (m^3/h unit)
W_{belt} is the width of the conveyor belt of the aggregate (m)
PW_{belt} is the power of motor for the conveyor belt (kW)
N_{cement} is the number times of the cement filling that calculated by $\frac{V_{truck} * Cc}{M_{cement}}$ where N_{cement} is round up
V_{truck} is a truck volume (m^3)
Cc is a cement content in (1 m^3) of concrete
M_{cement} is a capacity of the cement weight scale of batch plant (kg)
$N_{aggregate}$ is the number times of the aggregate filling that calculated by $\frac{V_{truck} * A_w}{M_{aggregate}}$ where $N_{aggregate}$ is round up.
Aw is an aggregate content (kg) in 1 m^3 of concrete
$M_{aggregate}$ is the capacity of the aggregate weight scale of batch plant (kg)
D is the distance between the truck mixer location and the project site (km)
S_{truck} is the speed of the truck mixer (km/hr)
T is the traffic situation.
$N_{truck\ mixers}$ it is the number of truck mixers that available in one hour.

The final model was created to estimate WRMC productivity by using SPSS. Table 3. summarizes the factors that are employed in developing the MLR equation for WRMC which comprises the weights and the standard error.

Table 3. The summary of WRMC model using MLR.

Factor	Unstandardized Coefficients	
	Beta (β)	Std. Error
Constant	7.408	25.431
W_{belt}	-3.292	28.973
P_{Wbelt}	2.189	0.590
N_{mixer}	0.139	1.257
D	-2.167	0.600
S_{truck}	0.043	0.175
T	0.393	4.964
$No._{truck\ mixers}$	2.656	0.985

Eq. 8 represents the final model used to estimate WRMC productivity by MLR.

$$\text{WRMC Productivity} = -3.29W_{belt} + 2.19PW_{belt} + 0.139N_{mixer} - 2.167D + 0.043\ S_{truck} - 0.393T + 2.656N_{truck\ mixers} - 7.408 \quad (8)$$

Where
N_{mixer} is the number times of the loading the mixing unit for one truck mixer that calculated by:
$N_{mixer} = \frac{V_{truck}}{V_{mixer}}$ where N_{mixer} is round up.
V_{mixer} is a mixer capacity (m^3).
D is the distance between the truck mixer location and the project site (km).

IOP Conf. Series: Materials Science and Engineering **901** (2020) 012020 doi:10.1088/1757-899X/901/1/012020

3.2. ANN Model

The simulation of the artificial neural network using NEUFRAME V.4 software was employed in this study that works underlying the mathematical formulas and supervised learning. After the training of DRMC algorithm using ANN successfully, the optimal model of estimating the value of the productivity was determined from the sigmoid transfer function as shown in Eq. 9.

$$\text{DRMC Productivity} = \frac{P_{Rang}}{1+e^{(-5.58\tanh k+2.98)}} + P_{min} \quad (9)$$

Where

P_{Range} is the difference between maximum and minimum values of actual productivity which used in training the network which equal to 85.

P_{min} is the minimum value of actual productivity which used in training the network which equal to 20.

Note before using the equation (k), all the input of variables should be changed between 0 to 1 utilizing the data ranges in the training of ANN model with the Eq. 10 applied.

$$x_n = \frac{x - x_{min}}{x_{max} - x_{min}} \quad (10)$$

$$k = [-1.2 + 1.78\text{Wbelt} + 0.042\text{Pbelt} - 0.024\text{Ncement} - 0.058\text{Naggregate} - 0.024\text{D} + 0.0068\text{Struck} + 0.007\text{T} + 0.0366\text{No. truck}] \quad (11)$$

- N_{cement}, $N_{aggregate}$, W_{belt}, P_{Wbelt}, D, S_{truck}, T, No. of truck mixers.

The optimum Eq. 12 of the wet ready mixed concrete (WRMC) productivity was a sigmoid transfer function type that was established during the training process of the artificial neural network (ANN).

$$\text{WRMC Productivity} = \frac{P_{Rang}}{1+e^{(-8.11\tanh k+4.14)}} + P_{min} \quad (12)$$

Where

P_{Range} is the difference between the maximum and the minimum values of the actual productivity which is used in training the network which equal to 70.

P_{min} is the minimum value of the actual productivity which is used in training the network which equals to 20.

Note before using the equation (k), all the input of variables should be changed between 0 to 1 utilizing the data ranges in the training of ANN model with Eq. 13 and Eq. 14 applied.

$$x_n = \frac{x - x_{min}}{x_{max} - x_{min}} \quad (13)$$

$$K = -0.54 - 0.28W_{belt} + 0.09P_{belt} + 0.025N_{mixer} - 0.076D - 0.003S_{truck} - 0.063T + 0.117\text{No. truck} \quad (14)$$

3.3. SVM Model

The training data were used to develop the support vector machine (SVM) model using (WEKA 3.9.1©2016) software. Two models were created; The first model was employed to estimate the dry ready mixed concrete (DRMC) and the second one was used to estimate the wet ready mixed concrete (WRMC). Once the training of the algorithm finished successfully, the optimum equation was found at poly kernel function, whereas C and Epsilon parameters were equal to 0.3 and 0.001 respectively. Eq. 15 represents the best of the DRMC Algorithm using SVM.

$$\text{DRMC Productivity} = 14.77W_{belt} + 2.04\,PW_{belt} - 0.0215N_{cement} - 0.145\,N_{aggregate} - 1.091\,D + 0.2248\,S_{truck} - 2.7768\,T + 1.1423\,N_{truck\ mixers} + 11.9934 \quad (15)$$

Eq. 16 represents the optimum model of WRMC productivity after completing the training processes.

$$\text{WRMC Productivity} = 10.0688W_{belt} + 1.447PW_{belt} - 0.904N_{mixer} - 1.9998D + 0.1\,S_{truck} - 4.7914\,T + 2.5222N_{truck\ mixers} + 22.183 \quad (16)$$

4. Performance and Validation of Model

The validation and performance of the model were satisfied using three common statistical parameters which are used to measure the quantification of the error which are; Mean absolute error (MAE), Root mean squared error (RMSE), and the correlation (r) between the actual and the predicted values of the productivity for validation the data. The validation of the data comprises fifteen batch plant which are not used in the training process. Table 4. presents the performance of DRMC productivity and table 5. shows the performance of WRMC productivity with the different types of modeling techniques.

Table 4. The performance of DRMC productivity.

Performance measure	MLR	ANN	SVM
R	0.91	0.87	0.99
(MAE)	15.46	12.66	6.14
(RMSE)	17.98	17.82	6.50

Table 5. The performance of WRMC productivity.

Performance measure	MLR	ANN	SVM
R	0.52	0.70	0.99
(MAE)	31.84	25.04	8.87
(RMSE)	45.92	36.23	12.45

Figures 5 and 6 present the comparison between the actual productivity and the predicted values calculated using the three modeling techniques. Figures 5 and 6 show that the SVM model is the best followed by the neural network while the regression results are the least because the regression technique is based on the principle of data preservation as for the neural network technique is only acceptable when it is within training limits, as opposed to SVM technique that exceeds the shortage in the two techniques above.

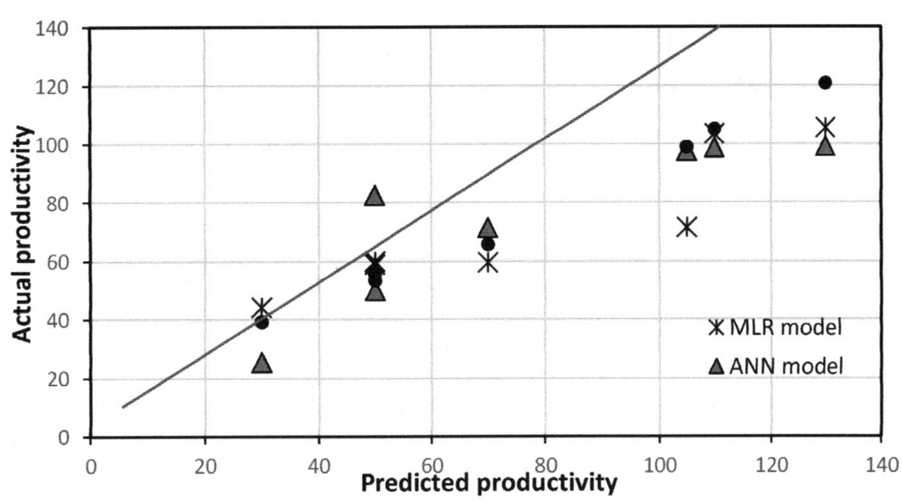

Figure 5. Predicted and Actual DRMC productivity.

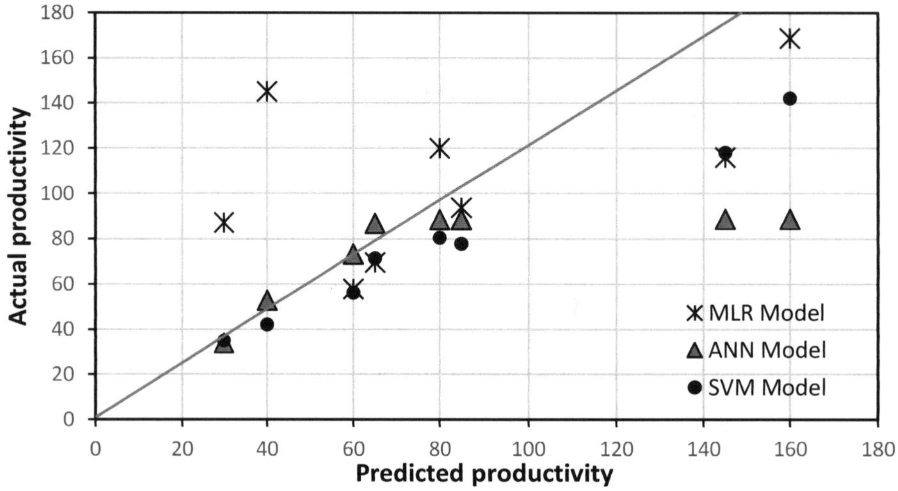

Figure 6. Predicted and Actual WRMC productivity.

5. Conclusion

After obtaining the predicted model of productivity for both the dry ready mixed concrete DRMC and the wet ready mixed concrete WRMC successfully, the conclusion can be summarized as below:

- MLR model presented good results with the validation of training dataset, but poor results with the external dataset. Therefore, the lack of data division into training and testing technique to obtain the minimum risk rate during the process of obtaining the mathematical model
- ANN model showed satisfying results of the validation dataset with a range of the training dataset and poor results with the data that exceeded the range of the training.
- SVM model produced good results for both the training and the validation dataset, best generalization, where the SVM model was the optimum.
- Generally, the results of the artificial intelligence techniques were more précised than those calculated by the conventional techniques.

References

[1] A. Sheikh, M. Lakshmipathy, and A. Prakash, Application of queuing theory for effective equipment utilization and maximization of productivity in construction management, International Journal of Applied Engineering Research, 11(8) 2016 5664-5672.

[2] Productivity Commission Report, Australian Productivity Commission, Reform of Building Regulation, Research Report, 2004.

[3] H. Fan, S. AbouRizk, and H. Kim, Building Intelligent Applications for Construction Equipment Management, Computing in Civil Engineering, 2007 192–199.

[4] O. Tatari, and M. J. Skibniewski, Integrated Agent-Based Construction Equipment Management: Conceptual Design Integrated Agent-Based Construction Equipment Management, Journal of Civil Engineering and Management, XII 2006 231-236.

[5] R.L. Peurifoy, C.J. Schexnayder, and A. Shapira, Construction Planning, Equipment, and Methods, Seventh Edition Robert, 7th, Texas, 2006.

[6] K.R. Kinner, D.C. Montgomery, G.C. Runger, J.W. Fowler, D.R. McCarville, T.R. Rhoads, and J.D. Stanley, Multivariate statistical methods for modeling and analysis of wafer probe test data, IEEE Transactions on Semiconductor Manufacturing, 15(4) 2002 523-530.

[7] D. Kriesel, A Brief Introduction to Neural Networks, Germany, University of Bonn, 2005.

[8] S. Akkol, Comparison of Artificial Neural Network and Multiple Linear Regression for Prediction of Live Weight in Hair Goats, Conference 8th of The Eastern Mediterranean Region of the International Biometric Society, Turkey, 27(1) 2015 21-29

[9] A.J. Smola, Regression estimation with support vector learning machines, Master's thesis submitted to the department of physics, Universitat Munchen, 1996.
[10] Y.B. Dibike, S. Velickov, D. Solomatine, and M.B. Abbott, Model induction with support vector machines: introduction and applications, Journal of Computing in Civil Engineering, 15(3) 2001 208-216
[11] P. Samui, Support vector machine applied to settlement of shallow foundations on cohesionless soils, Computers and Geotechnics, 35 2008 419–427.
[12] V. Vapnik, The Nature of Statistical Learning Theory, springer, 2nd Ed, 1999.
[13] P. Samui, V.R. Mandla, A. Krishna, and T. Teja, Prediction of rainfall using support vector machine and relevance vector machine, Earth Science India, 4(IV) 2011 188-200.
[14] A.J. Smola and B. Scholkopf, A tutorial on support vector regression, Statistics and Computing, 14 2004 199-222.
[15] B. Boser, I. Guyon, and V. Vapnik, a training algorithm for optimal margin classifiers. Fifth Annual Workshop on Computational Learning Theory, pages 144-152, Pittsburgh, ACM. 1992.
[16] P. Samui, and D. Kim, Utilization of support vector machine for prediction of fracture parameters of concrete, Comput. Concrete, 9(3) 2012 215-226.

Employ 6D-BIM Model Features for Buildings Sustainability Assessment

Hussain M. Habib[1,a*] and Kadhim R. Erzaij[1,b]

[1]Civil Engineering Department, University of Baghdad, Baghdad, Iraq.
[a]h.habib1001@coeng.uobaghdad.edu.iq, [b]kadhim1969@yahoo.com

Abstract. To translate sustainable concepts into sustainable structure, there is a require a collaborative work and technology to be innovated, such as BIM, to connect and organize different levels of industry e.g. decision-makers, contractors, economists, architects, urban planners, construction supplies and a series of urban planning and strategic infrastructure for operate, manage and maintain the facilities. This paper will investigate the BIM benefits as a project management tool, its effectiveness in sustainable decision making, also the benefit for the local industry key stakeholders by encouraging the BIM use as a project management tool to produce a sustainable building project. This paper presented a basic idea about the Building Information Model (BIM) technique and explain the levels of BIM and its dimension, the potential of BIM in sustainability design and the role of BIM in sustainability aspects had been discussed. The role of BIM in the planning and design stages explained the BIM role in sustainability rating system and grant credits in classification processes. Processing of creating 6D had been detailly explaining starting from the framework (ISO 1440) which adopted in the processing of establishing 6D complete system, phases of this system, and development steps from 3D to 4D, 4D to 5D and 5D to 6D. The 6D model features in the sustainability buildings had been explained in this research. As a result, many conclusions have been reached, the most important is there are still significant constraints in adopting of the integration approach between the (BIM) technology with sustainability in Iraqi buildings projects, both are facing great challenges in terms of economic, technical and the awareness level of government and society.

Keywords: Sustainability, BIM, 6D Model, LOD, Life Cycle Costing

1. Introduction

Building information modeling (BIM) is an approach of innovative new to building design, construction, and management, had been introduced by Autodesk in 2002. BIM is an integrated and technically coordinated process to create intelligent and informative 3D representation for stakeholders and the designers to take the design decision, build and run the projects in a cost-effect and sustainability manner [1]. It is a set of interactive policies, techniques, and processes that generate methodology to manage the project basic design and data in digital format through the building life cycle [2], is using 3D models along with additional information and data, like as schedule-related data (4D) and the component pricing and cost or budgeting information (5D). Recently the 6D of CAD is proposed to be the sustainability dimension in BIM technique as a tool for life cycle sustainability assessment and decision-makers and designers' approach to take the informed decision related to sustainability design facility management.

2. Levels of Development (LOD) in BIM

Level of Development (LOD) describes the level of completeness to which a Model Element is developed. LOD is often referred to in Building Information Modeling (BIM) delivery documents. It is an essential parameter for describing digital content in the context of BIM. It is an important tool for determining the exchange of information during the life cycle of a building [3]. The levels of BIM had been determining in three levels as described in Table 1, but only level one currently using by most users [1].

Table 1. Levels of Details (LOD) in BIM [3].

Level	Character	BIM
0	Unmanaged CAD, in 2D, data interchange by paper or electronically.	No.
1	Managed CAD, in 2D or 3D format in Common Data Environment (CDE) provided by using a collaborative tool with a unified strategy to the structure and format of data. Content information is created in level 1 using standard methods of data structures and storing (CAD standards) in standard formats to be transferred between varying CAD applications	Yes
2	BIM Level 2 is a managed BIM environment that contains intelligent BIM forms maintained in separate disciplines (specialized models), sharing and formatting by using a structured approach and integrated using a special or customized design program for the design, project management, and maintenance purposes.	Yes
3	Level 3 BIM is an integrated and collaborative BIM system that is made by Web Services for concurrent Building Information using open codes Without interoperability problems and expansionary BIM applications to project life cycle management.	Yes

2.1.BIM Dimensions

There is no consensus among experts on the number of dimensions in BIM that can be utilities in the construction industry because it is continuously expanding into new additions that meet the construction requirements in various phases of construction and contribute greatly to reducing project uncertainty. Table 2 contains explanations for 7 of BIM dimensions.

Table 2. BIM Dimensions adapted by authors [4].

No.	BIM Dimension
3D	The main difference between BIM 3D and Assisted Computer 3D (CAD) is that the latter describes the building in separate 3D views such as plans, sections, and heights that require scanning the entire document. They are described in aspects such as distances, columns, beams, and walls [5].
4D	4D is realized when linking 3D data to a management program makes the users able to visualize the entire project process and a detailed analysis of the project phases requirements and the owner or client benefits from the process conception, the contractor benefits from risk management and prevention. The 4D BIM effective management can help to visualize confliction before it happens, thus resulted in money and time saving on site
5D	To develop BIM 5D, the quantity data and cost plan are linked to the BIM data set; the quantitative and qualitative measurement is automatically taken from the 5D model, which is connected to the cost of the data and establishes a dynamic cost plan.

6D	The emergence of 6D has been strongly associated with an increase in global concern for sustainability resulting from the awareness of the seriousness of sustainability issues within the global context, and in all its economic, social and environmental aspects that are influential in the construction industry. Respectively, 6D can affect the other dimensions of BIM, because the characteristics of sustainability are likely to change over time and have a wide-ranging impact [5]. Designers have taken advantage of the integration of sustainability into the BIM model in achieving important objectives in assessing and achieving building performance in terms of sustainability such as carbon emissions and energy analysis. It helps to select the appropriate decisions in the early phases of the project and test alternatives and compare them because it enables easier integration of the elements of sustainability such as daylight, natural and mechanical ventilation, water efficiency, heat effect and renewable energies in the early stages of the building project. Finally, it allows the validation of design decisions [6].
7D	The seventh dimension (7D), Facility Management (FM), and life cycle costs (LCC) are called 7D BIM data. When integrate FM processes with the BIM models using and data. thus, provide significant financial savings and improved maintenance operation in the long-term. The application of BIM in the building's operation leads to many capacities and challenges. Many scientists, which use BIM in FM, appear to provide significant potential support [7].

There is no agreement on what on the sixth dimension should constitute the experts argue that due to the significance of sustainability should be in this dimension[4]. Dimensions 6 and 7 of BIM evolve so far and their understanding. Some researchers argued that BIM can unleash any dimensions number, describing them as nD modeling, noting that BIM model applications are multiple and can be at any dimension number in the next few years [9].

2.2.BIM in Sustainability Areas

In the design stages and prior construction stages, the most important decisions on sustainable design alternatives will be made by using the potential of BIM [10]. BIM technique may address many sustainability areas as explained in Table 3.

Table 3. BIM potential in the sustainability areas [9].

No.	Sustainability Area	The benefit of BIM using
1	Building orientation	Decide the best orientation for energy costs reduction
2	Building massing	for building form analyzing and building envelope optimizing
3	Day-lighting analysis	lighting and cooling loads reduction.
4	Water harvesting	water demands reduction in a building
5	Energy modeling	Reducing energy requirements and renewable energy choices analyzing to contribute to lower costs energy
6	Sustainable materials	reducing material needs and using recycled materials
7	Site and logistics management	waste and carbon footprints reduction

2.3.BIM Roles in Sustainability Aspects

To improve the overall building design and performance, BIM can play a large role in its ability to provide the necessary information. It offers an ideal benefit of sustainable building design through an Integrated Project Delivery (IPD) as well as design optimization. BIM supports solutions that can significantly contribute to identifying the best alternatives to reduce energy consumption and resource consumption [10]. The great importance of this technology in model development lies in combination visualization and performance simulations readily available for decision making through the rapid application of groups in the BIM model (enabled by the BIM platform). This BIM project is in fact due to its current role in building sustainability as shown in Table 4 [12].

2.4.BIM with Sustainability Rating Systems

In recent years, BIM's ability to provide a classification of sustainable building systems has been studied [22]. They were trying to have integration between BIM and sustainable building classification certificate, an argument that the Autodesk Revit BIM tool can evaluate thirteen credits and one requirement directly for the LEED rating system. Some studies argue that BIM applications adopting is assessing additional seventeen credits and two basic requirements in LEED [23].

Table 4. BIM role in sustainability aspects.

	Role of BIM in sustainability Aspects
Environmental	BIM's contributions to integrated project implementation and design improvement are well recognized as important in reducing the use of materials and energy. It gives users the ability to perform full lifecycle assessments and use the data of lifecycle assessment to run comparative analyses for different design options. By using BIM software, the designers will be able to move from models and environmental impact calculations to "rule the experiment" to real-time assessments at crucial moments, and easily, to produce understandably and customizable data graphics [12].
Economic	When using BIM technique to estimate the project cost generates rapid cost feedback throughout the building lifecycle [13]. BIM facilitates the process of cost estimation simpler and more accurate, the design of the digital model requires the architect and engineer cooperation, BIM provides more detail within lesser time and expenses [14]. BIM software has the potential to perform the quantity take-offs and pricing necessary for cost estimating while recognizing actual materials used in construction. The cost estimation accuracy is often positively related to the available volume of project information [15].
Social	There are some indirect means added by BIM to social sustainability. Such as a centered indoor localization algorithm, developed by BIM for supporting the response in building fire emergency [16]. The critical path method was also used by BIM to simulate the effect of overall temperature and humidity on a building project [17]. BIM proposed a system for forecasting air pollutant concentration while the construction process [17]. Finally, BIM use in building safety [18]. BIM may address some issues related to social sustainability in the built environment by proposing a MIB-based quality inspection process on building safety issues. But there is still a big gap that BIM needs to bridge in this dimension.

2.5.BIM in Stages of Sustainability Planning and Design

The basic decisions on sustainability, energy consumption, and the environment are taken at the planning and design phases are the point. It is indisputable that as soon the accurate and informed design decisions been taken as possible can help develop a more efficient and cost-effective sustainable design [10][5]. BIM analysis tools can rapidly assess different design options to help the design team for choosing the best green designs and make informed decisions [23]. These analyses provide more understand to the planners for their design's performance on the environmental aspects, the building efficiency, and the occupants. In the traditional projects, the design system makes project stakeholders and designers be much lower efficiency in early design decisions making because of the conceptual difficulties [24].

In the past few years, the developments in technology fields and the increasing use of BIM have made the use of BIM applications make design decisions more sustainable. The BIM applications

IOP Conf. Series: Materials Science and Engineering **901** (2020) 012021 doi:10.1088/1757-899X/901/1/012021

supporting the sustainable construction in the planning and design phases were identified and classified in seven areas [25].

1) Evaluation of the direction of construction (choosing a suitable direction can reduce energy consumption);
2) Analysis of the building structure and improvement of the building envelope in equal transparency ratio manner.
3) Analysis of daylight.
4) Water harvesting assessment for building water needs reducing.
5) Energy performance modeling for energy demands to reduce and to analyze renewable energy options that help to lower costs of the energy.
6) 6. Study the requirements of sustainable materials and then of replacing recycled materials;
7) Reduce waste and carbon footprints through appropriate site design and logistics management.

3. Procedure of Conceptually 6d Model

The motivation to establish a 6D system comes from the existing sustainability rating systems can't provide quick and reliable support for the design decision. The major purpose of this model is the assistance for design and a post-construction assessment tool. it can automatically evaluate lifecycle build-up estimates. It includes 3D design, 4D (schedule), 5D (cost) and 6D (sustainability). Estimate the sustainability performance for the options of the alternative design, the assessment included the building life cycle phases starting from material production through construction and operation phases to maintenance, demolition and disposal in the end [4]. The design of a complete 6D system can achieve the automatically building life cycle sustainability assessment and has a major contribution to help design buildings and support decision-making are:

- Quantities can be derived automatically from a 5D CAD model.
- A lifecycle cost analysis can be provided.
- Provide an assessment of the sustainability of the life cycle.
- Comparison of impacts of environmental, social and economic in different design
- Help designers and developers in making more informed decisions for sustainability designs, all of the buildings life cycle stages should be included in support for design decisions [24].
- Government departments will also be able to develop a sustainable building performance database and setting minimum standards for sustainability.

By providing an easy and quick sustainability evaluating to the design phase and facilitating the performance standards and database development, it hopes that more sustainable buildings will be executed in the future. This requires the coordination of efforts among the competent public bodies, including academic, technical, governmental, legislative and supervisory.

3.1. Phases of 6D CAD Model

According to ISO 14040 for LCA requirements (Figure 1), developing a 6D MODEL will consist of three phases, named:

1) The goal and scope definition,
2) The 6D model development for inventory analysis automatically and the assessment of the impacts.
3) Results interpretation.

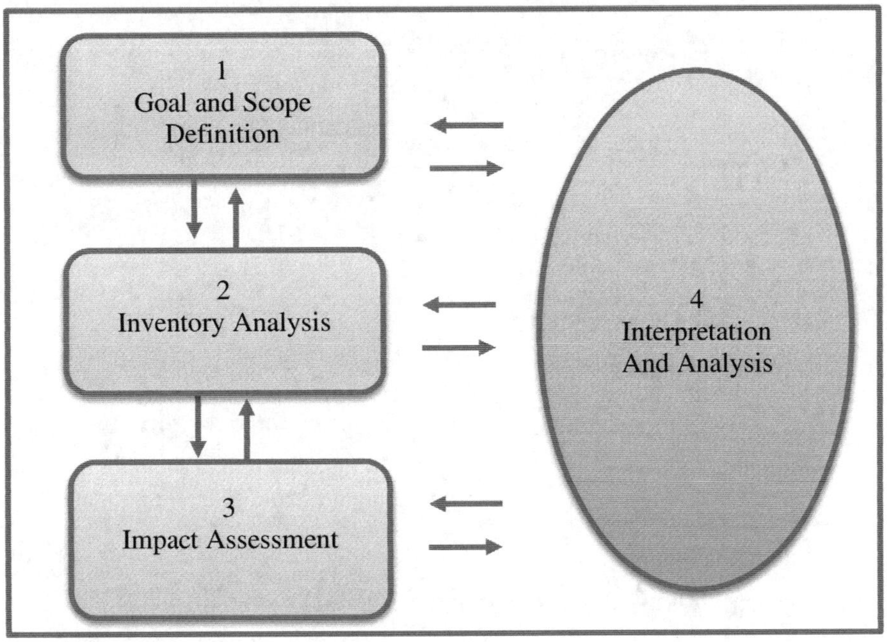

Figure 1. Stages of life cycle assessment as defined by ISO 14040 [25].

3.2. First Phase: Goal and Scope Definition

The goal is developing an integrated 6D model able to conduct the buildings life cycle sustainability assessment automatically to help buildings designers and support decisionmakers. All the buildings life cycle stages should be included in the vision of the three sustainability aspects (Environmental, Social, and Economic).

3.3. Second Phase: Inventory Analysis and Impact Assessment

Some software such as SimaPro (LCA tool) and(EcoInvent LCA (an embodied database) will be used At this stage [26]. The life cycle inventory data and estimate the impact results for a particular commodity unit will construct the database, including construction products [27]. For example, (according to some advanced means, such as Ecological Ecarcity 1997 or Environmental Indicator 1999) by the database can calculate the inventory data and can the assessment environmental impact against 1 cubic meter of the building block or 1 concrete cubic meter and calculate the amount of material used in the building [26]. To complete this phase, the impacts of the other two sustainability aspects (social and economic) must be assessed. The 6D system must contain three modules Input unit, core unit, and output unit Figure (2), all necessary data will be collected in the input unit of the model. These include:

- Based on 3D object design model, which may be created using, for example, Revit or Autodesk Architecture
- 4D schedules can be developed with, for example, Primavera or Microsoft Project.
- Site and location data, it may use to conduct the requirements for the heating and cooling calculation.
- Assumptions of life service recurring material servicing and maintenance costs must be included in the assessment. For example, repainting is usually performed in 10 years, but the carpet tiles need to replace every 8 years, etc.

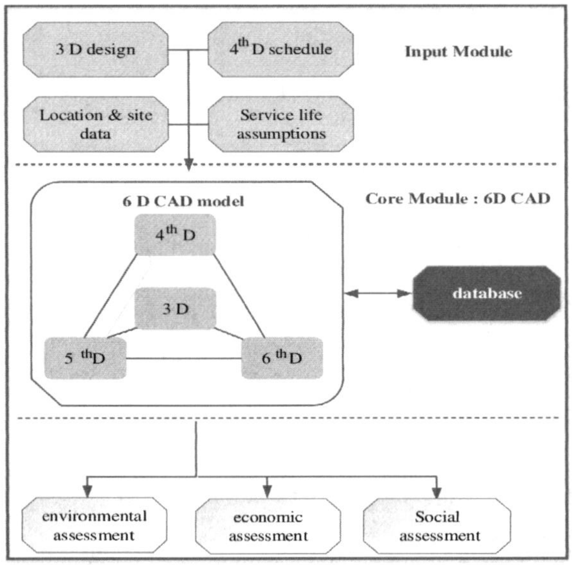

Figure 2. Overview of the 6D CAD model [4].

4. Steps for 6D Model

4.1. Step One. From 3D to 4D
For equipment information, labor, and materials in the temporary work, table 4D should be linked to the 3D design, this can be automatically linked to any unique object ID and activity chart. When it connected, the 4D model becomes available by some software, such as the Autodesk Naviswork program [28].

4.2. Step Two. From 4D to 5D
In traditional methods can automatically calculate the permanent works quantities in the design with bills of quantities (BQs). And can verified These quantities with those measured according to shown in the BQ. and the priced BQ can derive rates of each item, or the quantity surveying firm provide the cost database. This gives the permanent works cost. but project cost besides the permanent works comprise the preliminaries costs too, including site staff, temporary works, plants, etc. 4D can derive most items of the preliminaries according to the included method of construction.

4.3. Step Three. Life Cycle Costing
The aim of this step is the life cycle cost determining. the user model is assigning the default building life as 25 years or they can adjust it to suit their standards. The operational cost consists of several types of costs such as energy for lighting, ventilation, cooling, heating, and electricity for electrical appliances. Some programs can be commercially available to use for simulate annual energy usage such as TRNSYS or EnergyPlus after obtaining the energy cost of the processed companies[31]. Future energy costs should be considered at an appropriate rate of interest, as well as the cost of maintenance and the replacing materials cost [4]. The age of components or equipment has a shorter lifecycle than the building lifetime should be determined for calculating the replacing cost. For the demolition and disposal cost, we can use the building size and the appropriate cost database to estimate the demolition cost at the end of life. The cost of disposal includes the transport the destroyed construction waste to landfill sites and associated taxes.

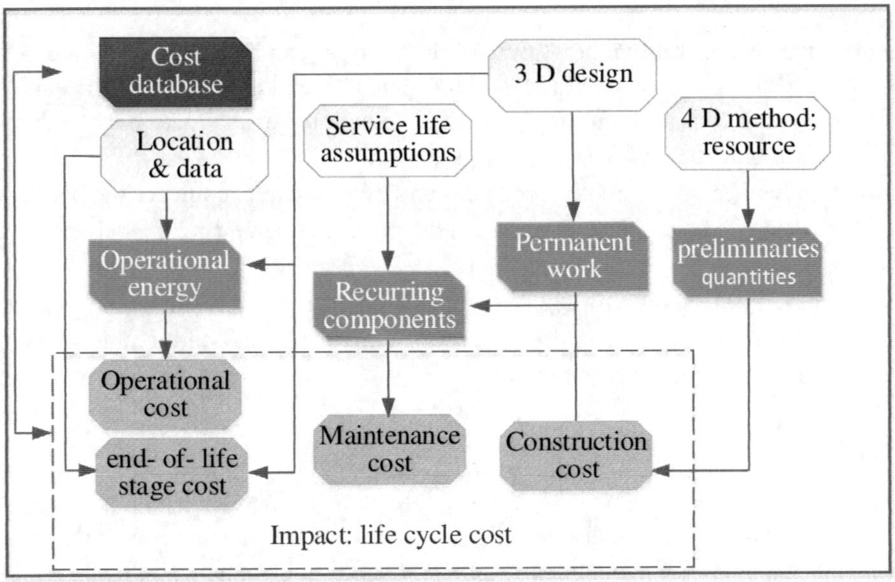

Figure 3. Conceptual map of economic assessment framework.

4.4. Fourth step. From 5D to 6D

Since the economic aspects been addressed in step three of the model procedure, this step is assessing only the environmental and social impacts of the building life cycle (Figure 4). For this purpose, the quantities derived from the 4Dh construction schedules are exported to EIA system, as SimaPro software. The program will calculate the construction materials environmental impacts. The 4D schedule provides information about the worker's number in each construction activity. which means employment opportunities and a significant impact social aspect.

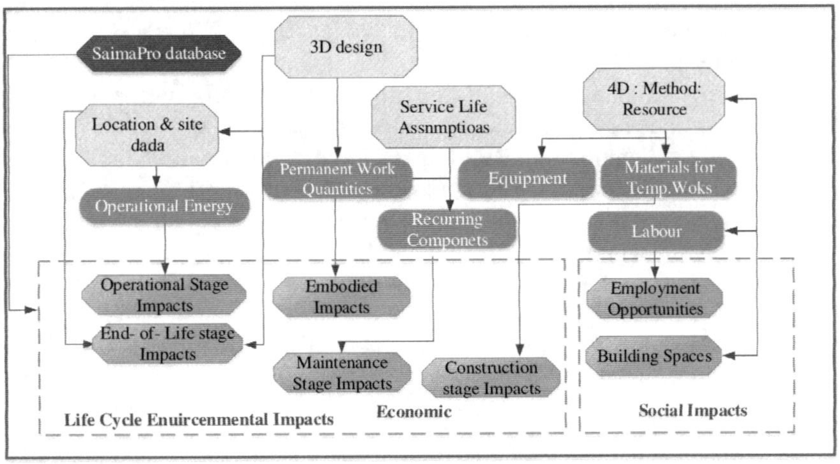

Figure 4. Conceptual map of environmental and social assessment framework.

4.5. Third phase: Interpretation of the Output Module

The environmental impacts which it been derived from the second phase, many different categories are included, such as, resource consumption, ecosystem spoiling, energy use, and each category comprises the number of sub-categories, some of the qualitative and other quantitative they cannot be aggregated, such these environmental impacts, the economic and social aspects. this problem was a motivation for developing the analysis tool can assess the impact of the interrelationship between the environmental, economic, and social [8].

5. 6D Model Features

Besides providing the sustainability index, some of the analyses can be produced by the 6D model:

1) The performance comparison for a certain standard of building (e.g. the air conditioning electricity annual consumption), to declared or to national or international standards. Forms of bar charts or tables can be used to present the results.
2) Comparison between standard performances with, for example, compare the performance of the embodied energy and the operation performance, maintenance, and life-cycle energy performance.
3) The store data ability and performance comparing of different alternatives in the same function, such as, conducting comparison for the cost of individual glass and double glass windows, or compare steel and concrete structure, etc.

6. Conclusions

One of the most important constraints of the sustainability rating system is the inability to provide rapid and reliable design decision support, therefor it is creating a CAD 6D model that can assess the sustainability of a building automatically. will help stakeholders and designers to make more informed decisions at the design stage. This research presents a description of a conceptional model in detail. For much more sustainability performance in the building need to a database created for performance standards.

Recommendation

Despite the great benefit of the integration of BIM technology in sustainability assessment for construction projects, the process of transition in this field should be considered and not quick and sudden and according to a roadmap involving all relevant institutions (academic, technical and economic) and supported by government procedures in both legislative and regulator

References

[1] C.N. Preece, M.T. Shafiq, Z.I. Syed, C.M.M. Isa, and H. Golizadez, Towards BIM enabled sustainable urban developments in the UAE, In Proceedings of Sustainable Built Environment Conference 2016 in Cairo 332-345.

[2] B. Succar, Automation in Construction Building information modelling framework : A research and delivery foundation for industry stakeholders, Autom. Constr., 18(3) 2009 357–375.

[3] R. Volk, J. Stengel, and F. Schultmann, Building information modeling (BIM) for existing buildings - Literature review and future needs, Autom. Constr., 38(March) 2014 109–127.

[4] P. Yung and X. Wang, A 6D CAD model for the automatic assessment of building sustainability, Int. J. Adv. Robot. Syst., 11(1) 2014 1–8.

[5] G. Kapogiannis, M. Gaterell, and E. Oulasoglou, Identifying uncertainties toward sustainable projects, Procedia Eng., 118 2015 1077–1085.

[6] K.D. Wong and Q. Fan, Building information modelling (BIM) for sustainable building design, Facilities, 31(3) 2013 138–157.

[7] B. Becerik-Gerber, A. M. Asce, F. Jazizadeh, N. Li, and G. Calis, "Application Areas and Data Requirements for BIM-Enabled Facilities Management," vol. 138, no. March, pp. 431–442, 2012.

[8] C. Fu, G. Aouad, A. Lee, A. Mashall-Ponting, and S. Wu, IFC model viewer to support nD model application, Automation in Construction, 15(2) 2006 178-185.

[9] S. Azhar, Building information modeling (BIM): Trends, benefits, risks, and challenges for the AEC industry, Leadership and management in engineering, 11(3) 2011 241-252.

[10] S.L. Shareef and H. Altan, Building sustainability rating systems in the Middle East, In Proceedings of the Institution of Civil Engineers-Engineering Sustainability, 170(6) 2016 283-293.

[11] T. Ahmad, A. Aibinu, and M. Jamaluddin, BIM-based iterative tool for sustainable building design : A conceptual framework, Procedia Eng., 180 2017 782–792.

[12] F. Jalaei and A. Jrade, Integrating building information modeling (BIM) and LEED system at the conceptual design stage of sustainable buildings, Sustain. Cities Soc., 18 2015 95–107.
[13] A. Sattineni and R.H. Bradford, Estimating with BIM: A survey of US construction companies, Proceedings of the 28th ISARC, Seoul, Korea, 2011 564-569.
[14] S. Azhar, M. Khalfan, and T. Maqsood, Building information modelling (BIM): now and beyond, Construction Economics and Building, 12(4) 2012 15-28.
[15] G. Kim, J. Yoon, S. An, H. Cho, and K. Kang, Neural network model incorporating a genetic algorithm in estimating construction costs, Building and Environment, 39(11) 2004 1333–1340.
[16] N. Li, B. Becerik-Gerber, B. Krishnamachari, and L. Soibelman, Automation in construction a BIM centered indoor localization algorithm to support building fire emergency response operations, Autom. Constr., 42 2014 78–89.
[17] Y. Shan and P. M. Goodrum, Integration of building information modeling and critical path method schedules to simulate the impact of temperature and humidity at the project level, 2014 295–319.
[18] S. Park, and I. Kim, Bim-based quality control for safety issues in the design and construction phases, Archnet-Ijar, 9(3) 2015.
[19] L.C. Bank, M. McCarthy, B.P. Thompson, and C.C. Menassa, Integrating BIM with system dynamics as a decision-making framework for sustainable building design and operation, In Proceedings of the First International Conference on Sustainable Urbanization (ICSU) 2010.
[20] S. Barnes and D. Castro-Lacouture, BIM-enabled integrated optimization tool for LEED decisions, In Computing in Civil Engineering, 2009 258-268.
[21] S. Azhar, A. Nadeem, J.Y. Mok, and B.H. Leung, Building Information Modeling (BIM): A new paradigm for visual interactive modeling and simulation for construction projects, In Proc., First International Conference on Construction in Developing Countries, 1 2008(August) 435-46.
[22] T. Ahmad and M.J. Thaheem, Economic sustainability assessment of residential buildings: A dedicated assessment framework and implications for BIM, Sustainable cities and society, 38 2018 476-491.
[23] E. Krygiel and B. Nies, Green BIM: Successful Sustainable Design with Building Information Modeling, John Wiley & Sons (2008).
[24] A. Haapio and P. Viitaniemi, A critical review of building environmental assessment tools, Environ. Impact Assess. Rev., 28(7) 2008 469–482.
[25] M.A. Rosen and H.A. Kishawy, Sustainable manufacturing and design: concepts, practices and needs, Sustainability, 4(2) 2012 154–174.
[26] https://simapro.com/
[27] R. Frischknecht, N. Jungbluth, H.J.Althaus, C. Bauer, G. Doka, R. Dones, and Y. Loerincik, Implementation of life cycle impact assessment methods, ecoinvent report No. 3, 2007.
[28] A. Russell, S. Staub-french, N. Tran, and W. Wong, Automation in construction visualizing high-rise building construction strategies using linear scheduling and 4D CAD, Autom. Constr., 18(2) 2009 219–236.
[29] D.B. Crawley, L.K. Lawrie, F.C.Winkelmann, W.F. Buhl, Y.J. Huang, C.O. Pedersen, and J. Glazer, EnergyPlus: creating a new-generation building energy simulation program, Energy and Buildings, 33(4) 2001 319-331.
[30] D. Crawley and I. Aho, Building environmental assessment methods: applications and development trends, Building Research & Information, 27(4-5) 1999 300-308.

Modeling of Estimate at Completion in Earned Value

Sabreen Mortadha Baqer[1,a*] and Sedqi Esmaeel Rezouki[1,b]

[1]Civil Engineering Department, University of Baghdad, Baghdad, Iraq.
[a]sabreen9013@gmail.com, [b]sedki_razqi@yahoo.com

Abstract. This paper will explain the concept of Earned value management and definition of the influenced factor, earned worth management (EVM) is an associate trade commonplace for observance the performance of in progress comes. The performance baseline is about up within the designing section to live any time and price deviations throughout project execution. supported this progress, associate estimate at completion (EAC) is forecasted, investigate a wide range of factors impact on earned value in different countries, especially in Iraq and the region to reduce cost project in estimate to complete and estimate at complete, which helps decision-makers to make the right decision to reduce costs, achieve planned duration and taking into account other aspects.

Keywords: Earned value management; Estimate at completion; Project risk; factors affecting, Cost and schedule.

1. Introduction

Earned value management (EVM) is one of the potential measurement method of performance and evaluating the success of a construction project [1]. It addresses the 3 performance key areas: scope, schedule and budget [2], additionally earned price management (EVM) addresses several project management areas together with project organizing, planning, scheduling, budgeting, accounting, analyzing, coverage and variance watching, and so, it's doable to forecast value or schedule overruns and other important factors such as quality, risk, safety and social factors) in an early stage of the construction project [3].

Many applications could be used in different projects to demonstrate the advantages of Earned value management (EVM) in extracting additional information from data about the status, trend, and future project schedule performance and associated cost. Earned Value Management (EVM) can be defined as a management technique for project performance monitoring [4]. EVM integrates scope, price, and schedule management underneath an equivalent framework and it provides performance variances and indexes which permit managers to discover over-costs and delays. Moreover, underneath this technique, new real knowledge generated throughout the project run time is employed to explain trends for the longer-term project total price and finishing date. The potential benefit of using it is to estimate and address the delay factors in the construction projects [5] so its application to the Iraqi projects may have a significant impact in the case of conducting adaptive and adjustments commensurate with the local scope of Iraqi projects.

2. Methods of Analysis

2.1.The Use of EVM on Iraqi Delaying Project

Many Iraqi construction projects have not been completed within the duration and cost planned and this affects the objectives of the project significantly in terms of cost, time, and quality. The using EVM will have a positive impact to mitigate that risk because the EVM will indicate the Deviation in the early stages of the project, and will explain the basic equation of EVM.

2.2.Earned Value Management (EVM)

It is a project management technique for objectively measuring project performance and progress as shown in Eq. (1):

EV= Budget at Completion (BAC) * Actual Complete % (1)

2.3.Cost Performance Index (CPI)

It is a life of the money effectiveness and potency of a project. It represents the number of completed work for each unit of value pay.

CPI = (EV/AC) (2)

Where CPI greater than one is favorable (under budget):

< 1 means the price of finishing the work is above planned (bad);

= 1 means the price of finishing the work is true on set up (good);

> 1 implies that the value of finishing the work is a smaller amount than planned (good or typically bad).

Having a high CPI (in some cases, terribly high is simply one.2) could mean that the arrange was too conservative, and therefore a high range could not be smart, because the CPI is being measured against a poor baseline

2.3.1. Schedule Performance Index (SPI)

It is a measure of schedule efficiency, expressed as the ratio of earned value to planned value (PV).

SPI = (EV/PV) (3)

It can be concluded the following with the above formula: SPI > 1 More work has been completed than the planned (ahead of schedule); SPI < 1 Less work has been completed than the planned (behind schedule), and SPI = 1 Work is being completed at about the same rate as planned (on time).

2.3.2. Estimate to Complete (ETC)

ETC is that the estimate to finish the remaining works of the project. ETC should be supported objective measures of the outstanding work remaining, generally supported the measures or estimates accustomed to produce the first planned price (PV) profile, together with any changes to predict performance supported historical performance, actions being taken to boost performance, or acknowledgment of degraded performance. While algebraically, ETC ought to ne'er be computed mistreatment either EAC or AC. within the following equation,

ETC = ((BAC-EV)/CPI) (4)

ETC is that the variable, EAC is that the variable and AC are fastened supported expenditures thus far. ETC should be reportable in truth to replicate the project team estimate to complete the outstanding work. If ETC pushes EAC to exceed BAC, then project management skills are used to either suggest performance enhancements or scope modification, however, ne'er force ETC to convey the "correct" answer so EAC = BAC. Managing project activities to stay the project among budget may be the activity of a human factor, not a function

2.4.Budget at Completion (BAC)

The entire planned worth (PV or BCWS) at the tip of the project. If a project incorporates a management reserve (MR), it's usually not enclosed within the BAC, and severally, within the performance mensuration baseline.

2.5.Estimate at completion (EAC)

Projects throughout their execution part might deviate from their original planned cost; so, the project team forecasts the EAC that differs from the budget at completion (BAC) and relies entirely on the present performance of the project [6]. Though a variety of studies supported customary formulae and alternative models are offered to forecast EAC, that one offers the most effective answer is that the practitioner's quandary. Within the face of no specific guideline concerning the most effective model for many correct results, the selection is left with analysts and project managers to determine the EAC formula to be used for foretelling [7].

The sole attainable intuitive recommendation is obtainable by the project conditions and level of experience possessed by the project management team [8]. EAC formulae have supported the mix of many information components, that embody the budgeted price of labor schedule (BCWS), budgeted price of labor performed (BCWP), and the actual price of labor performed (ACWP). In line with PMI (2013), 3 of the foremost common

EAC foretelling techniques for the estimate to complete (ETC) are mentioned within the following. The first methodology uses the particular price till the information date and predicts that the remainder of the project is going to be dead at the planned rate. The mathematical illustration of this methodology is given in Eq5.

$$EAC=AC+ETC \quad (5)$$

$$EAC=AC+(BAC-EV) \quad (6)$$

The second methodology considers the present price-performance index (CPI) of the project and assumes that the remainder of the project is going to be completed at the present rate, as mathematically expressed in Eq7.

$$EAC = \frac{\mathbf{BAC}}{\mathbf{CPI}} \quad (7)$$

The third statement equation considers each SPI Associate in Nursingd CPI of the project and is usually helpful once the schedule of the project is an impacting issue. This methodology will be varied by distributing a complimentary weight to every index as per the project manager's call. However, the total of those weights sometimes equates to one [9]. The mathematical illustration of this methodology is given in atomic weight.

$$EAC=ETC+AC \quad (8)$$

2.6.Variance at completion (VAC)

The vacation may be a forecast of what the variance, specifically the price Variance (CV), is going to be upon the completion of the project. it's the dimensions of the expected cost or underrun. In several things, the project manager should request further funding as early as potential, or a minimum of report the potential for associate overrun. The vacation represents the dimensions of this request.

$$VAC= BAC-EAC \quad (9)$$

3. Types of Regression

3.1. Linear Regression

Linear regression assumed that there's a linear relationship between variables. There are unit 2 kinds of statistical regression; easy linear regression and multilinear regressions. When there's quite one variable quantity the multi statistical regression is employed (MLR).

3.2. Nonlinear Regression

One of the foremost blessings in exploitation nonlinear regression is that the broad varies of functions that maybe work. Several scientific or physical processes area unit inherently nonlinear; so, to aim a linear work of such a model would essentially cause poor results [10]. Nonlinear regression conjointly shares the characteristic with statistical regression of getting economical use of knowledge. That's nonlinear regression (NLR) will provide smart estimates of the unknown parameters within the model exploitation comparatively little knowledge sets [11]

4. Methodology

4.1. Developing Six Equations Index by Using NLR

The Leven Berg-Marquardt technique is used to develop the NLR equations. This technique is based on inserting variables in a nonlinear equation built according to some values of equation parameters and checked by the "coefficient of determination" test. Based on the checklist results, the independent variables for the cost index that have the most significant impact on the estimate at a complete model of projects are treated to fit the analysis mechanism so they become as shown in Table 1.

On the other hand, the first equation for the cost index as shown in Eq. (10), Table 2 shows the best values of the parameters of the equation in this research

$$COI = C_1{*}\ I_1 + C_2{*}\ I_2^{C3} + C_4{*}\ I_3 + C_5{*}\ I_4 + C_6{*}\ I_5^{C7} + C_8{*}\ I_6 + C_9{*}I_7^{C10} + C_{11}{*}\ I_8^{C12} + C_{13}{*}\ I_9 \quad (10)$$

Where COI is the cost index.

Table 1. Input Variable for Cost Index (COI).

Items	Symbol
Policy and requirements of the client	I_1
Policy and experience of the company	I_2
Price changes (marital, labor, equipment)	I_3
Errors in estimated quantities.	I_4
Funding and factors affecting it: provide financial liquidity to contractor	I_5
Funding and factors affecting it: The Company has access to banking facilities	I_6
Funding: grants or loans	I_7
Funding: governments financing	I_8
Funding: private financing (private banks)	I_9

Table 2. Parameters Estimates for Cost Index (COI).

Parameter	C_1	C_2	C_3	C_4	C_5	C_6	C_7	C_8	C_9	C_{10}	C_{11}	C_{12}	C_{13}
Estimate	-0.185	0.414	1.508	0.326	0.229	0.338	0.336	-0.072	-0.035	0.164	0.023	1.233	0.048

Furthermore, based on the checklist results, the independent variables for the time index that have the most significant impact on the estimate at a complete model of projects are treated to fit the analysis mechanism so they become as shown in Table 3.

Table 3. Input Variable for Time Index (TII).

Items	Symbol
Rise production capacity through: Increasing labor force	I_1
Rise production capacity through: Increase machines and equipment	I_2
Rise production capacity through: Increase the number of working hours	I_3
Rise production capacity through: Raising the skill of the workforce	I_4
Quick completion of work	I_5
Effect of the management method on the delay of completion of the project	I_6
The effect of the (economic situation) factor on the delayed completion of the project	I_7
Availability of money liquidity and exchange rate	I_8

Moreover, the second equation for the time index as shown in Eq. (11), Table 4 shows the best values of the parameters of the equation in this research.

$$TII = T_1 * I_1 + T_2 * I_2^{T_3} + T_4 * I_3^{T_5} + T_6 * I_4 + T_7 * I_5 + T_8 * I_6^{T_9} + T_{10} * I_7 + T_{11} * I_8^{T12} \quad (11)$$

Where TII is the time index.

Table 4. Parameters Estimates for Time Index (TII).

Parameter	T_1	T_2	T_3	T_4	T_5	T_6	T_7	T_8	T_9	T_{10}	T_{11}	T_{12}
Estimate	-0.059	-20.177	0 .006	1.753	0.061	0.468	0.474	0.12	5.264	-0.03	18.472	0.00022

Otherwise, based on the checklist results, the independent variables for the quality index that have the most significant impact on the estimate at a complete model of projects are treated to fit the analysis mechanism so they become as shown in Table 5.

Table 5. Input Variable for Quality Index (QUI).

Items	Symbol
Reduce the cost of acquisition and use the raw materials.	I_1
Contribute to strengthening the competitiveness of the institution	I_2
Matching materials to specifications	I_3
matching work to quality specifications	I_4
Does the organization conduct internal audits periodically to determine the conformity of the quality management system to ISO 9001 – 2015	I_5
Does the employer take the necessary procedures to remove the causes of nonconformity and prevent its recurrence?	I_6

The third equation for the quality index as shown in Eq. (12):

$$QUI = Q_1 * I_1 + Q_2 * I_2^{Q3} + Q_4 * I_3^{Q}{}_5 + Q_6 * I_4^{Q}{}_7 + Q_8 * I_5 + Q_9 * I_6 \quad (12)$$

Where QUI is the Quality index and the best values of the parameters of the equation in this research are. Q_1= 0.275; Q_2= 0.179; Q_3=0.906; Q_4= 0.067; Q_5=0.87; Q_6=0.295; Q_7=0.972; Q_8= 0.153; Q_9=0.116. Based on the checklist results, the independent variables for the risk index that have the most signify cannot impact on the estimate at the complete model of projects are treated to fit the analysis mechanism so they become as shown in Table 6.

Table 6. Input Variable for Risk Index (RSI).

Items	Symbol
The quality of the land is not appropriate in terms of its ability to withstand construction or because of its other characteristics.	I_1
Lack of infrastructure connected to the place of execution such as roads or inadequate place of implementation.	I_2
Delay payments to local companies from the employer	I_3
Financing cut off unexpectedly by local companies	I_4
The high prices of materials necessary to carry out civil works in construction sites due to economic conditions	I_5
Difficulty accessing project sites due to security situation	I_6

The fourth equation for the risk index as shown in Eq. (13):

$$RSI = R_1 * I_1^{R2} + R_3 * I_2 + R_4 * I_3^{R5} + R_6 * I_4 + R_7 * I_5^{R8} + R_9 * I_6 \quad (13)$$

Where RSI is Risk index and the best values of the parameters of the equation in this research are R_1=0.231; R_2=1.007; R_3=0.194 ; R_4=0.075; R_5=1.243; R_6=0.187 ; R_7=0.257 ; R_8=1.033; R_9=0.138. Based on the checklist results, the independent variables for the safety index that have the most significant impact on the estimate at a complete model of projects are treated to fit the analysis mechanism so they become as shown in Table 7.

Table 7. Input variable for safety index (SAI).

Items	Symbol
Provide adequate personal safety equipment (gloves, glasses, helmets, etc.) on-site	I_1
Does the use of new machines affect the provision of appropriate training.	I_2
Do accidents occur due to lack of safety standards in construction companies	I_3
The safety officer, if any, or the implementing contractor shall be responsible for the laying down of guiding plates by the approved design of the Civil Defense, indicating to the site's employees or non-employees the risk of the site according to the nature of the executed works.	I_4
Are buildings adjacent to the drilling work supported if there is a potential for impact on these works?	I_5

The fifth equation for the safety index as shown in Eq. (14):

$$SAI = S_1 * I_1 S_2 + S_3 * I_2 S_4 + S_5 * I_3 S_6 + S_7 * I_4 + S_8 * I_5 S_9 \quad (14)$$

Where SAI is Safety Index and the best values of the parameters of the equation in this research are S_1=0.383; S_2=1.600; S_3=0.102; S_4=;1.266 ; S_5=5.344; S_6=-10.450; S_7=0.173; S_8=0.501; S_9=1.438. Based on the checklist results, the independent variables for the social index that have the most significant impact on the estimate at a complete model of projects are treated to fit the analysis mechanism, so they become as shown in Table 8.

Table 8. Input variable for social index (SOI).

Items	Project size	Reliance on local labor	The impact of social customs and traditions on the region
Symbol	I1	I2	I3

The sixth equation for the social index as shown in Eq. (15):

$$SOI = A_1 * I_1 + A_2 * I_2^{A3} + A_4 * I_3^{A5} \quad (15)$$

Where SOI is Social Index and the best values of the parameters of the equation in this research are A_1=0.302; A_2=0.191; A_3=1.000; A_4=0.455; A_5=1.180

4.2. Developing Model Estimate at Completion

After the six equations have been developed, the effect of each factor was taken for (14) case study by invited selected professional site engineers and experienced engineers to assess the importance of each factor in the project to calculate the value of six equations index.

$$ETC^* = ((BAC-EV)/(CPI + SPI + CI + TI + QI + RSI + SI + SOI) \quad (16)$$

$$EAC^* = ETC^* + AC \quad (17)$$

$$VAC^* = BAC - EAC^* \quad (18)$$

Where:
ETC*: Improvement Estimate to Complete,
EAC*: Improvement Estimate at Complete,
VAC*: Improvement Variance at Completion.

Table 9 shows the calculation of estimate at completion for (15) case studies before improvement the results illustrate high deviation in estimate at completion of projects through cost project index and schedule project index by using the Eqs. (1) to (6), assuming the project scenario when it is over budget and behind schedule will remain until the project complete as well for another project status. It's noted that most of Iraq projects suffer from lagging and high deviations in cost and time from what is planned in the contract and we in this research strive to address and reduce these deviations by measuring the performance of the project at all stages and improve the calculations of the final costs and taking into account all things that affect cost deviations through improved performance of final cost calculations. As well as Table 10 shows the calculation of estimate at complete for (15) case studies.

Table 9. Estimate at complete for (15) case studies.

Project	Contract Value	Actual Cost (AC)	Contract Duration	Actual Duration	% of Completed	Planned Value (PV)
P.1	2,863,770,000	2,870,346,030	540	472	86.50	2,503,147,111
P.2	6,500,000,000	6,529,791,000	330	290	92.00	5,712,121,212
P.3	3,472,344,810	3,480,170,810	420	380	77.50	3,141,645,304
P.4	25,114,681,760	19,589,451,773	1003	990	71.00	24,789,167,440
P.5	27,477,372,000	7,040,055,388	910	841	20.90	25,393,922,914
P.6	4,510,987,000	3,600,145,282	360	343	55.50%	4,297,968,169
P.7	4,899,000,000	4,899,000,000	540	495	82.50%	4,490,750,000
P.8	8,435,529,200	6,832,778,652	240	202	78.30%	7,099,903,743
P.9	1,736,360,000	980,125,830	420	360	49.00%	1,488,308,571
P.10	14,958,000,000	13,620,754,800	720	668	77.00%	13,877,700,000
P.11	572,368,411	681,118,409	240	280	85.23%	667,763,146
P.12	663,496,920	390,145,280	240	197	50.40%	544,620,389
P.13	590,191,920	301,258,362	240	200	33.90%	491,826,600
P.14	29,600,000,000	19,964,715,698	540	400	61.00%	21,925,925,926
P.15	125,699,508,800	90,780,940,500	900	436	48.00%	88,458,215,001

Table 9. Continued.

Project	Earned Value (EV)	CPI	SPI	ETC	EAC	VAC
P.1	2,477,161,050	0.86	0.99	447,973,080	3,318,319,110	-454,549,110
P.2	5,980,000,000	0.92	1.05	567,807,913	7,097,598,913	-597,598,913
P.3	2,691,067,228	0.77	0.86	1,010,372,171	4,490,542,981	-1,018,198,171
P.4	17,831,424,050	0.91	0.72	8,001,325,372	27,590,777,145	-2,476,095,385
P.5	5,742,770,748	0.82	0.23	26,644,420,153	33,684,475,541	-6,207,103,541
P.6	2,503,597,785	0.70	0.58	2,886,602,974	6,486,748,256	-1,975,761,256
P.7	4,041,675,000	0.83	0.90	1,039,181,818	5,938,181,818	-1,039,181,818
P.8	6,605,019,364	0.97	0.93	1,893,630,865	8,726,409,517	-290,880,317

P.9	850,816,400	0.87	0.57	1,020,130,966	2,000,256,796	-263,896,796
P.10	11,517,660,000	0.85	0.83	4,068,537,148	17,689,291,948	-2,731,291,948
P.11	487,829,597	0.72	0.73	118,034,951	799,153,360	-226,784,949
P.12	334,402,448	0.86	0.61	383,952,498	774,097,778	-110,600,858
P.13	200,075,061	0.66	0.41	587,409,373	888,667,735	-298,475,815
P.14	18,056,000,000	0.90	0.82	12,764,326,430	32,729,042,128	-3,129,042,128
P.15	60,335,764,224	0.66	0.68	98,346,018,875	189,126,959,375	-63,427,450,575

Table 10. Improvement estimate at complete for (15) case studies.

Project	CPI	SPI	COI	TII	QUI	RSI	SAI	SOI	ETC*	EAC*	VAC*
P.1	0.86	0.99	0.83	0.79	0.66	0.64	0.15	0.83	69,895,054	2,940,241,084	-76,471,084
P.2	0.92	1.05	0.85	0.94	0.77	1.00	0.44	0.85	80,884,348	6,610,675,348	-110,675,348
P.3	0.77	0.86	0.74	0.25	0.16	0.24	0.38	0.74	206,385,189	3,686,555,999	-214,211,189
P.4	0.91	0.72	0.65	0.16	0.22	0.20	0.30	0.65	1,947,198,512	21,536,650,285	3,578,031,475
P.5	0.82	0.23	0.48	0.28	0.20	0.23	0.19	0.48	7,818,581,904	14,858,637,292	12,618,734,708
P.6	0.70	0.58	0.80	0.32	0.39	0.30	0.35	0.80	522,731,053	4,122,876,335	388,110,665
P.7	0.83	0.90	0.38	0.84	0.70	0.64	0.29	0.38	170,775,490	5,069,775,490	-170,775,490
P.8	0.97	0.93	0.72	0.94	0.83	1.00	0.62	0.72	265,419,075	7,098,197,727	1,337,331,473
P.9	0.87	0.57	0.71	0.82	0.66	0.20	0.24	0.71	198,841,110	1,178,966,940	557,393,060
P.10	0.85	0.83	0.29	0.79	0.77	0.57	0.51	0.29	720,489,510	14,341,244,310	616,755,690
P.11	0.72	0.73	0.81	0.94	0.77	0.19	0.99	0.81	15,815,861	696,934,271	-124,565,860
P.12	0.86	0.61	0.60	0.25	0.16	0.30	0.35	0.60	94,635,017	484,780,297	178,716,623
P.13	0.66	0.41	0.90	1.01	0.63	0.93	0.40	0.90	72,232,356	373,490,718	216,701,202
P.14	0.90	0.82	0.46	0.72	0.94	0.59	0.13	0.46	2,182,024,039	22,146,739,737	7,453,260,263
P.15	0.66	0.68	0.22	0.17	0.25	0.19	0.13	0.22	26,129,712,162	116,910,652,662	8,788,856,138

Table 11 shows a comparison between estimate at complete before improvement and after improvement and residual value in cost.

Table 11. Comparison between EAC and EAC*.

No.	Project	EAC	EAC*	Residual Value
P.1	Construction of gate in Basra Sports City	3,318,319,110	2,940,241,084	-378,078,026
P.2	Construction of Olympic stadium in Najaf	7,097,598,913	6,610,675,348	-486,923,565
P.3	Construction of a closed hall in Saydiya	4,490,542,981	3,686,555,999	-803,986,982
P.4	Yarmouk Intersection/Mosul	27,590,777,145	21,536,650,285	-6,054,126,860
P.5	Muthanna Intersection	33,684,475,541	14,858,637,292	-18,825,838,248
P.6	Pavement the entrance of Rabia Nineveh	6,486,748,256	4,122,876,335	-2,363,871,921
P.7	Construction of a closed sports hall in Kirkuk	5,938,181,818	5,069,775,490	-868,406,328
P.8	Construction of Schools 12 classes in Salah Al din	8,726,409,517	7,098,197,727	-1,628,211,790
P.9	Construction of the General commission Taxes/Karbala	2,000,256,796	1,178,966,940	-821,289,855
P.10	Construction of a water purification unit for Basra Sports City	17,689,291,948	14,341,244,310	-3,348,047,638
P.11	Adnan Ghazwan School 18 classes	799,153,360	696,934,271	-102,219,090
P.12	Zuhawr Al-Iraq School 18 classes	774,097,778	484,780,297	-289,317,481
P.13	Sibawayh Al-Obeidi School 18 classes	888,667,735	373,490,718	-515,177,017
P.14	Al Shaiba Intersection/Basra	32,729,042,128	22,146,739,737	-10,582,302,391
P.15	Basra Sport City Street 20 km length	189,126,959,375	116,910,652,662	-72,216,306,713

5. Conclusions

The main conclusions are that:

- The Iraqi construction industry considered is delaying in developing modern methods that integrated relevant factor data with uncertainties for performance forecasting.
- Additional factors will greatly help project managers to better guess the EAC because it will be more comprehensive in the aspects affecting the project.
- Through the 15 projects studied in the fourth quarter, it is noticed the deviations in time and planned cost due to the neglect of many aspects of high impact on the projects .
- EVM found significant benefits in dealing with deviations, especially when considering other aspects such as (quality, health and safety, risk, and social) as well as key factors (cost and time).
- The original formula for the final cost calculations in EVM was developed by adding the effect of the factors, which led to the improvement of the value of the final cost calculations.
- The model developed in this study is practical and predicts reliable figures during the project construction period which helps stakeholders make better decisions.
- This research is an important addition in the world of knowledge, which aims to improve the appreciation of project objectives and increase confidence in expectations which can be verified statistically.

References

[1] Q.W. Fleming and J.M. Koppelman, Earned value management: Mitigating the risks associated with construction projects, Program Manager, 31(2) 2002 90–95.
[2] S.A. Mubarak, Construction project scheduling and control, Wiley, Hoboken, NJ, 2015.
[3] T. Narbaev and A. De Marco, Cost estimate at completion methods in construction projects, Proc., 2nd Int. Construction and Project Management Conf., IACSIT Press, Singapore, 2011 32–36.
[4] C.M. Popescu and C. Charoenngam, Project Planning, Scheduling, and Control in Construction: An Encyclopedia of Terms and Applications, Wiley, Hoboken, NJ, 1995.
[5] D. Hillson, Earned value management and risk management: A practical synergy, PMI 2004 Global Congress Proc., PMI, Prague, Czech Republic, Newton Square, PA, 2004 1–7.
[6] PMI (Project Management Institute), A guide to the project management body of knowledge, Newton Square, MA 2013.
[7] S.H. Iranmanesh and M. Zarezadeh, Application of artificial neural network to forecast actual cost of a project to improve earned value management system, In World congress on science, engineering and technology, 2008 240-243.
[8] J. Brandt, H. Thordal-Christensen, K. Vad, P.L. Gregersen, and D.B. Collinge, A pathogen-induced gene of barley encodes a protein showing high similarity to a protein kinase regulator, The Plant Journal, 2(5) 1992 815-820.
[9] T. Narbaev and A. De Marco, Combination of growth model and earned schedule to forecast project cost at completion, Journal of Construction Engineering and Management, 140(1) 2014 04013038.
[10] M. Berenson, D. Levine, K.A. Szabat, and T.C. Krehbiel, Basic business statistics: Concepts and applications. Pearson higher education AU, 2012.
[11] J. Adamowski, H. Fung Chan, S.O. Prasher, B. Ozga-Zielinski, and A. Sliusarieva, Comparison of multiple linear and nonlinear regression, autoregressive integrated moving average, artificial neural network, and wavelet artificial neural network methods for urban water demand forecasting in Montreal, Canada, Water Resources Research, 48(1) 2012.

The Factors Affecting on Earned Value Management

Sedqi Esmaeel Rezouki[1,a] and Sabreen Mortadha Baqer[1,b*]

[1]Civil Engineering Department, University of Baghdad, Baghdad, Iraq.
[a]sedki_razqi@yahoo.com, [b]sabreen9013@gmail.com

Abstract. Earned price management (EVM) is an associate degree business customary for observance the performance of in progress comes. The performance baseline is about up within the coming up with part to live any time and value deviations throughout project execution. EVM solely focuses on the project schedule (SPI) and value (CPI), and doesn't address alternative vital aspects of quality, Risk, safety and social factors. Therefore, this paper will explain the statistical study aims to know the most important factors that affect in implementation of EVM in a construction project in Iraq and it will be consisting of two main tools which are the assessment of the questionnaire and checklist to obtain real weights affecting the Iraqi construction projects. There is a need for understanding all affecting factors on estimate at completion in EVM. This will develop the equation of EVM. It will be necessary to study affecting factors to achieve major factors of EVM. Based on previous studies and interviews, 207 influential factors were identified. After the first questionnaire, the number was reduced to 73. Based on the interviews with experts, weights were determined and the most influential factors were 37 and were adopted as a checklist.

Keywords: Earned value, cost factors, time factors, quality factors, risk factors, safety factors, and social factors.

1. Introduction

The construction industry is mixed with uncertainty risk in several comes. The important part of the project is to complete among the calculable times, value with expected quality is that the predominant goals of the project management. However, essentially a variety of attributes varied |of varied |of assorted} varieties and a range of constraints involve from time to time that need to be managed with success by overcoming various obstacles. Completion of the development project involves many factors and attributes of varied varieties contribute to their extent of impact on the project. These factors were known by consultations with consultants and active engineers excluding reviewing literature. In this section, the most important indices affecting the EVM are shown in Fig. 1 will be discussed in detail and these factors are cost, time, quality, risk, safety, and social aspect and will explain each factor briefly during this chapter and the impact of each factor on EVM Thus improving the estimate at completion [1, 2].

2. Factors Affecting EVM

2.1.Cost factor

Cost management is bothered by the method of designing and dominant the budget of a project or business. It includes activities like coming up with, estimating, budgeting, financing, funding, managing, and dominant prices so the project will be completed inside the approved budget [3]. Price management covers the total life cycle of a project from the initial coming up with part towards activity the particular price performance and project completion. Several value estimating strategies is applied to predict what proportion it'll value to perform the project activities. The selection for the estimation methodology depends on the amount of data on the market, Analogous estimating mistreatment the particular value of previous, similar comes will function a basis for estimating this project [4]. Remaining uncertainties in estimates which will probably end in extra value is lined by reserving value (e.g. mistreatment step-up and contingencies).

They are several factors touching on the value of comes, however, the analysis can specialize in the foremost powerful factors within the estimate at completion in EVM, together with the policy and needs of the leader and company policy, quality assurance of processed materials, and Economic facet. Economical management of finances is that the key think about dominant the overrun of the project. A disciplined approach got to be adopted within the management of the budget and value of activities to ascertain the project cost. Most of the price studies reviewed suitably specialize in the differential cost of EVMS compliance and coverage. The "normal" prices of operating an internal control system aren't thought-about relevant as a result of they might be incurred within the absence of associate demand for an earned worth management system [5].

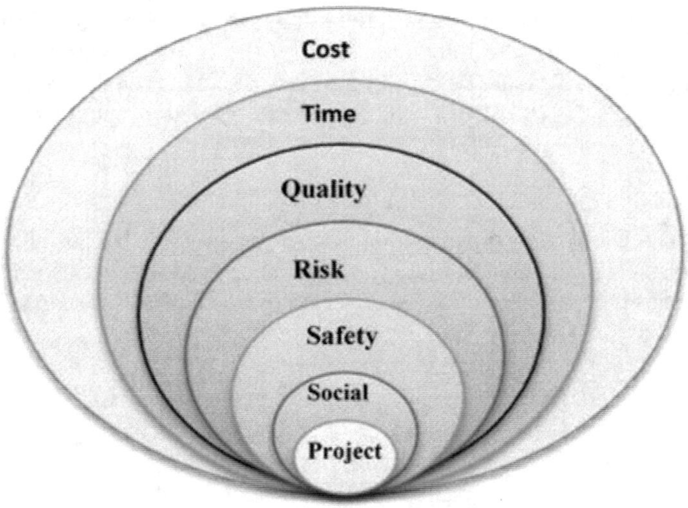

Figure 1. The affecting factors on EVM modified by the authors [1].

2.2.Schedule factor

Schedule management needs the active involvement of all project participants to stay up on concerning the standing of the project and any delays which will impact schedule performance. A Schedule Management arranges ought to contain all of the subsequent major elements:

- Baseline Schedule Development and Approval.
- Schedule Analysis and Coordination.
- Schedule Changes and Revisions.
- Schedule Delay Mitigation designing [6].

In the observation of the development business designing and programming could be a vital task to be performed by the old planners. However, the programing is usually applied either supported time or value and consequently termed to be value bound schedule and time-bound schedule betting on the kind of project [7]. Most of the time, the resources and their accessibility play an important role, and therefore value bound programming is most generally used and once circumstances secure for the completion of the project at intervals the time, then all activities are back worked to fulfill the completion time. Consequently, the resources are allotted, leveled, and optimized for the most effective utilization of resources further as project time length. However, in observe the programing could be an advanced task involving variable attributes and also the assessment of their impact on the project [8].

2.3.Risk factor

Project risk management is important to the success of the project because it deals with uncertainty in two situations: the first is the threat (negative impact) the second is the opportunities (positive impact). This issue includes a group of various sub-factors betting on their importance, that depends on the character of the project, as well as the standard of the land in terms of suitableness and talent to resist construction, the presence of groundwater, the shortage of infrastructure resulting in the location and monetary aspects like delays of advances by the leader and wages of staff for businesses, high material prices…. etc.

EVM is an element of Project Time Management and Project price Management information areas, whereas Project Risk Management could be a different information space with no apparent relation with the previous ones, though the basics of EVM square measure influenced by risk management [9]. This analysis seeks to clarify the mechanisms of this relation, to boost this relationship, a variety of things moving the EVM were enclosed as index (RSI) in the equation of estimate at completion. PRM and EVM have similar objectives, i.e., providing info regarding the factors moving the project performance. However, no framework exists for the interconnection of the results obtained from every technique. Because the same issues square measure being self-addressed by each tool, there's a great deal of interest in combining their results. If the PMB of a project consists of well-equated risk factors, uncertainties will be expected and a more robust EAC will be forecasted [10].

2.4.Quality factor

Quality management in construction comes could be an important issue because it is meant to make sure that the development project has been completed among the desired specifications and planned with the constrains (time and cost) [11]. Quality could be a very important consider EVM. The target of Quality earned price (QEV) is to live the projectability to deliver the standard necessities outlined by the project's neutral, throughout the project execution. Therefore, the standard is taken into account in the third part of earned price Management [12]. It focuses on providing pictures of the project potency to deliver the project quality necessities supported the used time and spent value (actual costs) [13].

Project management aims to deliver comes on time, with the united scope consistent with mere needs, inside the planned budget. This scope, schedule, and value constraints are known as triple constraint or project triangle. The project triangle implicitly covers the standard and assumes that scope suggests that scope with quality. It's not therefore applicable to bundle scope with quality in contrast to the opposite forms of comes. Even the task is completed, and scope target is achieved, it would need several iterations on the task to repair the bugs, improve the options and then on. They're tightly connected however not precisely along, from the close relationship between quality and EVM, we found several factors that have a high impact on the calculations of the estimate at completion in EVM. These factors include analyzing the costs of quality and its applications. Planning, controlling, and improving strategies based on quality work, conforming materials to specifications, conforming work to quality specifications… etc.

2.5.Safety Factors

Safety management is a crucial management perform and desires to be controlled expeditiously on a project. Earned worth technique (EVM) is one in every of the popular strategies employed in the dominant construction project. Safety statistics ought to be reported within the same manner as price, schedule, and quality. The authors found the need to consider the impact of safety factors in the calculations of the estimate at completion in EVM The most important factors are personal safety (the provision of safety equipment), lack of training when using new machinery, providing emergency exits in buildings, the development of an escape map indicating the exits, and the development of signs by the Civil Defense…etc. Several researchers have emphasized the importance of safety. According to [14], safety must be an integral part of the company's procedures. It is noted that many construction companies follow the 'Zero accident' or 'Zero incidence' policy [15].

2.6.Social Factor
The social aspect has recently become one of the influential aspects of projects as it must pay attention to the culture of society, customs, and traditions, and this interest is evident in international companies in their activity outside their countries. In contraction projects, the user and occupier are going to be either public or personal generally. The society contains users further as project participants. The support of the society is extremely crucial within the project completion. Society as an entire could object, protest, and hinder the development method in obtaining clearances and moving the lads, materials, and machinery. Within the case of the general public comes the land acquisition and their cooperation is extremely a lot of needed [1].

3. Methodology

3.1. Data Collection
It will be collecting the data by two stages as following:

3.2. Open Questionnaire and Interviews
In such cases of the research type questionnaire survey is an effective method to seek a large sample size for quantitative data analysis. The researcher based on previous studies and interviews with experts to discuss the main affecting factors. who develop and add factors to the factors that the researcher already has, 207 influential factors were identified. Table 1 will describe the characteristics of the experts.

Table 1. The characteristics of the experts.

Scientific Qualification	No.	Specializations	Experience
Ph.D.	6	Civil engineers	
MSc	5	Civil engineers	More than 20 years
BSc	7	Civil engineers	

3.3. Closed Questionnaire
After the completion of the questionnaire form, it has been distributed to the sample, Research sample consists of A group of engineers, project managers, and heads of departments in government departments that perform or finance the implementation of projects or who have a scientific and practical experience in this field. the questionnaire divided into two parts, the first part is about personal information and the second part to identify the influence of the affecting factors by using a Likert scale. Based on analysis results of Questionnaire, and according to relative important index (RII) value, the experience cousins the value of (0.7) will be the threshold, therefore, any factor has RII over (0.7) will be concerned and any factor has RII less than (0.7) will be neglected. Finally, the researcher developed 73 factors that had been identified as the most affecting factors. Then the second round of interviews had been connected to filtration and prioritize the factor resulted in a shorter list consist of 37 factors.

3.4. Distribution of the Questionnaire
The questionnaire offers an exact formulation of the items, suitability of the area to be measured, and getting the arbitrators validity. The form has been distributed to the target sample that consists of one hundred respondents. The distribution method is in 2 ways:

- The form is directly distributed to the participants, wherever the research worker offers straightforward information regarding the study and its objectives. This technique is taken into account as a perfect within the follow-up of form and to induce results that adjust to reality. The direct meeting technique includes sixty-six respondents; they kind half of 1 mile from the entire variety of the participants.
- The form is indirectly distributed to the participants through the web, wherever it's distributed ten forms. The research worker conjointly offered straightforward clarification regarding the study

and its objectives. the total form forms that square measure collected during this method square measure solely vi forms, and solely five of them don't come back.

Thus, the study gets sixty-six complete and proper forms from the entire variety that is one hundred. The two-part written form was the most instrument of knowledge assortment. half one wanted general data regarding the non-public characteristics of respondents, like qualification profile, category of their organizations, and years of expertise, amongst others. partially 2, the respondents were requested to rank these factors and criteria for achievement. The on top of literature review in previous chapters provides the theoretical basis for developing the analytical framework for this study according to its vital in keeping with a five-point Likert scale (1–5) where:

1) There is no effect
2) A failing grade
3) Median degree
4) High degree
5) Very high degree

The analysis of the factors and its importance done by using (SPSS statically program), using the resulting information in building the mathematical model and software to improve the EVM implementation Extract the weights of success factors and criteria for success deduced from questionnaire survey which is built on a proposed system to check and improve EVM implantation. Table 2 shows the number of questionnaires.

Table 2. The number of questionnaires.

Description	Total number	Percent, %
The total number of questionnaires distributed	100	100
Collected questionnaire	66	66
Questionnaire without response	34	34

3.5. The Second Part (The Influences of The Main Axes)

The second part of the questionnaire consists of six axes: cost, time, quality, risk, safety and social from this part the researcher has been identifying the effect of all factors on the projects in Iraq.

3.6. The Third Part (Study Axes)

The third part (Study axes) of the questionnaire consists of six axes, the first (The cost axis) consists of 29 factors, and the second (The time axis) consists of 22 factors, and the third (The quality axis) is made up of two-part first case there is a system of quality follow and consists of 23 factors and second case there is no quality system consists of 11 factors, and the fourth (The risk axis) consists of 38, and fifth (The safety axis) consists of 40 factors, and sixth (The social axis) consists of 13 factors. Questionnaires were distributed to some governmental institutions specialized in the creation and project management and The National Investment Commission, Services Department of the martyr's Foundation, The Ministry of Education, The Ministry of higher education and The Ministry of water resources in Iraq, the Ministry of Construction and Housing, the University of Baghdad University of Technology, and the University of ALrafdain), and after that was the explanation thorough and take advantage from the sample in the development of the form and answer all their questions were selected 100 Engineer (identified by the Supreme departments), of all ages and disciplinary, managerial positions, of both gender, then received 76 forms and there were 10 forms are not clear and a section of questions unanswered about or answered unrealistic so it has been deleted from the accounts and we relied on 66 any form by the response to the amount of 70%. The researcher has been analyzed the data, and identify the honesty and reliability of the sample. Then show how extract Weights and importance of the factors and calculated (mean, standard deviation the Relative important index Alpha Z Distribution).

3.7. Statistical Analysis of the Questionnaire Data

Data collected from surveys were analyzed using various statistical techniques including descriptive statistics, Z test, reliability analysis. SPSS for Windows version 25 and MS Excel were the basic software tools used to analyze raw data. Before presenting the results of statistical analyzes in the following chapters, the minimum criteria and justification for applying these tests are discussed in the following sections. Reliability and validity take into account because the most significant methodology conditions for analysis tools style. Therefore, it should be provided the dependableness and validity of the form before any applied math analyses for knowledge.

4. Validity and Reliability

4.1. Validity

It is one of the necessary conditions for building the tests, standards and honesty demonstrates the extent to measure the paragraphs of the phenomenon to be measured, and that the best way to measure honesty is the virtual honesty, which is to present the paragraphs of the scale on a group of experts to rule on its validity, was achieved sincerity ostensibly scale of through the presentation of paragraphs to a group of specialists in the field of construction project management as all the paragraphs and the factors and criteria listed in the questionnaire belong to the field.

4.2. Reliability

It is a measure of consistency in the results as giving the same results after its application twice in two different times on individuals themselves, stability was measured as flowing: Coefficient (alpha) for Internal Consistency: The Alpha coefficient provides us with a good grade in most situations and adopt this method on the consistency of the individual performance of a paragraph to another and extract stability according to this method was the use of all forms search of (66) form, Table 3 shows the values of reliability and validity according to Cronbach's alpha for each factor in the questionnaire.

Table 3. Reliability and validity Statistics of factors groups.

No.		Factor	No. of Item	Reliability*	Validity
1		Cost	43	0.929	0.963
2		Time	22	0.891	0.943
3	Quality	not found system	25	0.969	0.978 0.984
		Found system	12	0.925	0.962
4		Risk	38	0.926	0.962
5		Safety	52	0.969	0.984
5		Social	13	0.851	0.922

*(Cronbach's Alpha)

4.3. Likert Scale

A psychology response scale chiefly employed in questionnaires to induce participant's preferences or degrees of necessary the issue on the EVM application. The respondent's square measure asked to point their assessment to the impartment level for every issue by exploitation Associate in Nursing ordinal scale [16]. Likert Scale was Associate in Nursing ascendant order to make sure the participant review all choice to elite their rate in additional accurse. The design of form is predicated on it the solution of every question that is one in all 5 choices. So, the Likert Scale Quintet is employed. it's typically coming into values (weights) as within the Table 4 [17].

Table 4. Likert scale Quintet weights.

Opinion	There is no effect	A failing grade	median degree	high degree	very high degree
Weight	1	2	3	4	5

4.4. Z Test

This statistical examination is useful in ascertaining the sample that participated in the questionnaire, it has a degree of dependability to be used to predict the results of the research in general. The value of (Z calculated) for each category was found using the following equations:

$$\bar{x} = \frac{\sum_{i=1}^{k} xi\,.\,fi}{\sum_{i=1}^{k} fi} \quad (1)$$

where:

x̄: Arithmetic Mean

xi : Degree of the criterion's importance

fi: The Frequency

Calculate the standard deviation for respondents for each criterion based on the following equation:

$$S = \left[\sum_{i=1}^{k} (xi - \bar{x})^2 . fi / \left(\sum_{i=1}^{k} fi\right)\right]^{1/2} \quad (2)$$

where:

S: Standard deviation

X: Arithmetic Mean

Xi: Degree of the criterion importance

fi: The Frequency

$$Z = \frac{\bar{x}}{\frac{S}{\sqrt{n}}} \quad (3)$$

where:

X: Arithmetic Mean

S: Standard deviation

n: Size of the sample

If Z calculated > Z tabular, then accept the results of the questionnaire

If Z calculated < Z tabular, then reject the results of the questionnaire.

4.5. The Relative Importance Index (RII)

The relative importance index could be a statistical procedure to see the ranking of various factors. A formula is employed to rank the foremost cogent factors on EVM supported frequency of incidence as known by the specialists. The ratio index formula is shown below:

$$RII = \sum W/A*N \quad (4)$$

where W = weight given to each factor by the respondents; A = highest weight, i.e., 3 in this case; and N = total range of respondents [18]. The issue with the very best rank indicates that it's the most frequency of incidence on the EVM whereas the issue with all-time low rank indicates that it's the smallest amount frequency of incidence on the EVM. By victimization SPSS program (version 25), the mean of all factors are calculated, and RII are calculated by the equation No.4. The researcher adopted RII value (0.7) as a limit between that influential and non-influential factor on EVM and it achieved the expert's consensus.

4.6. Design and Development of the Assessment Checklist

After conducted statically analysis and review the results, a checklist has been developing, it consists of six-axis and 37 factors. The authors made another interview that reduced factors to 37 factors as shown in Table 5.

IOP Conf. Series: Materials Science and Engineering **901** (2020) 012023 doi:10.1088/1757-899X/901/1/012023

Table 5. Axes and factors of statical characters.

Category	No.	Factors	Mean	RII
Cost	1	Policy and requirements of the client	3.55	0.709
	2	Policy and experience of the company	3.89	0.779
	3	Price changes (marital, labor, equipment)	3.67	0.733
	4	Errors in estimated quantities.	3.52	0.703
	5	Funding and factors affecting it: provide financial liquidity to contractor	3.98	0.797
	6	Funding and factors affecting it: The Company has access to banking facilities	3.64	0.727
	7	Funding: grants or loans	3.61	0.721
	8	Funding: governments financing	3.56	0.712
	9	Funding: private financing(private banks)	3.52	0.705
Time	1	Rise production capacity through: Increasing labor force	3.77	0.755
	2	Rise production capacity through: Increase machines and equipment	3.79	0.758
	3	Rise production capacity through: Increase the number of working hours	3.64	0.727
	4	Rise production capacity through: Raising the skill of the workforce	3.79	0.758
	5	Quick completion of work	3.91	0.782
	6	Effect of the management method on the delay of completion of the project	3.56	0.712
	7	The effect of the (economic situation) factor on the delayed completion of the project	3.61	0.721
	8	Availability of money liquidity and exchange rate	3.71	0.742
Quality	1	Reduce the cost of acquisition and use the raw materials.	3.56	0.712
	2	Contribute to strengthening the competitiveness of the institution	3.52	0.700
	3	Matching materials to specifications	3.65	0.730
	4	matching work to quality specifications	3.58	0.715
	5	Does the organization conduct internal audits periodically to determine the conformity of the quality management system to ISO 9001 - 2015?	3.58	0.712
	6	Does the employer take the necessary procedures to remove the causes of nonconformity and prevent its recurrence?	3.61	0.712
Risk	1	The quality of the land is not appropriate in terms of its ability to withstand construction or because of its other characteristics.	3.83	0.767
	2	Lack of infrastructure connected to the place of execution such as roads or inadequate place of implementation.	3.89	0.779
	3	Delay payments to local companies from the employer	3.73	0.745
	4	Financing cut off unexpectedly by local companies	3.73	0.745
	5	The high prices of materials necessary to carry out civil works in construction sites due to economic conditions	3.82	0.764
	6	Difficulty accessing project sites due to security situation	3.65	0.730
Safety	1	Provide adequate personal safety equipment (gloves, glasses, helmets, etc.) on-site	3.68	0.736
	2	Does the use of new machines affect the provision of appropriate training?	3.52	0.703
	3	Do accidents occur due to lack of safety standards in construction companies	3.50	0.700
	4	The safety officer, if any, or the implementing contractor shall be responsible for the laying down of guiding plates by the approved design of the Civil Defense, indicating to the site's employees or non-employees the risk of the site according to the nature of the executed works.	3.52	0.703
	5	Are buildings adjacent to the drilling work supported if there is a potential for impact on these works?	3.79	0.758
Social	1	Project size	3.68	0.736
	2	Reliance on local labor	3.65	0.730
	3	The impact of social customs and traditions on the region	3.65	0.730

5. Conclusions

The conclusions from the current study can be summarized by the following points:

- Definition of the factors of the participants in the following techniques: open interviews with experts and the questionnaire respondents.
- The researcher conducted an open interview with the experts to assist in the construction of the questionnaire.
- The questionnaire consists of 207 factors; these factors have a significant impact improve EVM to the results of interviews with experts.
- Perform a set of statistical operations on the questionnaire results to find reliability and type of data distribution. Compute the mean and Z Distribution and RII to see the trend of answers and the statistical significance respectively.
- Once the applied mathematics analyses are finished, the investigator mentions the explanations that have a big impact on rising EVM, that area unit seventy-three factors.

References

[1] R. Vidhyasri and R.Sivagamasundari, Assessment of Influencing Factors in Construction Project Scheduling, International Journal of Applied Engineering Research, 13(8) 2018 5693-5699.
[2] T. Narbaev and A. De Marco, Cost estimate at completion methods in construction projects, Proc., 2nd Int. Construction and Project Management Conf., IACSIT Press, Singapore, 2011 32–36.
[3] P. Raykar and A.N. Ghadge, Analysing the Critical factors influencing the time overrun and cost overrun in Construction Projects." International Journal of Engineering Research, 5(1) 2016 21-25.
[4] A.D. Marco, Cost and schedule monitoring of industrial building projects: case study, Journal of Management in Engineering, 135 2009 853-862.
[5] W.H. Lipke, Earned Schedule, Lulu Publishing, 2009.
[6] W. Lipke, Connecting earned value to the schedule, The Measurable News, 1 2004 6–16.
[7] W.H. Lipke, Schedule is Different, The Measurable News, Summer 2003 31-34.
[8] B.G. Hwang, X. Zhao, and S.Y. Ng, Identifying the critical factors affecting schedule performance of public housing projects, Habitat International, 38, 2013 214-221.
[9] V.K. Diamantas, Earned value management under risk, Int. J. Project Organization and Management, 3(3/4,) 2011 335-351.
[10] D. Hillson, Earned Value Management and Risk Management: A Practical Synergy, PMI 2004 Global Congress Proceedings, 2004 1–7.
[11] N. Banaitiene and A. Banaitis, Risk management in construction projects, Risk Management–Current Issues and Challenges, 2012 429-448.
[12] [S.D. Marcelo, Quality: The Third Element of Earned Value Management, 2010.
[13] Pinar Efe," Quality Integrated Earned Value Management",2015.
[14] W.A. Stanton and J.H. Willenbrock, Conceptual framework for computer-based construction safety control. Journal of Construction Engineering and Management, 116(3) 1990 383-398.
[15] K.N. Jha, Evaluation of Construction Projects in Terms of Cost, Schedule and Safety Performances, In Conference Paper, September 2012.
[16] F. Yao and W.E. Johns, A HYCOM modeling study of the Persian Gulf: 2. Formation and export of Persian Gulf Water, Journal of Geophysical Research: Oceans, 115(C11) 2010.
[17] A.M. Abdul-Fattah, D. Lechuga-Ballesteros, D.S. Kalonia, ad M.J. Pikal, The impact of drying method and formulation on the physical properties and stability of methionyl human growth hormone in the amorphous solid state, Journal of pharmaceutical sciences, 97(1) 2008 163-184.
[18] L. Muhwezi, J. Acai, and G. Otim, An assessment of the factors causing delays on building construction projects in Uganda. International Journal of construction engineering and management, 3(1) 2014 13-23.

Using BIM for Optimizing the Upgrading Cost to Convert the Traditional Buildings to Sustainable Buildings in Iraq

Sarah Ahmed Abdulrasool[1,a*], Ahmed Mohammed Raoof Mahjoob[1,b]

[1]Department of Civil engineering, University of Baghdad, Baghdad Iraq
[a]s.abdulrasool1001@coeng.uobaghdad.edu.iq, [b]ahmed.mahjoob@coeng.uobaghdad.edu.iq

Abstract. A high percentage of existing buildings in Iraq are traditional buildings, yet there is approximately no such green building in Baghdad or other governorates. Most of these buildings require urgent upgrading to increase their performance (operationally, economically, and environmentally), also the building owners looking for identifying and implementing many of the green building measures to reduce the operational and maintenance costs of their buildings. The decision-makers need to support the possibility of achieving sustainable measures of existing building rating systems such as LEED or BREEAM, and that would require an optimization model. The goal of this study is to maximize the building's operational efficiency, health, and comfort of the building to its occupants through minimizing upgrading cost and reducing the environmental impacts of the building. An existing building will be used as a case study to illustrate the optimization process and demonstrate the plans and processes required to achieve the green building measures, the capability of reducing the required upgrade costs. For this type of problem, the building drawings are represented Using Building information modeling (BIM) (Revit program), and the life cycle costing and environmental impact can be found using analytical tools (ONE CLICK LCA STUDENT VERSION online software). The results show that up to (15.4 and 23.2) % reduction in discounted and nominal costs respectively could be achieved, the CO_2 emissions also had decreased from 34233 Ton to 19299 Ton.

Keywords: BIM, sustainability, LEED, Pearl, ONE CLICK LCA STUDENT VERSION

1. Introduction

From studies, one can notice that around 30% of the total energy uses in modern countries comes from the energy uses in existing buildings and it may induce that almost 30% of carbon emissions are caused by the energy of existing buildings. The energy consumption of existing buildings in some modern countries has exceeded those in the industrial and transportation sectors. Therefore, it is of urgent need to find tools and methods focus on the retrofitting of existing buildings (which forms a high percentage in Iraq) not only from the economical aspect but also from the social and environmental aspects. It's also required to minimize the temperature rises and the effects of its climatic changes in the future while reducing the upgrading cost [1]. BIM is considered to be a revolutionary technique and process which transforms the method of designing, planning, constructing, and managing the buildings. It also can serve as an approach to help the construction industry to create new thoughts and practices. According to Smith [2], BIM digitally representing the physical and functional features of a facility. BIM offers a mutual knowledge source for information of a facility and forms a rational foundation for decisions through the life-cycle from the inception onwards. So, BIM is a clever technique for designing, producing, visualizing, and lifecycle managing of buildings. BIM's software combines multi-dimensional conception with the complete scientific databases. This facility the joint design and the facility management among the project team members.

IOP Conf. Series: Materials Science and Engineering **901** (2020) 012024 doi:10.1088/1757-899X/901/1/012024

Some researchers have faith that BIM is a typical change rather than a representation of two-dimensional building design, collecting, and management of the life cycle [3]. BIM processes are well-known for constructing new structures, but a high percentage of existing structures are not maintained, updated, or deconstructed using BIM so far. The huge benefits of effective resource management inspired research to overcome worries of structure circumstances and lack of documentation in existing projects. Because of the quick developments in BIM research, relevant stakeholders require a recent overview of implementing BIM and researches in existing structures. The results show that the BIM application in existing structures is so frightened because of the following challenges: (1) hard process of modeling and conversion from structure data captured to BIM semantics, (2) hard updating of information in BIM software and (3) handling with the inexact data, objects, and relationships in BIM that occurs in existing structures [4]. The greatest capacity of BIM is its nature of integration among different construction processes, as it is relating to a sustainable world from the front of the project, until operation and maintenance.

2. Green Building Rating

For buildings under LEED® certification, numerous LEED® credits need the drawings to be presented, to achieve the requirements for the credit. Even though that most of the drawings could be done by traditional CAD programs, BIM programs deliver it more effectively as a portion of the building information modeling with the benefit of parametric change technique, this technique coordinates alterations and keeps consistency at any time. [5] In every construction project, the cost plays an important role. Lifecycle cost (LCC) means determining the entire cost of the structure over the expected life of it, companies with the cost of operation and maintenance. Alternative modern techniques can be used to improve calculating LCC without much changes in the structure. The lifecycle cost can be determined at any stage of the building, which may assist the decision-makers in providing the necessary financial information for maintenance, improvement, and construction facilities [6].

The Net present value (NPV) is the most suitable method for calculating the life cycle cost (LCC) since it considers the money's time value. This method calculates the sum of all the annual and non-annual costs comprising investment, replacement, operation, maintenance, and the resale value. [7] NPV is shown in Eq.1

$$NPV = C+R+A+M-S \tag{1}$$

Where
C: is the investment cost
R: is the Replacement costs
A: is the annual operating, maintaining, and repairing costs.
M: is the Non-annual operating, maintaining, and repairing costs.
S: is the Resale value.

To determine the lifecycle cost there may be difficulties in collecting the necessary information about the initial and future expenditures like the original acquisition, current, overheads, operating, maintaining, discounts, critical, escalating, and end of life costs [7]. One Click LCA student version calculates LCC depending on Eq.1

3. Measures and Technologies for Existing Buildings

The following are some techniques used to provide efficient lighting with lower energy consumption:

3.1. Energy Efficient Lighting

A. Fluorescent lamps consist of: (1) tabular lamp and (2) ballast, for more efficient energy, one can use smaller diameter of tabular lamps, or use the electronic ballast [8].
B. Induction lighting as a technique to gain more durable lighting sources used mainly in high ceilings or in very cold temperatures, it may stay valid for 100000 hours, however, the main problem is the relatively high initial cost. Another solution is using Lighting Emitting Diodes

(LED) for high efficiency and long-life lighting by converting electricity to light. [9]. It could stay valid for 50000 hours.

C. Motion-activated lighting is a technique to save energy by activating lights when is needed using infrared sensors to sense the motion of a heat source in the place as soon as the MAL system detects a motion in the place it turns lights bulbs on, and they remain on as the movement still detected. After the last motion, the system will be kept in detection mode for a short time, then it turns off automatically [10].

3.2. HVAC@ R Systems

LEED verified not to use chlorofluorocarbon (CFC)-based refrigerants HVAC&R systems. When these refrigerants are used in the building, it should minimize the annual leakage to equal or less than 5%, and the total leakage to less than 30% of the refrigerant's charge, not to use refrigerants, or use refrigerants of zero ozone depletion potential (ODP) and less than 50 global warming potential (GWP) [11]. Pearl has also required to verify that all gaseous fire defeat systems installed in the project site must have zero ozone depletion potential (ODP). CO_2 monitors should be installed in all densely occupied spaces. Rooms of area less than 14 square meters are exempt. These monitors should be above the floor between 900 to 1800 mm. The system should be designed to create a visual alarm to the operator if the concentration of differential CO_2 in the zone rises 15% or more above that comparable to the minimum required outdoor air rate [12]

3.3. Thermal Pane Glass

Typically, windows cause 3 to 10% heat loss greater than the heat loss in walls, as a result, careful selection of windows type is important to reduce energy consumption in buildings. Thermal Pane Glass provides better insulation than the regular pane glass, it has a separate double or triple pane of glass and air space to reduce the loss or gain of the heat flux. The thermal conductivity of the glass and the frame is the Base of the insulation of windows, the metal frame's insulation can get better by installing a thermal break to split the frame into two parts. Fourier's heat low is used to describe the windows overall insulation [13].

3.4. Photovoltaic System

Solar systems are renewable and sustainable techniques using the solar energy to produce electricity, save energy up to 100%. This system utilizes photovoltaic cells made of semiconductor materials that have a sensitivity to the sunlight. With an area of 1 to 100 cm^2 for each cell, the cells are connected and form a module with an area of 0.5 to 2 m^2, by combining modules a photovoltaic array is created to provide electricity [14]. The photovoltaic arrays are either separated on the ground or installed on the roof of the building. Another important factor affecting the output of the system is the orientation of the panels which depends on the season and altitude of the building to get the maximum power by tilting the panels in a specific angle to face the sunlight [15]. There are two types of solar photovoltaic systems: grid-connected and stand-alone systems.

3.5. Solar Heating Systems

Using the sunlight energy in the process of heating water is another technique to conserve energy and can be used in many purposes in the building, the hot water amount's provided by this system changing according to the type and size of the system and the climate conditions, a backup system of water heating could be installed and used in case the sunlight energy is not enough to meet the demand or when the sunlight is absent [16].

The system of solar water heating is categorized as active uses electrical pumps and valves, and passive depends on the natural convection. This system is also classified into direct and indirect according to whether the water is heated in collectors directly or by using a mechanism to exchange heat, this system provides a clean and sustainable renewable energy source, reduces the consumption of fossil fuel and greenhouse gas emissions, and reduces the energy cost [16].

3.6. Water-Saving Plumbing Fixtures
Reducing indoor water consumption by using water-efficient urinals, toilets, and faucets is another technique to enhance the efficiency of buildings. For example, using a dual flush technique in toilets that uses compressed air in flushing, the air is compressed in the dual flush vessel while it is filled with water, the approximate water consumption per flush by this technique is about 0.8 to 1.1 gallons and 1.6 gallons for liquid and solid wastes respectively. One can upgrade the manual faucets by installing flow restrictors on the head of the faucet to reduce the flow of water to about 0.5 to 2.5 gallons in a minute, or by replacing the faucet head with flow aerators which add air to the stream of water [17]. Pearl specified the requirement to color all pipes that contain recycled water to distinguish them easily from other pipes, to reduce the water demands through irrigation strategies, plant selection, and developing the reclaimed water use.

Ensuring that 100% of the demand for outdoor irrigation is served by the use of Exterior Water Allowance and that a recycled water mainline loop for reclaimed water is installed. The reclaimed water use is not available to allow for future change from potable to reclaimed water for outdoor irrigation demands. Pearl also suggested rainwater management system can collect and treat at least 90% of rainwater and the treating can accomplish the minimum standards of quality control:

- Remove 80% of total suspended solids (TSS), 95% of litter (gross pollutants, >1 mm), at least 90% of hydrocarbons and use of oil interceptors or suitable paving for parking for more than 4 lines.
- Use either Non-structural solutions for rainwater management contain sustainable urban drainage systems (SUDS) for example ponds, Vegetable granules, wetlands, etc.

Or structural solutions for rainwater management using structures for example pipes, tanks, concrete channels, etc [12].

3.7. Indoor Air Quality
LEED had forbidden smoking inside the building, and forbid smoking out of the building too except in designated areas for smoking which should be located at least 7.5 m from the entries, operable windows, and outdoor air intakes. It also forbids smoking out of the property line in case the space is used for business [11]. Pearl requires the following:

- Use low emission paints and coatings to enhance the desirability of the building to improve the health of occupants.
- Provide permanent access systems of at least 3 meters in the main direction of movement to eliminate the particles and dirt entering the building at the outside entrances regularly used.
- Establish that interior noise levels are of no more than 50 dB in the space to be employed.
- Pearl has demonstrated to use no Asbestos Containing Materials (ACMs) in the improvement and to remove all ACMs from renovated buildings.
- All insulation materials have a Zero Ozone Depleting Potential, and equal or less than five low Global Warming Potential
- Replace chlorine-based materials like Chlorinated polyethylene (CPE), Polyvinyl Chloride (PVC), Polychloroprene, Chlorosulfonated polyethylene (CSPE), and Chlorinated polyvinyl chloride (CPVC), with more green alternatives [12].

4. Building Description
The case study will be applied to a building that lies in Aletaifia, Baghdad, Iraq, the building is of (463.75) m^2 and 8 floors (including the ground level and the basement) as shown in Fig. 1 and Fig. 2 the building contains 90 small offices most of them are medical clinics, the details of the building are:

1) The building is 26.5 m width, 17.5 m length;
2) It contains 8 floors (ground floor and basement with 4 m high each and 6 floors with 3 m high each);

3) A 662 m^2 of 12 mm thick glass façade;
4) 33 columns of (300 mm * 450 mm) for each floor, finished with gypsum boards;
5) 7407 m^2 of walls, the external face of the building (1843.47 m^2) is finished with cement plaster and the inner face (and all the internal walls) are all finished with gypsum boards;
6) 90 rooms ranging in areas from 20 m^2 to 27 m^2;
7) 90 wooden doors of 1 m *2 m, and 90 aluminum doors of 0.8 m *2 m;
8) 14 windows (3*1.8 m), 7 windows (2.5*1.8 m), 7 windows (1.5*1.8 m), 7 windows (1*1.8 m), 21 windows (0.915*1.8 m), 21 windows (0.915*0.61 m), and 69 windows (0.61*0.61 m), the total area of windows is 539.77 m^2;
9) The building consumes 767 CFL (23 W) lights and 90 (1.5 Ton) split units;

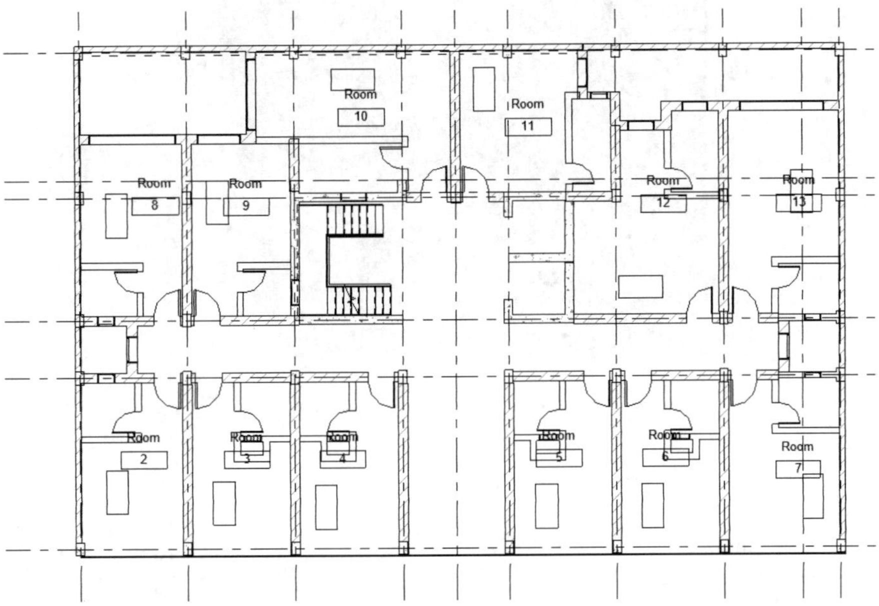

Figure 1. Plan of the case study building (researcher).

5. Methodology

First, the researcher gets the case study's drawings (hard copy) and built the model by using Revit (all the small details are important and can make a difference in the process of optimization) many errors occurred through building the model, but they all were solved and the model had been built successfully, as shown in figure (2), after that the researcher exported the model to the LCC software which is ONE CLICK LCA STUDENT VERSION, (can compute LCA and other environmental impact issues) then the real data of the building from the materials used to the annual consumption of electricity and water were collected and entered to the program after that the researcher run the program and got the LCA and LCC results (discounted and nominal) of the building before starting the optimization process, the researcher examined more sustainable alternatives to get the best set of alternatives, till got the set number 17 which had the less LCA and LCC results. Fig. 3 and Fig. 5 show the LCC results for the building before and after retrofitting respectively. The X-axis presents the LCC components and the Y-axis presents the percent of these components to the total LCC. Fig. 4 and Fig. 6 show the LCA results for the building before and after retrofitting respectively. The X-axis presents the LCA components and the Y-axis presents the percent of these components to the total LCA.

ONE CLICK LCA STUDENT VERSION is a Building Life Cycle software that calculates Life Cycle Assessment (LCA), Life Cycle Costing (LCC), and other impacts of the environment such as the Carbon footprint in like few minutes as a result to the simplicity and the advanced automation, this software is a web-based software used for sustainable Building credits, Low Carbon buildings, Eco-design, and Infrastructures. It is accommodating with LEED, BREEAM, and about 40 other sustainable rating systems, the user does not need to be an LCA or LCC expert [18]. The LCC discounted and nominal of the building before any retrofit were 6042275940 IQD and 17221281230 IQD respectively. LCC discounted and nominal of alternatives are shown in (table 1) in Iraqi Dinar (IQD), the negative signs in rows (5, 12, and 14) means additional costs will be used if those alternatives were applied.

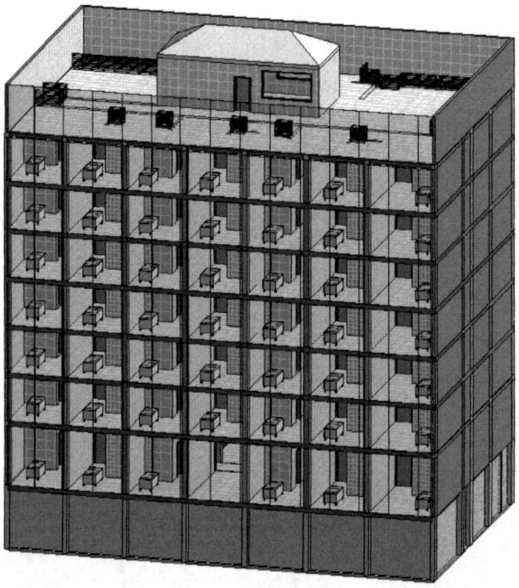

Figure 2. The case study building after rendering (researcher).

Table 1. LCC discounted and nominal of the alternatives

No.	Technique	LCC dis. [IQD]	% reduction	LCC nom. [IQD]	% reduction
1	Solar heater	5987510455	0.9	17035953880	1.076
2	LED lights	5918686093	2	16801608360	2.4
3	PV panels	5986086655	0.93	16917827510	1.76
4	PV (on grid)	5771432480	4.478	16313363086	5.27
5	Solar façade	6195695724	-2.5	16846602360	2.175
6	Solar façade (on grid)	5667142663	6.2	15275608884	11.29
7	Inverter AC	5636495167	6.72	15826188080	8.1
8	Central AC	5854734438	3.1	16385309890	4.85
9	VRF AC	5618978010	7	15694193740	8.867
10	Water saving	6026922624	0.25	16078887807	6.63
11	Wall painting	5537567762	8.35	15837795510	8
12	Marble tiles	6343176603	-4.98	18154702270	-5.4
13	Vinyl carpet	5990194910	0.86	17288118960	-0.38
14	Granite tiles	6164604075	-2	17311205450	-0.5
15	Ceiling painting	5812555564	3.8	16591580987	3.65
16	Insulation	6024080984	0.3	16779628398	2.56
17	Best alternative	5111379097	15.4	13225352521	23.2

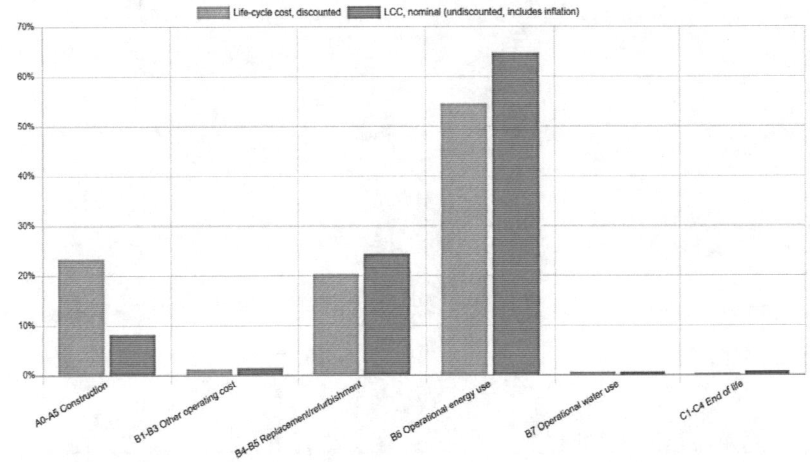

Figure 3. LCC results for the building before any retrofit.

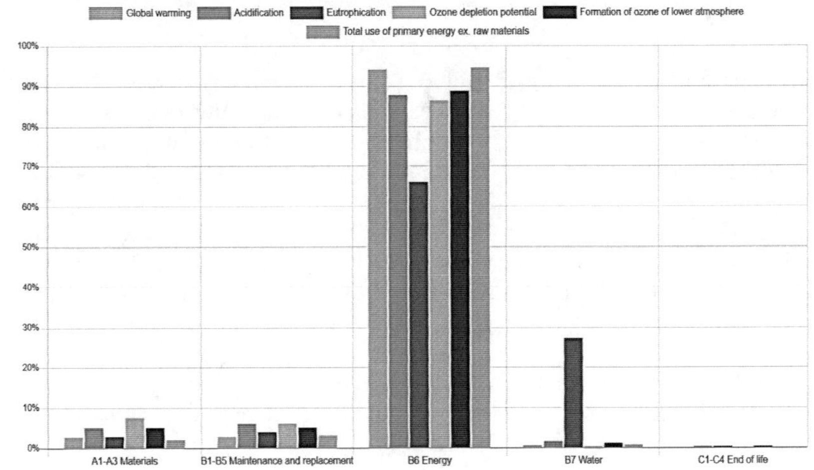

Figure 4. LCA results for the building before the retrofit process.

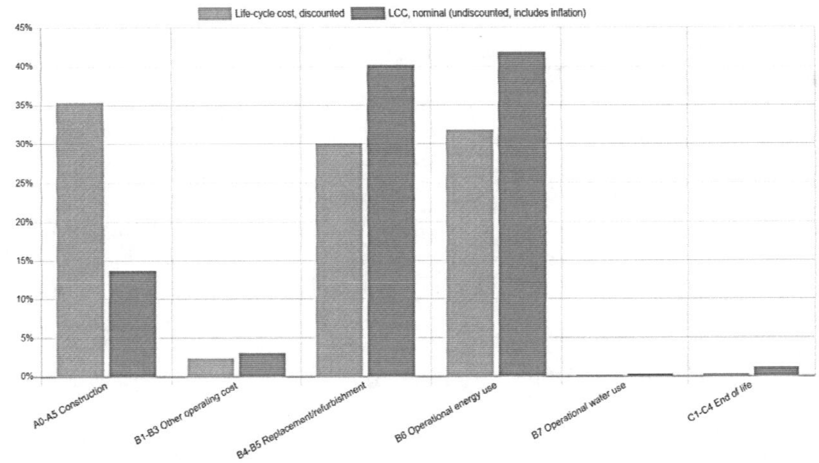

Figure 5. LCC results for the building after the retrofit process.

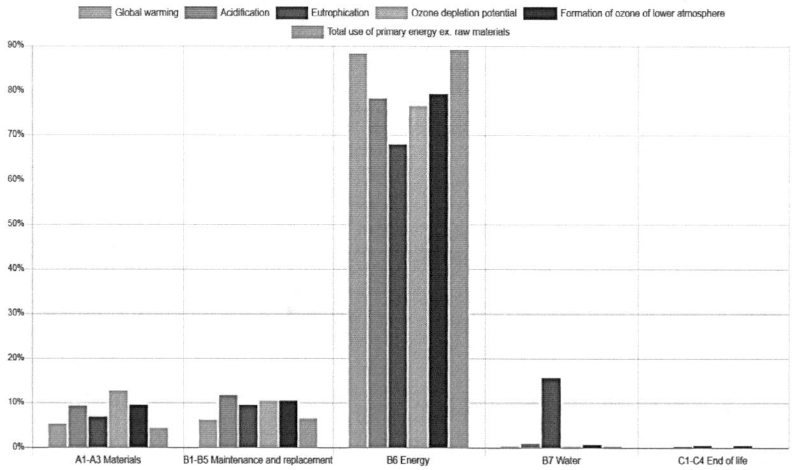

Figure 6. The results of LCA for the building after the retrofit process.

6. Conclusion

This research motivates using BIM to retrofit (upgrade) existing buildings, and it shows many reasons for what this technique is not used mainly for retrofit existing buildings. Verified some measures for the upgrading process, results of Table 1, by comparing LCC dis. and LCC nom. of the alternatives with those of the building before retrofit which are (6042275940 IQD and 17221281230 IQD respectively) and shown that saving in costs can reach up to 15.4 % and 23.2 % in discounted and nominal calculations respectively by adopting the set 17 which is the best alternatives, as shown in Table 1. At the same time, the CO_2 emissions had decreased from 34233 Ton to 19299 Ton as calculated by One Click LCA student version. The measures used in the optimum case include using solar water heater, LED lights, VRF inverter air condition, insulation for external walls, paint the internal walls with light colors, insulation for roof and basement, porcelain for other floors, paint the ceiling with light colors, double glass for the façade, photovoltaic panels in the roof and on the façade, using water-saving toilets, and greywater recycling system. The results also show that even if the initial cost increased, the total life cycle costing will be reduced to a reasonable percent by reducing the costs of operation, maintenance, energy, and end of life.

References

[1] B.C.M. Leung, Greening existing buildings [GEB] strategies. Energy Reports 4 (2018) 159- 206
[2] D. Smith, An introduction to building information modelling (BIM). Journal of Building Information Modeling (Fall 2007) 12-14.
[3] P. Russell and D. Elger, The meaning of BIM. Architecture in Computro, 26th eCAADe Conference Proceedings (2008) 531-536.
[4] R. Volk, J. Stengel, and F. Schultmann, Building Information Modeling (BIM) for existing buildings-Literature review and future needs. Automation in Construction 38 (2014) 109-127.
[5] S. Azhar, W. A. Carlton, D. Olsen, and I Ahmad, Building information modeling for sustainable design and LEED® rating analysis. Automation in Construction 20(2) (2011) 217-224.
[6] N.N. Kale, D. Joshi, and R. Menon, Life cycle cost analysis of commercial buildings with energy efficient approach. Perspectives in Science 8(2016) 452-454.
[7] A. S. Kshirsagar, M. A. El-Gafy, and T. S. Abdelhamid, Suitability of life cycle cost analysis (LCCA) as asset management tools for institutional buildings. Journal of Facilities Management, 8(3) (2010) 162-178.
[8] LRC, Fluorescent. Lighting Research Center, (2013), http://www.lrc.rpi.edu/resources/publications/lpbh/062Fluorescent.pdf

[9] AITI, About Induction. American Induction Technologies, Inc. (2008), http://www.americaninduction.com/IntroInduction/Introduction.html
[10] Makerere University, Summary of Computer Systems Engineering Projects - Motion Activated Light. Makerere University, (2009). http://mak.ac.ug/documents/FCITCSEProjectDescriptions.pdf
[11] LEED v4 for building operations and maintenance, U.S. Green Building Council, Washington DC, (2018).
[12] Pearl Building Rating System: Design & Construction, Abu Dhabi Urban Planning Council, (2010), Version 1.0
[13] ENERGY, Energy-Efficient Windows, U.S. Department of Energy, (2012). http://energy.gov/energysaver/articles/energy-efficient-windows .
[14] L. Oikkonen, J. Paatero, T. Carlsson, and P. Lund. Advanced energy systems: Photovoltaic energy. Helsinki University of Technology, (2005).
[15] C.R. Landau. Optimum Orientation of Solar Panels. The National Renewable Energy Laboratory, (2012). http://www.solarpaneltilt.com/
[16] NCSC, Passive and active solar domestic hot water systems. Raleigh. North Carolina Solar Center and North Carolina Department of Administration, (2002). http://infohouse.p2ric.org/ref/49/48019.pdf
[17] GAO. Water-Efficient Plumbing Fixtures Reduce Water Consumption and Wastewater Flows. Washington, D.C. (2000).
[18] One-Click LCA, Helsinki: Bionova Ltd, (2015).
[19] A. Mahjoob and M.M. Abed, Assessing BIM integration with sustainable requirement for buildings construction. International Journal of Science and Research (IJSR) 6(2015) 2258-2265.

Influence of Modern Processes of Life to The Architecture and Environmental Problems of Urbanization-Cities of Kazakhstan

Semenyuk Olga[1, a] *, Khvan Yelena[1, b], Chekayeva Rahima[1, c], Sadykova Sara[1, d], and Toibazarova Elmira[1, e]

[1]L.N. Gumilyov Eurasian National University, Nur-Sultan, Kazakhstan
[a]Ons_31@mail.ru, bkhvan-72@mail.ru, crahima.chekaeva@mail.ru, dsadykova_ssh@enu.kz, [e]Elmira2008_t@mail.ru

Abstract. Two directions of research were formed in the theory of architecture: The first is devoted to the issues of architectural design. In the second study the features of the influence of architecture on people. This is a two-way process of human interaction with the environment, it allows you to study in detail in the future, not only the design of an architectural object, but also the state of people interacting with it. On the one hand, a person changes the environment, and on the other reacts to its impact and influence. And this process is an integral part of human life, because the environment affects his emotional state, mental and physiological processes. This underscores the importance of studying the impact of the surrounding urban space on people, regulating the level of comfortable living by analyzing the ecological state of the environment and the quality of buildings being built. Life processes can be formulated as: spiritual, psychological, physiological, mental, bioenergetic, which affect morphogenesis in architecture. When describing an architectural object, we use such criteria as the functionality of the object, the aesthetics of the construction cost, and the modernity of the architectural solution of the building.

Keywords: Urbanization, ecology, architecture, resources, ecology.

1. Introduction

The modern city is a complex socio-economic organism, formulated demographic, diverse economy, and geographical relations with the surrounding economic space and the natural environment. Describing the economic development of the region, it should be noted that every year there is an activation of various sectors of the economy, in particular the construction industry, the development of small and medium-sized businesses. All this leads to an inevitable increase in the negative impact on the environment since all industrial enterprises that place their emissions, discharges, and waste in the natural environment are sources of its man-made pollution. Therefore, the sustainable development of the capital and other cities of Northern Kazakhstan is largely determined not only by the economic potential, the availability of land, human and other resources, but also by the state of the urban ecosystem. Living in a city, especially in industrial areas, it is difficult to talk about a favorable environmental situation. Landfills, chemical waste, automobile exhaust, industrial and industrial pollution of water and air, deforestation, the prohibition of fishing and hunting-all this directly entails environmental problems.

2. Environmental Engineering in Architectural Design

Architectural ecology as the ecological basis of architectural design includes environmental aspects of design, creating a healthy environment inside and around buildings, landscape architecture, improving the environment perceived by the senses, protecting the environment by architectural means, and social and spatial control over the environment and people [1]. Architectural ecology takes into account the ecological features of interaction between architectural objects and nature and the social and environmental needs of residents. Architectural ecology aims to bring people closer to nature, to save them from the monotony of urban space, lack of physical activity, and the correct distribution of the population on the territory. The distribution should be no more than 100 people per 1 ha, in the construction of neighborhoods for 30 thousand People with a ratio of low-rise and multi-storey construction in the ratio of 7: 3. It is necessary to preserve 50% of the urban space for natural areas and green spaces, to isolate the population from transport routes, to create conditions for people to communicate.

As the urban environment continues to be clean, permaculture opportunities are expanding, which significantly improves the appearance of buildings. Architectural and environmental solutions of buildings and engineering structures will be developed in the direction of supporting the animal world, which will improve the quality of the environment provided to people. One of the interesting areas may be the support of biodiversity, including the provision of conditions for small birds and animals in the city [2]. One of the directions of accounting for human ecology is to ensure communication between urban residents through architecture, including communication between neighbors in houses, within blocks, in cozy and well-maintained courtyards. The task of architectural ecology is continuous environmental education and education of urban residents with the help of a beautiful and healthy architectural and landscape environment [3]. Man and society interact with the environment that affects people. In this process, the personalization of the living space is necessary. For the permanent education of residents through architectural and landscape ecology, it is necessary to create a beautiful and healthy, clean, and favorable environment for the senses (vision, hearing, smell, touch). The participation of residents in maintaining such an environment contributes to their interest in it.

3. Environmental Problems of Urbanization of Nur-Sultan and Ways to Solve Them

The city of Nur Sultan is located in an arid zone and is characterized by limited water resources. The city's waterway is the Yessil river, which flows in its southern part. This reservoir is used as the main source of centralized water supply to the population, its flow is formed almost exclusively by meltwater. Within the city, the Yessil river receives two small tributaries-the Sarybulak and Akbulak streams. The main source of drinking water for the city is the reservoir on the Yesil river with a water reserve of 67.2 million m^3/year. The capital's reservoir is almost the only source to meet the domestic, drinking, and industrial needs of the city with a water reserve of 67.2 million m^3/year (design capacity of 410.9 million m^3).

The problems of ensuring environmentally safe and healthy living conditions of the population, rational nature management, and resource conservation are now directly related to the strengthening of man-made impact on the environment. In the context of further development of industries and agriculture, during the period of intensive development of Nur-Sultan and the growth of the capital's population, the need for energy and water resources, transport, housing and communal services is objectively increasing. All this leads to an increase in emissions into the atmosphere, the accumulation of industrial and household waste that must be recycled, disposed of and recycled. Atmospheric air is one of the main and most important components of the environment, a condition that significantly affects the global and regional climate systems. The state of atmospheric air in the city of Nur-Sultan is controlled at 7 stationary posts of RSE Kazhydromet. The state of air pollution was assessed based on the results of the analysis and processing of air samples taken at stationary observation posts. The main quality criteria are the values of maximum permissible concentrations (MPC) of pollutants in the air of populated areas.

In the 1st half of 2018, according to the stationary observation network, the level of air pollution was estimated as increased, it was determined by the values of SI = 4 (increased). The city's air is most polluted with sulfur dioxide. The volume of emissions of pollutants into the atmosphere from stationary sources amounted to 27.261 thousand tons, in the first half of 2015 - 21.922 thousand tons. The expected total emissions of pollutants in the 1st half of 2016 is 5,339 thousand tons more than in the same period of 2015 or 17% as a percentage. The reason is the increase in the volume of burned fuel at the TPP-2 of Astana-Energo JSC after the introduction of new capacities. As in any region, Nur Sultan has several environmental problems, based on which a register of environmental problems is compiled, which includes procedures for identifying, assessing and updating environmental problems to effectively plan the environment and constantly improve the environmental performance of the city. One of them is the increase in emissions of pollutants into the air from road transport; air pollution in winter from sources of thermal energy and Autonomous heat supply; improvement and cleaning of the Yessil river and its tributaries Akbulak and Sarybulak; lack of alternative sources of thermal energy. The problem of monitoring the environment and the ecology of the urban environment attracts the most attention. Today, this task is the most important and complex and is the subject of modern scientific research.

Atmospheric air is one of the main and most important components of the environment, a condition that significantly affects the global and regional climate systems. The state of atmospheric air in the city of Nur-Sultan is controlled at 7 stationary posts of RSE "Kazhydromet". The state of air pollution was assessed by analyzing and processing air samples taken at stationary observation posts. The main quality criteria are the values of maximum permissible concentrations (MPC) of pollutants in the air of populated areas. To improve the environmental situation in the framework of the action plan, TPP-1 will install environmental equipment, and by the end of 2019 will upgrade four boiler units at TPP-2. Initiated by The Department of natural resources. raids were carried out in the private sector, which resulted in the conclusion that it is unacceptable to use tires, plastics, and tires for heating, in some homes, low-quality coal is used for heating. In this case, an examination is carried out. But the environmental measures taken are not enough, and the situation shows that it is necessary to reconsider approaches to environmental conservation. It is necessary to propose to tighten environmental legislation, as well as to restructure the work of all existing structures in the city.

4. Environmental Problems of Urbanization in Northern Kazakhstan and Ways to Solve Them

An environmental problem is a change in the natural environment as a result of human impact or natural disasters that lead to disruption of the structure and functioning of nature.

In the Kostanay region there are the following environmental problems:

1) Lack of a complex of treatment facilities for biological treatment in the city of Kostanay. Sewage treatment facilities are presented in the form of earthen sedimentation tanks, which include three parallel maps measuring 182 x 87 m, fenced with dams that work alternately, the total area of which is 107.6 thousand m^2. Start of operation-1966. The currently used wastewater treatment technology does not meet modern requirements for wastewater treatment and allows cleaning only partially with mechanical impurities, which is confirmed by the results of chemical analyses in conditions of a constant excess of MPC. With the introduction of treatment facilities, the release of pollutants into the environment will be significantly reduced. Due to the high cost, the issue of construction of a biological wastewater treatment plant in the city of Kostanay was included in the Register of environmental problems, in the List of priority investment projects (programs) for 2007-2010 in the Kostanay region with funding from the Republican budget in the amount of 6300 million tenge.
2) Lack of storm sewer with a complex of treatment facilities in the city of Kostanay. Kostanay is located in the watershed of the Tobol river, which has a drinking and drinking purpose [4]. To eliminate water pollution, it is necessary to organize a system of interception, collection, and treatment of polluted stormwater (Fig. 1).
3) The absence of a complex for processing solid waste in the city of Kostanay. The absence of a solid waste processing facility in Kostanay leads to an increase in waste generation and contamination of land resources. Following the register of environmental problems in the Kostanay region, it is necessary to attract investment from the national budget for the construction of priority facilities, namely, the introduction of a complex for processing solid household waste with a capacity of 100 thousand tons per year [5].

4) Historical orphan pollution-storage of industrial effluents of the former copper-ammonia plant in Kostanay. The drive is located in the water protection zone of the Tobol River. The copper smelter ceased operations in the 80's. Excess MPC for copper is recorded at the location of the drive. There is a threat of pollution of the Tobol River. To eliminate pollution, it is necessary to Finance it from the national budget.
5) There are no fixed posts for monitoring air pollution in the cities of Rudny and Zhitikara. The need for monitoring in these cities is due to the increased technological load on the part of mining enterprises. According to the RD, a stationary pollution control point is created at the rate of 1 post per 50 thousand inhabitants. Currently, there are only 2 posts in the region in the city of Kostanay.
6) Inefficient operation of treatment facilities in cities due to high wear and tear. It requires reconstruction, modernization, and major repairs of Sosv, which requires a lot of money. State-owned utilities, which are on the balance sheet of the CBS, do not have sufficient financial resources to carry out these works. Sosv was built in the 70s and is on the balance of public utilities, which do not have sufficient financial resources for their repair and modernization. Due to the high wear and tear, the efficiency of Sosv is reduced annually.

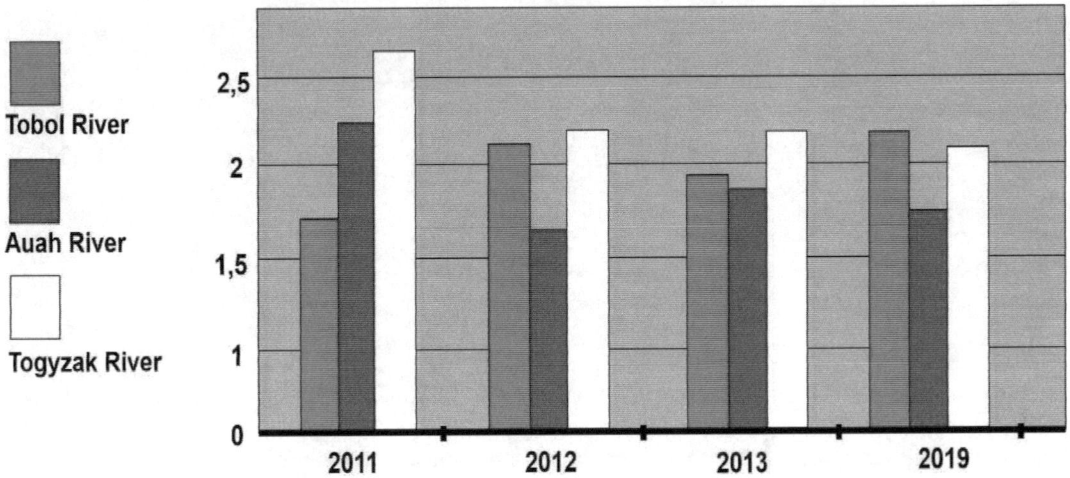

Figure 1. Polluted water index in Kostanay city.

To solve environmental problems, a register of environmental problems of the Kostanay region was developed. To solve the environmental problems of the construction of priority facilities, it is necessary to attract investment from the national and local budgets. The city of Petropavlovsk has a very high index of atmospheric pollution (Table 1). In 2018, 6 266 cases of hydrogen sulfide emissions above the maximum permissible standards were registered in Petropavlovsk. Petropavlovsk was named the dirtiest city in Kazakhstan. The Department of ecology of the North Kazakhstan region maintains a quarterly newsletter - an index of atmospheric pollution, which is affected by either a large number of emissions or an extremely high volume of emissions. According to this indicator, Petropavlovsk is on a par with Temirtau in terms of air pollution.

The source of harmful vapors in Petropavlovsk is the wastewater accumulator, which is located in the city. Residents of the city call this area toadstool because of the strong smell of hydrogen sulfide. There are rotting processes, and it is necessary to clean the bottom sediment. When the ice leaves in the spring, the process of hydrogen sulfide release begins, and observation posts operating in automatic online mode begin to signal this. To date, 11 cases of severe and one case of extremely high air pollution with hydrogen sulfide have been detected in Petropavlovsk. Every year in the city of Temirtau falls almost 300 thousand tons of industrial emissions, in winter there is almost no white snow, people complain that they are suffocating and try to appear less often on the street. The environmental problem in Temirtau gave rise to other problems: the outflow of the population from the city, the growth of cancer, dissatisfaction with the work of the city administration, and, as a result, social stress. The last reason to raise the topic of urban ecology was the black snow that fell in the city in January 2018.

Table 1. Characteristics of air pollution in the city of Petropavlovsk.

Impurity	Average hazard of pollution		Maximum single hazard of pollution		The number of cases of exceeding the Threshold Limit Values (TLV)	
	Mg/m^3	Multiplicity of exceeding the Threshold Limit Values	Mg/m^3	Multiplicity of exceeding the Threshold Limit Values	≥ TLV	≥ 5 TLV
Suspended substances	0.085	0.568	0.100	0.200		
Suspended particles, PM-2.5	0.013		0.513			
Suspended particles, PM-10	0.012		0.588			
Sulfur dioxide	0.008	0.153	0.351	0.702		
Sulfates	0.009		0.010	0.001		
Carbon monoxide	0.989	0.330	7.000	1.400	4	
Nitrogen dioxide	0.032	0.789	0.080	0.941		
Nitrogen oxide	0.001	0.014	0.066	0.165		
Ozone	0.057	1.896	0.108	0.675	-	-
Sulphuretted hydrogen	0.003	-	0.074	9.275	149	32
Phenol	0.002	0.489	0.003	0.300		
Formaldehyde	0.004	1.368	0.007	0.200		
Ammonia	0.002	0.044	0.069	0.346		
Methane	1.252		1.886			

To improve the environmental situation in the city of Pavlodar, green zones are being actively created. Regular checks are carried out for compliance with environmental legislation. The potential danger is the free economic zone, which is being built near Pavlodar. For the sake of profit, the authorities forbid environmentalists to revoke licenses from enterprises that violate environmental requirements. However, the Department of ecology in Pavlodar is characterized by increased activity. Due to its activity, the Kazakhstan electrolysis plant was not able to work actively, as this could lead to the release of pollutants in increased volumes. At the same time, the company has clear objectives that will help it maintain turnover and allow it to work if the concentration of harmful emissions remains within the normal range. The city and the entire region are characterized by the problem of disposal of mercury-containing lamps. Now special containers are being installed throughout the territory to help solve the problem. The city and the entire region are characterized by the problem of disposal of mercury-containing lamps. Now special containers are being installed throughout the territory to help solve the problem.

5. Principles of Ecological Architecture

The following principles of ecological architecture can be formed in the process of research of ecological problems of urbanization:

1) The principle of energy conservation. We are talking about a new design and construction of structures in such a way as to minimize the need for consumption of thermal energy for heating or, conversely, cooling.
2) The principle of reducing the volume of new construction. At all times, people used old buildings or just materials from their disassembly to erect new buildings. Thus, the builders of the Abbey of St. Alban in England at one time adapted the bricks from the ruins of the Roman city of Verulanum to work. In the practice of Russian and Scandinavian wooden architecture,

old healthy beams and rafters were often singled out, removed from previous buildings, and assembled in another building. The roof builders of medieval Europe did the same. But in the mid-20th century, a completely different approach seemed to prevail - developers convinced city officials and private investors that it was cheaper and more efficient to break everything down and build from scratch. It is not always cheaper, rarely more efficient, but certainly easier.

3) The principle of using renewable energy sources. To save money and be environmentally friendly, green architecture uses solar panels and solar energy storage devices. Also, on the Southside of the building, there is a large area of Windows, which in total gives up to 80% savings on heating and hot water [6].
4) The principle of respect for the resident. This is a significant change in the approach to the functioning of the building, when both the Builder and the architect, and the owner see the building is not a living machine, and corporate property, in the maintenance of which each resident plays a huge role. This role cannot be significant in the case of multi-storey standard multi-storey blocks with an unchanged configuration of walls and roofs, where the role of residents is solely to prevent vandalism. But in the reconstruction of old buildings, the role of future residents can be extremely great.
5) The principle of respect for place and region. This is a special position of consciousness, most represented by Eastern philosophy, in which merging with the natural environment, an infinite view of it from time immemorial was considered the highest value. In General, European consciousness has long cultivated a different attitude to nature - it was considered exclusively as a resource, as well as an object of purposeful activity of people. The appearance of green architecture gives a hint that humanity has finally begun to think about the harmful effects on nature and is gradually trying to learn to coexist with it in harmony [7].

6. Conclusions

The environmental situation in Kazakhstan is of serious concern to many residents. The growth of mining production has led to the release of huge amounts of toxic waste. The lack of effective cleaning systems in industrial plants contributes to air pollution. Associated gas is burned at oil and gas enterprises, which leads to soot emissions. Road transport runs on low-quality fuel, and the country's water resources are subject to degradation. Currently, the government of Kazakhstan seeks to participate in various environmental organizations, which should help the country to correct the situation in the future. Despite the difficult environmental situation and the almost complete absence of clean cities in the country, Kazakhstan shows that every citizen must fight for the environment. When discussing the problems of the environmental situation in general, we are talking not only about the overall global picture, but also about the actions of one person. Until people receive proper environmental education, these issues will remain without special attention, until everyone understands the importance of maintaining a favorable environmental situation, and it is impossible to avoid the inevitable disaster [8]. Defenders of animals and nature do not give up trying to achieve good results: rallies against deforestation, self-cleaning of forests and rivers from volunteers, as well as the construction of reserves have a positive effect [9]. Issues and problems of ecological construction in the recent past have not been a priority in Kazakhstan. But during the crisis, it became apparent that green buildings consume less expensive resources. Green construction began to develop actively in Kazakhstan with a change of priorities. Green building can not only solve the long-term problems of global climate change, but also change life and the local environment right now to meet the most pressing human needs.

References

[1] D. Zh. Markovich, Social Ecology-M, Education, 1991. - 176 p.
[2] N.A. Agajanyan and B.A. Nikityuk, Polunin N. N. human Ecology and integrative anthropology. M.-Astrakhan, 1996 224.
[3] N.A. Agajanyan and V.I. Torshin, Human Ecology, selected lectures-M, Ecocenter, 1994 225.
[4] P.A. Kazantsev, Fundamentals of ecological architecture and design, Guide. Vladivostok: Publishing house of the far Eastern state technical University, 2012 17-19.

[5] L.A. Danilenko and L.A. Malyshkina, Environmental protection and rational use of natural resources, Neftyanoe Khozyaistvo, 9 1997 72-74.
[6] Dilara Aronova, The newspaper "Kostanay news", January 17, 2013
[7] Sergey Shevchenko, The newspaper "Kostanay news", July 28, 2012
[8] F.A. Novikov, Architecture and architecture, Publication stereotype, URSS edition, 2014 107-113.
[9] E.V. Girusov, S.N. Bobylev, S.N. Novoselov, and N.V. Chepurnyh, Ecology and Economics of nature management. YuNITI-Dana, Moscow, 2007 591.

IOP Conf. Series: Materials Science and Engineering **901** (2020) 012026 doi:10.1088/1757-899X/901/1/012026

Developing Systems Engineering for Sustainable Infrastructure Projects

Khamael R. Obead[1,a*] and Meervat R. Wali[1,b]

[1]Civil Engineering Department, University of Baghdad, Baghdad, Iraq
[a]khamael.riyadh@yahoo.com, [b]meervat.r@coeng.uobaghdad.edu.iq

Abstract. Globally, Sustainability is very quickly becoming a fundamental requirement of the construction industry as it delivers its projects; whether buildings or infrastructures. Throughout more than two decades, many modeling schemes, evaluation tools, and rating systems have been introduced en route to realizing sustainable construction. Many of these, however, lack consensus on evaluation criteria, a robust scientific model that captures the logic behind their sustainability performance evaluation, and therefore experience discrepancies between rated results and actual performance. Moreover, very few of the evaluation tools available satisfactorily address infrastructure projects. The research introduces a system engineering model that abstracts the environment, the construction product, and its production system as three interacting systems that exchange materials, energy, and information. The model utilizes this setup to capture and quantify essential flows exchanged between such three systems, to evaluate sustainability. The research walks through the development of a generic case of the model, and then demonstrates its utility in evaluating the sustainability performance of civil infrastructure projects. The developed model will address an identified gap within the current body of knowledge by considering infrastructure projects. Through the ability to simulate different scenarios, the model will enable identifying which activities, products, and processes impact the environment more, and hence potential areas for optimization and improvement.

Keywords: Infrastructure, sustainable Infrastructure, systems engineering, realizing of systems.

1. Introduction

This paper is part of a series intended as brief introductions to the application of systems engineering approaches to infrastructure projects. It was developed by the International Council on Systems Engineering (INCOSE) Infrastructure Working Group in the interest of aiding industry [1-3]. Another characteristic in the construction of infrastructure projects has increased the complexity and number of stakeholders. Unlike other types of projects, construction of Infrastructure projects is often funded by government agencies. In many cases, there are multiple funding agencies at different levels of government. Add to this the many groups affected by the project, from local communities to utilities and transit systems that must be crossed or relocated, along with potential land-use restrictions and environmental impacts, and the requirements list grows exponentially.

The codes and standards that form part of the requirements add another layer of complexity in the construction of infrastructure projects. the construction of Infrastructure projects can cross into different jurisdictions, leading to varying and sometimes conflicting requirements for the different physical locations of packages of work [4]. There are several unique aspects of systems engineering on large infrastructure construction projects. One difference is that the system elements are often completed by different firms under separate contracts, and often at different times. This makes it even more difficult to integrate the work (as it's done across projects), and to complete the verification and validation that all of the requirements have been met (as each contract is only a part of a larger system).

Design-Build projects have some unique characteristics of their own that have the potential to increase project risk. On design-build projects, there is a shift in responsibility for the quality of the work, including the development of final requirements and specifications, away from the agency and towards the contractor. The design is initiated by one firm but completed by another: this handoff increases the potential for misinterpreted requirements or incomplete incorporation of requirements. Internal integration is much more complex, as the work is broken down into much smaller packages to enable a more continuous construction effort [5].

2. Systems Engineering Fundamentals

Systems engineering is an interdisciplinary approach and means to enable the realization of successful systems. It focuses on defining customer needs and required functionality early in the development cycle, documenting requirements, and conducting design synthesis and system validation. SE considers both the business and technical needs of customers to provide a quality product that meets user needs. Systems engineering involves multidisciplinary teams working together to define project requirements, identify alternative designs, select the optimal solution, build the product, conduct testing to ensure the product meets stakeholder needs and requirements, and operate and maintain the product through its lifecycle [6]. Many of these steps are undertaken on traditional design projects. The systems engineering approach differs in a few fundamental aspects:

- Requirements are thoroughly defined, refined, and allocated to lower-level system elements, assigned to suppliers and subcontractors, and tracked, through the testing and verification phase of the work.
- The product is viewed from a system perspective: links between requirements are identified so that changes to any one area can be analyzed for impacts to the requirements for the system as a whole.
- The entire product lifecycle is considered from the start of the project, including the disposal of the final product.

3. Process of System Engineering

Figure 1 provides a simple context diagram showing design engineering as a process. The design engineering process gathers inputs from the specifications and system requirements, identifies and specifies all systems, structures, and major components to be designed and produces an integrated design with interface requirements and descriptions of the system elements [7]. From a system engineering perspective, effective design management requires the following activities to be well executed:

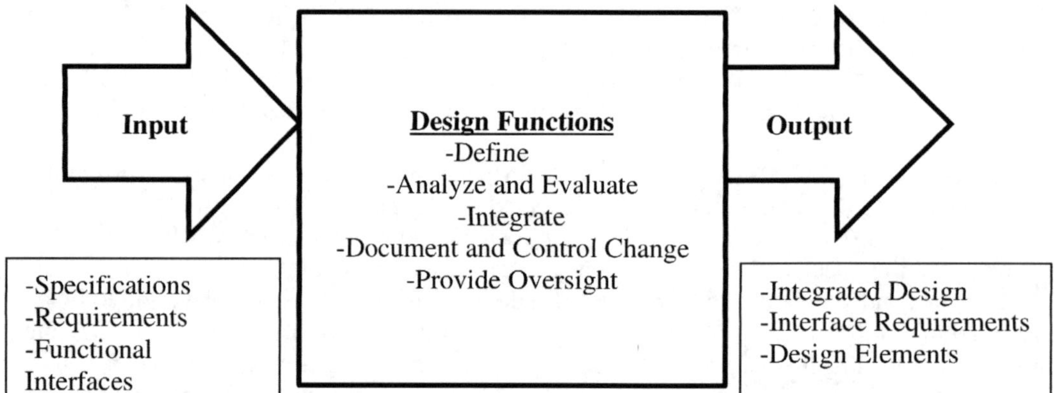

Figure 1. Design SE process.

3.1. Define
Describe a high-level system consistent with stakeholder requirements and specifications. Refine and allocate the system requirements to lower-level system elements as appropriate. Identify interfaces between system elements, and between the system and other external systems [7].

3.2. Analyze and Evaluate
Analyze the system as a whole to determine how best to achieve user and client requirements. Establish criteria for selecting from among the design alternatives. Evaluate design alternatives and decide on the best option [7].

3.3. Integrate
Define and manage interfaces with external systems. Define the strategy for integrating system elements. Ensure the design elements are integrated throughout the effort, from preliminary engineering through final design.

3.4. Document and Control Change
Document the preliminary engineering phases and the baseline design. Establish traceability between requirements and system elements. Analyze design changes for their full impact on the integrated design and the requirements. Approve changes through a prescribed process and appropriate levels of authority. Understand the impact on the rest of the system before committing to the change. Ensure approved changes are propagated to all affected items [7].

3.5. Provide Oversight
Manage the resulting information and the processes being implemented. Develop metrics useful to show the quality of the design management activities.

4. Benefits of Systems Engineering

As infrastructure projects become more complex about the size of the systems, the number of requirements and stakeholders involved, and the sophistication of the associated technology, the risks of project failure increase [8]. The application of systems engineering can help ensure these projects are delivered on time and budget, and that they meet requirements. Systems engineering improves project outcomes by:

4.1. Improving System Quality
Systems engineering involves an increased focus on thorough planning at the start of the project, including identifying stakeholder needs, documenting how the desired system will operate, and establishing system requirements, including tracing all requirements to their source documents. It also provides a detailed approach for ensuring those requirements are tracked through the testing stage and verified in a disciplined manner, through the use of a requirements verification traceability matrix (RVTM).

4.2. Gaining Increased Stakeholder Participation
Systems engineering requires the involvement of all stakeholders throughout the project to ensure the project continues to meet stakeholder needs. This continuous involvement helps the project team identify and document any changes to the requirements as early in the project as possible. The use of a standard systems engineering project process helps stakeholders actively participate in the development of the system, and facilitates the involvement of new stakeholders by reducing the learning curve.

4.3. Ensuring Design Integration
The thorough and detailed documentation of requirements in a requirements matrix provides a good foundation for design integration. The matrix documents the connections from the source files, such as contract documents, down to very detailed requirements for smaller system elements. It also documents interfaces and traceability between requirements, both internal and external. It also helps document

IOP Conf. Series: Materials Science and Engineering **901** (2020) 012026 doi:10.1088/1757-899X/901/1/012026

changes in requirements, and enables the team to identify impacts of the changes on the requirements for other parts of the system.

4.4. Reducing Cost and Schedule Overruns

Combined with the disciplined project controls that form part of the systems engineering approach, the rigorous tracking of requirements helps reduce the risk of schedule delays and cost overruns.

5. Practical Study and Statistical Analysis

There is a need for understanding all criteria and sub-criteria that used in the implementation of sustainable infrastructure projects. This will support and develop managing sustainable projects. For implementing sustainable infrastructure it will be necessary to study criteria and sub-criteria to achieve major components of sustainability. the statistical study aims to know the most important factors that affect in implementation of sustainable infrastructure. For identifying criteria and sub-criteria for sustainable infrastructure the questionnaire has been developed to reach to the opinion of the experts through surveying. The questionnaire contains a list of criteria for sustainable infrastructure. In this study, a systematic approach has been taken to identify and analyze the criteria to implement sustainable infrastructure through literature review and interviews with experts and questionnaires.

The respondents were requested to rank these criteria and sub-criteria according to Likert scale that is the most well-known used approach to scaling responses in the research survey. In Likert scale, the (1= Does not affect and 5 = Very effect) The analysis of obtained data was done by using the Relative Importance Index (RII) method. The researcher was distributing 40 questionnaire forms and 30 forms were returned, The quantitative data collected was analyzed using (SPSS) a statistics program. 32 factors were considered to rank by the Relative Importance Index (RII) method. These factors are a category in seven components namely (Environmental side, Economic side, Social side, Importance of site selection, Energy efficiency, Water consumption efficiency, and Sustainable materials used). the rating of criteria and sub-criteria are taken out for all seven components deduced from the questionnaire [9-11].

5.1. Mean Score Ranking Technique

The responses of some collected data were analyzed by using the mean score ranking technique. The results are shown in Table 1, the rank options in descending order and it was calculated by using the formula below for mean score (MS):

$$\mathrm{MS} = \sum_{1}^{n} \frac{X1*S1+X2*S2+\cdots+Xn*Sn}{N} \tag{1}$$

5.2. Standard Deviation Technique

The Standard Deviation (SD) Measures the dispersion of the average. If the standard deviation is small, it indicates that it is close to the average, whereas if the standard deviation is large, it indicates that the answers were wide and not convergent. The results are shown in Table 1. The standard deviation was calculated according to the equation shown:

$$\sigma = \frac{\sqrt{\sum_{1}^{n}(\mathrm{xi}-\mathrm{x})^2}}{\mathrm{N}} \tag{2}$$

Where σ = standard deviation; x = mean score; xi = degree of importance; and N= total number of responses

6. Relative Importance Index (RII) Method

Through the questionnaire, the concept of importance is explained in a general way and it was comparable in preference and the definition of all factors was provided in the questionnaire. the questionnaire is designed in a way that will support in keeping of integrity and consistency in data. To ensure that the data were good in quality, a summary of the subject was written and the methodology of the research was explained to each respondent on the questionnaire. The results are shown in Table.1 The analysis of the data obtained was done by using the RII method by using the equation below:

$$\text{RII} = \sum_1^n (\text{X1} * \text{S1} + \text{X2} * \text{S2} + \cdots + \text{Xn} * \text{Sn})/(\text{A} * \text{N}) \quad (3)$$

Where RII = the Relative Importance Index; S = weights of factors; X = frequency of each factor; N = total number of respondents for each factor; and A = highest weight.

Table 1. The results of the questionnaire.

No.	Criteria	MS	SD	RII	Rank
	Environment side				
1	The importance of environmental standards to be considered in the design of sustainable infrastructure	4.17	0.648	0.83	2
2	Considerations of the sustainable impact of the project on safety and public health	4.27	0.691	0.85	1
	Economic side				
1	the impact of sustainable designs on reducing costs at all stages of the project life cycle	3.97	1.129	0.79	1
2	Appropriate economic systems and the encouragement of the State for such projects	3.90	0.995	0.78	2
3	the impact of sustainable project implementation on increasing costs	3.60	1.133	0.72	6
4	Take the cost of disposal of materials at the end of the service period in consideration	3.63	1.159	0.72	4
5	Sufficient profits through the operation of sustainable project	3.60	0.968	0.72	5
6	Capital recoverability during the operating period	3.60	1.221	0.72	7
7	the amount of the price or fees for the service or product provided by the sustainable project	3.90	0.995	0.78	3
	Social side				
1	the social impact on the formation of building designs for the community	3.43	1.073	0.68	4
2	security, social and political stability of the country	3.60	1.163	0.72	3
3	Understand and accept sustainable infrastructure ideas by society	3.63	1.098	0.72	2
4	Establish a reasonable and logical legal framework to accommodate the concept of sustainable infrastructure projects and ensure the rights of all parties involved in the project	3.67	1.028	0.73	1
	Importance of site selection				
1	Ease of linking infrastructure with existing infrastructure	4.07	0.828	0.81	1
2	Near the site of the main transport routes	4.03	0.718	0.80	3
3	Provide close transportation	3.93	0.980	0.78	4
4	The impact of location on ecosystems	4.07	0.944	0.81	2
	Energy efficiency				
1	The use of alternative and renewable energies in energy generation	4.17	0.592	0.83	1
2	The importance of using solar energy	4.10	0.759	0.82	2
3	Environmental techniques for cooling and heating	4.03	0.809	0.80	3
	Water consumption efficiency				
1	Low water consumption	4.10	1.029	0.82	1
2	Gray water recycling for watering gardens	3.90	1.062	0.78	3
3	Use of water from cooling and heating	3.63	1.326	0.72	6
4	Use of counters	3.67	1.093	0.73	5
5	Rainwater harvesting	3.40	1.163	0.68	7
6	The use of special techniques to determine if there are leaks in the water system	3.77	1.165	0.75	4
7	Develop designs that contribute to reducing water consumption	4.07	1.112	0.81	2
	Sustainable materials used				
1	Use of substances with little environmental impact	3.77	1.104	0.75	4
2	Use materials that can be recycled or used	4.00	1.017	0.80	1
3	The use of materials of high sustainability	3.90	1.029	0.78	2
4	Availability of sustainable resources	3.60	1.163	0.72	5
5	Consider the possibility of disposal of material after the end of service (reuse or recycling) taking into account the functional performance	3.80	1.186	0.76	3

From the results in Table 1, the following conclusions are drawn. Decision-makers should know the direct impact of sustainable management practices and spread awareness of sustainability, and the important impact of it is that the environmental factor is the most influential. This indicates that implementing sustainable management depends mainly on the desires of the shareholders themselves. Since the primary purpose of the project is to earn profits, the organization must realize that sustainable management will inevitably bring huge benefits to the organization, any organization can implement engineering systems that depend on sustainability and consider all criteria (environmental, social, economic, site selection, water efficiency, Energy efficiency and the use of sustainable materials) are all important factors that should not be overlooked. To enhance the implementation of sustainable management in institutions, it is necessary to enhance the administrative awareness of basic institutions in introducing engineering systems that depend on the application of sustainability in infrastructure.

7. Conclusion

Infrastructure projects are constantly evolving. The purpose of this paper is to introduce modern engineering systems that bring together relevant factors and project management with engineering systems covering all of the project life cycle to reduce delays or increase costs since all risks that affect the project are taken. Therefore, several major and sub-criteria have been defined which will greatly assist project managers in covering all factors, taking them into consideration, understanding infrastructure and engineering systems greatly, and knowing the sustainable factors that have a great link to sustainable infrastructure. Engineering systems help stakeholders monitor the project as a whole, even if there are secondary contracts to make paragraphs of the project. This paper is an important addition in the world of knowledge, which aims to improve project monitoring and increase confidence in expectations that have been calculated throughout the project life cycle.

References

[1] E.W. Aslaksen, Systems engineering and the construction industry, Requirements and Construction, 2005 1-4.
[2] California Department of Transportation, Division of Research and Innovation, SE Guidebook for ITS (2007).
[3] M.R. Emes, A. Smith, and L. Majanovic-Halburd, Systems for construction: lessons for the construction industry from experiences in spacecraft systems engineering, Intelligent Buildings International, 4(2) 2012.
[4] O. Hoehne, Entering a brave new world – applying systems engineering to American infrastructure projects – case study: the California high-speed train project, Proceedings of the 22nd Annual International Symposium, International Council on Systems Engineering, Rome, Italy, July 9-12, 2012. (Available through INCOSE)
[5] International Council on Systems Engineering, Systems Engineering Handbook: A Guide for System Life Cycle Processes and Activities (2011).
[6] International Council on Systems Engineering, Guide for the Application of Systems Engineering in Large Infrastructure Projects (2012).
[7] International Council on Systems Engineering, Systems Engineering in Transportation Projects: A Library of Case Studies (2013).
[8] S. Taylor, J. Wright, B. Mcintyre, and C. Anderson, Sustainable Infrastructure, Costa Rica, USA, 2017.
[9] International Association for Impact Assessment, IAIA, USA, 2017, http://www.iaia.org/.
[10] Risk Management Toolkit, MITRE, SE Process Office. Retrieved 8 September 2016.
[11] International Council on Systems Engineering, Systems Engineering in Transportation Projects: A Library of Case Studies (2013).

An Evaluation of Barriers obstructing the Applicability of Standard Bidding Documents (SBDs) in Public Procurement of Iraq

Sadeq AbdulHamza Hasan[1, a *] and Sawsan Rasheed Mohammed[1, b]

[1]Department of Civil Engineering, University of Baghdad, Baghdad, Iraq.
[a]sadeqah2004@yahoo.com, [b]civilsawsan2@gmail.com

Abstract. According to Order Number 87 of 2004 which had been issued by Dissolved Coalition Provisional Authority in Iraq, the Ministry of Planning MoP in the Iraqi Government started its efforts toward enhancing and developing the federal level of Public Procurement System by issuing a stream of procedures to the contractual entities (Standard Bidding Documents SBDs, procurement Planning Forms, Performance Indicators, Guide of Archiving Process… etc.). The financial crises which hit the international economy and the noted reduction in crude oil prices and the war against ISIS on the local level reflected an obvious shortage in funding process to the implementation of the project to the very minimums. At 2018 the Iraqi government allocated more funds to compensate the shortage in initial services projects at a time when the SBDs became mandatory to apply by all the public contractual entities, as a result, the local contractors were not able to meet the standard requirements and criteria of a modern experiment in the bidding process which led to more delay by re-announcing these projects. This research is to determine if the Iraqi contractual environment can be considered as healthy and supportive for the SBDs or not? Nine evaluation criteria for the SBDs projects were listed in the questionnaire targeted 100 private sector respondents and 130 public contractual entities, the mean comparison and relative important index RII were used to rank and match respondent agreement on assessing the importance mentioned criteria evaluation. The findings indicated that all respondents had agreed on ranking the average of annual revenues and financial liquidity as the top hurdles that need to be considered by MoP to ensure the success of the SBDs system.

Keywords: Standard Bidding Document SBDs, Private sector, enabling environment; barriers, MoP.

1. Introduction

The measurement of any successful construction project passes two stages (contracting and implementation), multi factors and criteria integration to achieve these stages that mean (Economy, Efficiency, accountability, good management, transparency, risk prevention and control, and fairness in contracting stage), (time, cost, quality in implementing stage). One of the most difficulties that might be facing the project is the keeping of harmony among the above-mentioned factors and criteria, thus these factors must be measured and valued towards integration among them. Delays in implementation schedule, budget overruns, poor quality, weak productivity development, and low customer satisfaction, are the statements that have been often used to characterize the construction industry, Many countries carried out modern experiments to explore the contractors' experiences on the best value of procurement to improve the methods of procuring future projects and learned from contractor's experiences in several projects, thus the result was using the new procedure of procurement [1] and development in business responsibility [2].

Taking Iraq as developing country, the MoP provide options for decision-makers to be considered when modernizing the Iraqi procurement regulations and to support the proper implementation in daily practice [3], Despite the efforts of the practitioners' and scholars in the field of construction, these difficulties are away from being settled. As in reference [4] noted the calls for change during the previous three decades were extremely various, rather than focusing on new apparatuses, advances or abilities, the analysis is regularly focused on the major working of the business. The better way to reduce the costs in any project is going through good procurement, impartial, consistent, and therefore reliable. Such an environment can directly expand the purchaser's options and opportunities as well as reduce the delays in implementation. it also reduces the financial complications in the project [5].

The procurement process is a standard practice of many contractual entities who are concerned in using the type of competitive bidding which holds the costs in its lowest level, Construction of industry projects participants have started meets that accepting the least price effort does not guarantee maximum amount [6] Contractor's competitiveness is usually used for ranking the contractors in a competitive tendering procedure, thus enhance competitiveness becomes a contractor's large strategy to sustain its business development. The topic of contractor's competitiveness has been catching research interests from many scholars [7]. The SBDs criteria like financial position, experience, key staff and management efforts are considered as valuable success factors pushing contractors to be more focused in competition but at the same time working in restricted resources [8]. Also, it is the major reason behind the failure in meeting its required standards. Proposal to avoid such problems is the medium bid awarding system that may be practiced by public agencies for the procurement of works [9]. 70% of the respondents considered the multi-parameter bidding according to this method, the study will be extra effective than lowest bidding method, this method sorted as the better one amongst various methods. research and SBDs suggest that the competitiveness of firms is equal too.

2. Competitiveness Procurement

In the early 19th century the UK started implementing the bidding documents, it was applied for the procurement of barracks during the Napoleonic wars [10,11]. The power of competitiveness defines the organization's capability in various markets to compete for business. The competitiveness of the Company is defined as the ability to produce design, marketing of products superior to those offered by competitors taking into consideration the non-price qualities and price as well [12]. According to the Competitiveness Policy in Untied State U.S. Council (1992), the competitiveness defined as the capability to output services and goods that meet the checking of international markets through citizen's standards earn of living for both high and sustainable over the long turn on [13]. The competitiveness of contractor's bidding defined as the difference between a percentage of the contractor's bid against the lowest bid.

Competitive bidding occurred where the acceptable bidding of the project is awarded to the bidder of the lowest price, which made it a fundamental part of the construction industry. This process of project delivery is designed to promote well competition to ensure the lowest price for the project. In the public sector, the principles are carefully designed to place all competent contractors on an equal foundation for competition, and carefully enforced to stop collusion among them as well as the illegal actions or unethical behaves by public officials [14]. However private owners may select to award contracts differently since they are not demanded by law such as the public agencies [15].

Since the bidding behavior is affected by many factors concerning both the features of the project and nonregular changed situations, the problem of bidding decision is highly unstructured. The decision of bidding always considered as very complex requiring a simultaneous estimate of a large number of extremely inter-variables related to arrive at a suitable decision [16] and [17] which lead to both determination of bid price and the bid decision are very important for the contractor. Public projects define as any establish for construction work that is funded by public budged, such as the sale of bonds or taxes [18] and most contracts between the contractor and the owner are extremely divided into different categories, each category has several variations that commonly determined by the kind of fees paid by the contractor [19] as shown in Fig. 1.

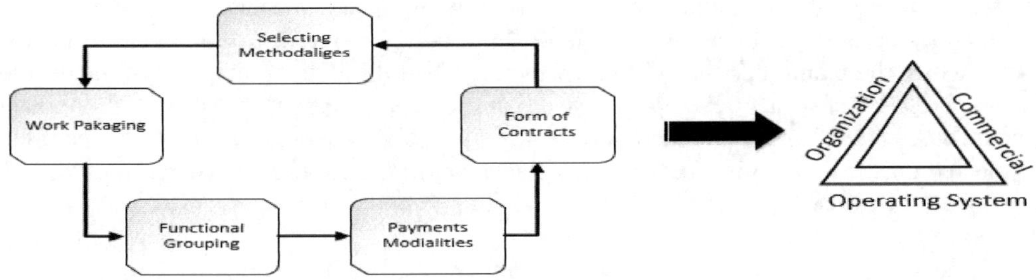

Figure 1. Relationship of the project procurement and delivery systems adapted from [20] and [21].

3. Standard Bidding Documents

According to the order of the dissolved temporary coalition authority # 87 which referred to recognizing that public contracts laws should conform to international standards of transparency, predictability, fairness of treatment; it also refers to provide a suitable mechanism to solve the disputes among contracted parties, and to be free from corruption and undue influence, and to create a system to procure goods and services at the best possible value for the government. Noting that the concept of full, fair, and open competition is of critical and high importance to the economy of Iraq and the goal of free trade among nations and building on the existing Iraqi laws in this field, including administrative instructions, and modernizing these instructions following best international practice. The Ministry of Planning as the representative of the GoI in this field is continuing to enhance the Public Procurement System at the federal level through issued international procedures for contractual entities like (standard Bidding Documents, Procurement Standard Process, Procurement Plan, key performance indicators, Procurement Records Management, Reporting on Procurement Activities and Procurement Department Webpage).

4. Factors Influencing of Procurement System

The accompanying criteria were utilized to look at client's prerequisites and specialists' inclinations for the exhibition of every acquisition technique as referred to in [22-26] recommend utilizing the accompanying criteria to build up a profile of the client's necessities speed (during both structure and development); sureness (cost and the stipulated time and learning of how much the client needs to pay at every period during the development stage); adaptability in pleasing plan changes; quality (temporary workers notoriety, style and trust in structure); multifaceted nature (client may determine specific subcontractor or build-ability investigation); chance assignment/evasion; duty (the culmination of the program, value, item quality, plan and development); value rivalry (covering such issues as an incentive for cash, support expenses, and focused offering); and questions and discretion, as shown in Fig. 2.

5. SBDs Contract Management

According to Seeley [27], contracting management is a framework whereby a primary temporary worker is delegated, either by exchange or in rivalry, and works intimately with the group of experts. [28] opined that "in an administration contract, the perpetual works are developed under a progression of development contracts put by the administration temporary worker after endorsement by the customer. "All physical developments will be implemented by sub-contractual workers chosen through focused offering, in this framework, for the most part, has the primary temporary worker called the administration contractual worker who gives the administration aptitude in the development of the venture for a charge. This Manager is designated at the initiation or better still practicality stage to join the customer's group of advisors, to help in preparing the program plan and site activities, He will oversee and co-ordinates the work bundles to individual sub-contractual workers and similarly gives on the site administration, plant and gear, comforts and so on for the work.

The expense paid to the administration contractual worker relies upon the nature and degree of the work done and not on the expense of the work. Notwithstanding, the board contracting framework is most suitable for enormous and complex activities that display specific issues that militate against the work of fixed value contract methods. Run of the mill instances of which are: Projects for which convoluted apparatus and/substantial gear are to be introduced simultaneously with the structure works. Projects for which the plan procedure will of need proceed all through the majority of the development time frames; Projects on which development issues are to such an extent that it is vital or alluring that the structure and supervisory crew incorporates a reasonably experienced structure, the contractual worker selected on such a premise, that his advantages are to a great extent synonymous with those of the business' expert advisors.

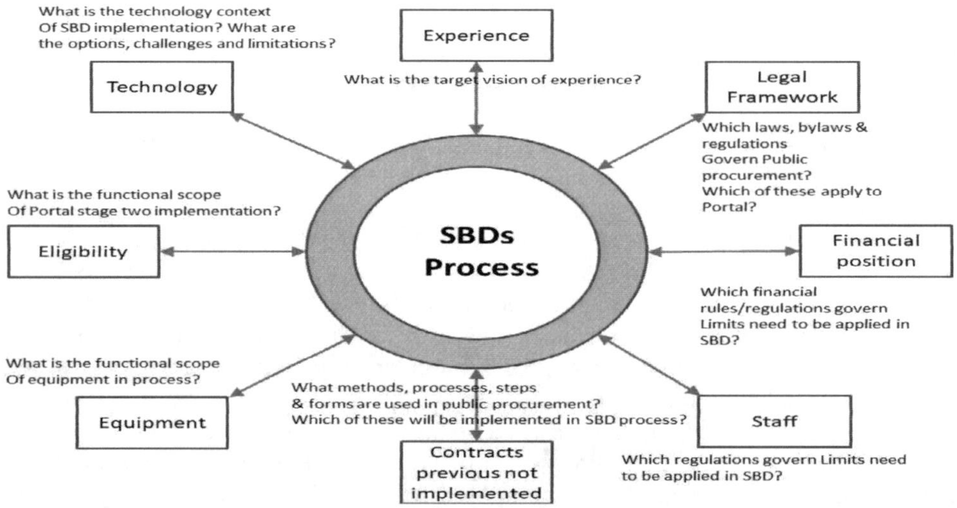

Figure 2. Procurement system, SBD Criteria of influence choice according SBDs.

5.1. Methodology

The information utilized for this investigation was gathered by two questionnaires, the first to determine important level factors criteria of SBDs and second of impact SBDs on the private sector to implement the projects, all around organized polls which were self-controlled to development experts engaged with the use of different acquisition frameworks in Iraqi provinces, Iraqi Contractors Federation was picked for the examination on account of its trademark as the real center point of local contracts for development exercises in Iraq, gathering and tabulating the answers to determine the relative importance index (RII) of all factors according to on the ranking given by the private sector, after the ranking, conclusions were created and support recommendations according to the findings of the study.

5.2. Assessment Factors that Influencing Decision for SBDs

The table of the factors identified by the author was the vision to be significant and comprehensive. it was the point of view of the author that the presented factors in this study were suitable in the context of Iraq. The survey adopted influencing factors that set that decision to bid from the survey of the developed questionnaire. All factors were gathered under six different groups, the first group is the "Financial Position which include Previous Financial Performance, Annual Revenues Average, Financial Resources", the second group is the "Experience which includes General Experience, Specialized Experience, Specialized Experience in main activities", the third group is "Previous Not implemented Contracts which included Previous Not implemented Contracts, Suspended Claims", the fourth group is the "staff leader characteristics", the fifth group is the Legal framework which included Nationality, Conflict of Interests, Disqualification by the Employer, Companies Owned by Government, Disqualification issued by the United Nations" and the sixth group is "Equipment that demonstrates his ownership or possession". A survey of a questionnaire that was distributed including these factors to contractors, with various classifications in the 15 Provinces in Iraq. to determine the grade of these factors based on classification given by the contractors, a technique relative importance index (RII) was adopted [29–30].

The formula used to calculate the RII for each factor is shown in Eq.1.

$$\text{RII}(\%) = \frac{5(n5) + 4(n4) + 3(n3) + 2(n2) + n1}{5n} * 100 \quad (1)$$

Where N denotes the overall number of responses received, n1, n2, n3, n4, and n5 state the respondent's number who selected 5, representing “extremely important”; 4, representing “highly important”; 3, representing “important”; respectively 2, representing “slightly important”; and 1, representing “not important”. The size of the model was specified for contractors classified according the first (the company must have a capital of (50000000000) five billions of IQ dinars and the company has successfully completed one contracting business Its cost is not less than (150000000000) fifteen billion IQ dinars), second (the company must have a capital of (40000000000) four billions of IQ dinars and the company has successfully completed one contracting business Its cost is not less than (120000000000) twelve billion IQ dinars), third (the company must have a capital of (30000000000) three billions of IQ dinars and the company has successfully completed one contracting business Its cost is not less than (80000000000) eight billion IQ dinars), fourth (the company must have a capital of (20000000000) two billions of IQ dinars and the company has successfully completed one contracting business of cost not less than (50000000000) fifth billion IQ dinars), fifth (the company must have a capital of (10000000000) one billions of IQ dinars and the company has successfully completed one contracting business of cost not less than (30000000000) three billion IQ dinars). the appropriate sample size was computed to be 42 contractors. A total of 100 responses were obtained.

5.3. Sample Size and Profile of the Respondents

A sum of 120 (one hundred twenty) polls were circulated and managed to the focused-on experts utilizing conclusive advantageous inspecting system out of which 100 numbers were returned and classified as reasonable for examination. The questionnaire distributed to the private sector (contractors) included three sections, the first section was allocated to general information to include all the data of the sample of the study, as well as the data of the subcontractors (company name, nationality of the company, the name of the ministry and the province, and general experience). The second section focuses on the general experience in constructions through the introduction of eight detailed questions in this field. The third and final section focuses on the specialized expertise in the SBDs. 37 detailed questions were asked, some of which require closed answers within three options (yes, no, unlike) the rest of the questions that require detailed answers according to the respondent's information. These questions were asked to obtain detailed data and at the same time to ascertain other things that may not be understood by the respondent in closed questions. In all measurement questionnaires, the triple scale was used exactly as in the public sector questionnaires for achieving a degree of homogeneity and thus the possibility of conducting a comparison between the two sectors. Each option was given a point and gave the research to the responses of the respondents in the surveyed sectors (yes, no, unlike), (3, 2, 1) respectively, with a response (3-1), and a hypothetical mean of (2) as shown in Table 1.

Table 1. Distribution and collection of questionnaires–private sector.

Target Company	Distributed	Collected
Class 1	40	36
Class 2	8	2
Class 3	23	20
Class 4	14	11
Class 5	35	31
Total	120	100

A reliability analysis was conducted to test the internal consistency of the study variable data using the Cronbach's alpha method based on internal consistency. Despite the data reliability, Cronbach's alpha method is also vital for internal validity incorrectly interpreting the relationships between Based on standard coefficient of Cronbach's alpha, the maximum value of reliability coefficient is 1.00 were for the output with a reliability coefficient of less than 0.6; the questionnaire in data collection is considered as not reliable and thus should be corrected or eliminated from the data collection. It is clear from the results in Table 2 that all stability coefficients of the research variables are very highly accepted in engineering and technical terms. The above result confirms the validity and stability of the data collection scale, as shown in Tables 2 and 3.

Table 2. Rang of reliability and its coefficient for Cronbach's alpha.

Coefficient of Cronbach's alpha	Reliability level
1.00	Very high
0.80 – 0.99	High
0.60 - 0.79	Moderate
Less than 0.59	Low

Table 3. Survey coefficient for Cronbach's alpha (researcher).

Cronbach's Alpha	Cronbach's Alpha Based on Standardized Items	No. of Items	Sector
0.996	0.970	171	Public
0.998	0.978	177	Private

6. Analysis and Finding

Table 4 shows the ranking of the 18 factors, arranged based on their respective group. Also, show the overall ranking of the factors. The three most important factors based on the respondents were "Annual Revenues Average", "Specialized Experience", and "General Experience". The least factors had in the four group "Legal framework". It is interesting to note that these factors primarily come under the Financial position. In other words, the most important factors influencing the private sector bidding decisions on SBDs. It can be consummated from the above results that private sector decisions to SBDs bid or not to bid are strongly influenced by the degree of criteria requirements especially the Annual Revenues Average as shown in Table 4.

Table 4. Factors arranged according to their respective ranks.

No.	Category	Factor	RII	OR (N)
1	Financial position	Previous Financial Performance	75.2	6
2		Annual Revenues Average	92.3	1
3		Financial Resources	78.8	4
4	Experience	General Experience	81.7	3
5		Specialized Experience	83.2	2
6		Experience in main activities	69.4	8
7	Previous Not implemented Contracts	Not implemented Contracts	67.1	9
8		Suspended Claims	63.5	11
9		Current commitments of projects	70.7	7
10	Legal framework	Nationality	22.4	18
11		Conflict of Interests	27.8	15
12		Disqualification by the Employer	26.3	16
13		Companies Owned by Government	61.6	13
14		Disqualification issued by the United Nations	24.9	17
15	Equipment	Ownership	62.3	12
16		possession	41.1	14
17	Leader Staff	Project manager	77.8	5
18		Other staff	64.6	10

7. Served of Private Sector

7.1. First Stage
The data presented in the questionnaire will be presented and analyzed. The opinions and responses of the sample surveyed by the study population represented by (contractors, subcontractors' representative by the engineers, legal and administrative staff). This sample was selected for research and analyzing its variables represented the private sector for construction projects in the Iraqi government, and the contractors dealing with the new contracting system (standard bidding documents) issued by the MoP, through which the referral of the construction project for companies which have technical, legal and financial criteria. In this stage of the analysis will be addressed the second phase of the private sector.

7.2. Second Stage
Represented by Iraqi contracting companies (private sector), Most local companies do not have the culture of standard bidding documents or sufficient knowledge to prepare a tender that matches the requirements of government agencies. Most companies cannot win a tender according to this new system. The obstacles of the private sector to work in this system through the identification of construction companies in Iraq for most of Iraq's provinces to study the most important obstacles and facilitate them to work as a permanent method for the private sector. For construction companies in Iraq regarding all provinces except Kurdistan Region Governorate KRG who Deale with the most important obstacles and facilitate the work as a permanent method for the private sector.

Where the local companies in the case of understanding the work on these documents can work in all countries of the world because it has an international procedure and is applied in most countries of the world. Moreover, the comprehensiveness of the study for most of the provinces of Iraq for reaching the goal of the research will give a strategic extent to the results of the research. For this purpose, the researcher used the triple Likert method, which is distributed from the highest weight, which is given (3) degrees to represent the answer field (large extent), to its lowest weight, which is given (1) With a single weight of (2) to represent the answer fields (medium). The frequency distribution tables for each user to use this type of table for statistical analysis to obtain the arithmetic mean, standard deviation, coefficient of variance CV. On the other side, the researcher adopted a satisfactory mean of (2) criterion to measure and evaluate the score obtained through the responses of the sample members of the study within the verbal estimate of the weights of the questionnaire form three, mathematical mean (2) is the average weights of the scale, After the statistical analysis and processing of the variables of the study of the public sector, 30 questions were designed for this purpose. The arithmetic mean, the standard deviations, and the coefficient variables were reached at the level of each question Mentioned.

8. Funding
More attention to the SBDs as well as the study of tender bid and contracts to avoid consequences and confusion that lead to financial obligations in implementation of standard bidding documents, the majority of local contractors do not use model techniques for contingencies' estimating while preparing their tenders according to the SBDs and no computerized tools use to manage the responsibilities. More attention needs to be paid to the SBDs as well as the study of tender (bid) and contracts to avoid sequences and confusion that lead to financial obligations in the implementation process of standard bidding documents. The majority of local contractors do not use model techniques for contingencies' estimating while preparing their tenders according to the SBDs and no computerized tools use to manage the responsibilities. In general, there is no experience in employment or commitment, evaluating and managing responsibilities of all sections in SBDs and Salary and wages paid by the contracting companies shall not be an obstacle to selection leadership, administrative, scientific and cognitive. There is a statistically significant relationship between (Analysis the Implementation of Standard Bidding Documents and their Impact on Contractors and Contracting Entities) in the public and private sectors. To arrive at a knowledge and determine the degree of strength of the correlation between the investigated variables, the statistic (t) will be used. The following are the most important results of the research according to the statistical analysis of the test hypothesis of the research mentioned. The Republic of

Iraq developed 19 main sets of standard bidding documents which standardize issued by the Ministry of Planning as shown in Table 5.

Table 5. Frequency distribution, arithmetic mean, standard deviation, and percentage of private sector.

Q	Large Extent	%	Medium Degree	%	Small Degree	%	Arithmetic mean	Standard Deviation	Coefficient of variation
1	93	0.93	2	0.02	5	0.05	1.74	0.88	0.50
2	90	0.9	2	0.02	8	0.08	1.81	0.67	0.38
3	68	0.68	18	0.18	14	0.14	1.87	0.92	0.50
4	70	0.7	24	0.24	6	0.06	1.96	0.68	0.35
5	78	0.78	18	0.18	4	0.04	1.93	0.87	0.86
6	78	0.78	10	0.1	12	0.12	2.00	0.77	0.41
7	52	0.52	21	0.21	27	0.27	2.11	0.77	0.36
8	20	0.2	77	0.77	3	0.03	2.08	0.83	0.42
9	57	0.57	41	0.41	2	0.02	1.93	0.87	0.50
10	38	0.38	21	0.21	41	0.41	1.96	0.51	0.26
11	23	0.23	56	0.56	21	0.21	1.88	0.89	0.47
12	32	0.32	13	0.13	55	0.55	1.99	0.83	0.42
13	30	0.3	19	0.19	51	0.51	2.02	0.80	0.40
14	27	0.27	24	0.24	49	0.49	1.69	0.86	0.51
15	22	0.22	61	0.61	17	0.17	1.76	0.85	0.48
16	27	0.27	22	0.22	51	0.51	2.05	0.63	0.31
17	26	0.26	16	0.16	58	0.58	1.78	0.85	0.48
18	33	0.33	36	0.36	31	0.31	1.79	0.88	0.49
19	34	0.34	31	0.31	35	0.35	1.77	0.91	0.53
20	34	0.34	20	0.2	46	0.46	2.02	0.67	0.33
21	11	0.11	74	0.74	15	0.15	1.97	0.90	0.45
22	34	0.34	25	0.25	41	0.41	2.55	0.54	0.21
23	39	0.39	30	0.3	31	0.31	2.17	0.44	0.19
24	35	0.35	40	0.4	25	0.25	2.25	0.85	0.38
25	34	0.34	33	0.33	33	0.33	2.66	0.68	0.26
26	34	0.34	25	0.25	41	0.41	2.74	0.52	0.19
27	21	0.21	54	0.54	25	0.25	2.64	0.59	0.23
28	37	0.37	13	0.13	50	0.5	2.53	0.73	0.28
29	15	0.15	51	0.51	34	0.34	2.82	0.55	0.19
30	29	0.29	16	0.16	55	0.55	2.88	0.45	0.15
					Overall Index				
	0.407		0.298		0.295		2.11	0.35	0.17

9. Research Limitations

There are multi limitation should be to be mentioned for a logical interpretation of the study findings. First, the majority of private-sector respondents were from companies under classification (1) are specialized in executing the infrastructure projects, in addition to Iraq Contractors Federation. Thus, the finding of this study is according to their practice with contractual entities; based on future studies may cover or consider the risk of SBDs during the implementation of projects. Second, due to the Shortage in funds for infrastructure projects after the decline in oil prices since 2014, the Government of Iraq reduced financing infrastructure projects so this procedure of SBDs simply implementing and shortage number during 2014-2018. Third, SBDs is newly procedure practiced in Iraq, and there is restricted experience with it.

Acknowledgments
This study is conducted with a great help, cooperation and support of the Ministry of Planning, Iraqi Contractors Federation and Ministry of Construction and Housing and Public Municipalities.

References

[1] E.S. Lesjo, P.A. Wondimu, and O. Lædre, Best Value Procurement from a Contractor Point of View, In Proc. 27th Annual Conference of the International Group for Lean Construction (IGLC). Dublin, Ireland, 3-5 Jul 2019 121-132

[2] A. Brekkhus, The construction industry has never had such a large share of GDP, 2017, http://www.bygg.no/article/1310113. (11.15, 2018).

[3] Improving Transparency within Government Procurement Procedures in Iraq OECD Benchmark Report https://www.oecd.org/gov/ethics/44736006.pdf

[4] T.C.L. Alves, C. Milberg, and K.D.Walsh, Exploring lean construction practice, research, and education, Engineering, Construction and Architectural Management, 19(5) 2012 512-525.

[5] L.T. Van, N.M. Sang, and N.T. Viet, A Conceptual Model of Delay Factors affecting Government Construction Projects, ARPN Journal of Science and Technology, 5(2) 2015 92-100.

[6] T.H. Khan, K.H. Road, R. Cantt, and A.Q. Khan, Effects of Lowest Bidding Bid Awarding System In Public Sector Construction Projects In Pakistan, 5(3), 2015 132–147.

[7] L.Y. Shen, W.S. Lu, Q.P. Shen, and H. Li, A computer-aided decision support system for assessing a contractor's competitiveness, Automation in Construction, 12 2003 577-587.

[8] W.S. Lu, L.Y. Shen, and M.C.H. Yam, Critical success factors for competitiveness of contractors: China study", Journal of Construction Engineering and Management, 134(12) 2008 972-982.

[9] S.K. Shrestha and C. Campus, Average Bid Method – An Alternative to Low Bid Method in Public Sector Construction Procurement in Nepal, 10(1) 2014 125–129.

[10] M. Williamson, O.D. Wilson, and R.M. Skitmore, Client abuses of the competitive tendering system: Some general principle and a case study, Journal of Construction Research 5(1) 2004 61-74.

[11] L.Y. Shen, W.S. Lu, and C.H.M. Yam, Contractor Key Competitiveness Indicators (KCIs): a China study", Journal of Construction Engineering and Management (ASCE), 132(4) 2006 416–424.

[12] K. Momaya and K. Selby, International competitiveness of the Canadian construction industry: a comparison with Japan and the United States, Canadian Journal of Civil Engineering, 25 1998 640-652.

[13] W.K. Fu, D.S. Drew, and H.P. Lo, Competitiveness of experienced and inexperienced contractors in bidding, ASCE Journal of Construction Engineering and Management, 129(4) 2003 388–395.

[14] J. Wells, The construction industry: Issues and strategies in developing countries: The World Bank, IBRD, Washington, 1984, (1985): 335-338.

[15] J. Sparks, A methodology for Estimating the Level of aggressiveness in Competitive Bidding Market, MSc, Blacksburg, Virginia 1999.

[16] D.K.H. Chua and D. Li, Key factors in bid reasoning model, Journal of Construction Engineering and Management, 126(5) 2000 349–357.

[17] A. Shash, Factors considered in tendering decisions by top UK contractors, Journal of Construction Management and Economics, 11 1993 111-118.

[18] N.B.H. Benjamin, Competitive Bidding: The Probability of winning, Journal of the Construction Division, ASCE, 98 (CO2), 1972 313-329.

[19] B.M. Samuels, Construction Law, Prentice-Hall, Inc. A Simon & Schuster Company, Englewood Cliffs, New Jersey 07632 (1996).

[20] M.M. Kumaraswamy, Industry development through creative project packaging and integrated management. Engineering, Construction and Architectural Management, 5(3) 1998 228-237.

[21] C. Thomsen, J. Darrington, D. Dunne, and W. Lichtig, Managing integrated project delivery, CMAA, McLean, VA (2010).

[22] J. Elfving, I. Tommelein I. and G. Ballard, Consequences of Competitive Bidding in Project-Based Production, Journal of Purchasing & Supply Management, 11 2005 173-181.

[23] P.E.D. Love, M.R. Skitmore, and G. Earl, Selecting an appropriate procurement method for the construction process: an empirical study, Construction Management and Economics 16 1998 221-233.
[24] National Economic Development Office (NEDO), Think About Building, NEDC, London, 1985.
[25] R.M. Skitmore and D.E. Marsden "Which procurement system? Towards a universal procurement selection technique", Construction Management and Economics, 6, 71Ð89, 1988.
[26] S. Singh, "Selection of appropriate project delivery system for building construction projects", Proc. CIB W-90 International Symposium on Building Economics and Construction Management, Sydney, Australia, pp. 469Ð480, 1990.
[27] F.Y.Y. Ling, S. L. Chan, E. Chong and L.P. Ee, Predicting performance of Design-Build and Design-Bid-Build projects. Journal of Construction Engineering and Management, 130 2004 75-83.
[28] [28] A.S. Oyegoke, Construction Management Contracting Systems in the UK and US Practices: Consulting and Contracting Perspectives, Journal of the Nigerian Institute of Quantity Surveyors, 34 2001 22-34.
[29] [29] S. Kometa, P. Olomolaiye, and F. Harris, Attributes of UK construction clients influencing project consultants'performance, Constr. Manag. Econ. 12 1994 433–443.
[30] [30] M. Kumaraswamy and D. Chan, Determinants of construction duration, Constr. Manag. Econ. 13 1995 209–217.

The Law of Decentralization and Its Impact on Construction Projects

Emad Essa Abed[1, a *] and Sawsan Rasheed Mohammed[1, b]

[1]Department of Civil Engineering, University of Baghdad, Baghdad, Iraq.
[a]emadessa2013@gmail.com, [b]civilsawsan2@gmail.com

Abstract. Decentralization is a democratic practice in which the Government allows ministries, governorates and districts to delegate their powers within the law. This is done in two parts: the first is the decentralization of the operational budget and the second the decentralization of the investment budget. This study aims to draw up a strategy and plan for decentralization in the governorates after the transfer of powers from the ministries. The researcher seeks to discover the success or failure of applying this idea and the importance of modern construction projects in the study of decentralization in these ministries in selected governorates. Transfer of powers This research is very important through the selection and timing, and is important in that this study is the first to target the impact of decentralization on the performance of construction projects. Despite the existence of previous studies and research on administrative decentralization, it did not address the impact of decentralization and its role in the process of implementation and completion of projects. The research methodology was divided into two parts, theoretical work and fieldwork. The researcher concluded that there is a need for decentralization in particularly in the governorates far from the central government. However, these conditions are subject to limitations and restrictions. They can only be applied after studying the presence of human cadres and providing expertise that deals with post-transfer jobs. The absence of Iraqi government legislation in this area and even if we assume there are laws, they have not been applied so far about the system of decentralization in Iraq in its entirety.

Keywords: Decentralization, investment budget, construction, problems.

1. Introduction

Over the past years, there has been increasing interest in the issue of decentralization with its political and administrative dimensions economic and financial. This concern was part of the trend towards expanding the participation of citizens and their role in the governance process, reducing the role of the State in production and direct management of its institutions, and giving the private sector and civil society institutions a greater role in the development process [1]. World Bank on World Development under many addresses such as decentralization and rethinking, making the country closer to people and transition to localization [2,3] there has also been interest in decentralization in the adoption of many countries in the world such as the Russian Federation and countries Eastern Europe, Qatar, and Bahrain that have taken a more or less decentralized approach to Decentralization at present, characterized by a view of decentralization. Economists have seen it They were influenced by neo-liberal ideas as a means of transferring power from the central state.

Advocates of political pluralism as a means of giving interest groups a space of freedom and the ability to organize and compete. The leaders of some dictatorships in Asian countries Africa is a democracy rather than a democracy at the national level and a safe way to gain more Legitimacy and support from units at lower levels, while Democratic politicians in developing countries saw them as a way to make the government more responsive to needs and priorities Local [4]. The purpose of this study is to develop a strategy and plan for decentralization in the provinces to transfer powers from the ministries and to address the large shortcomings in the performance of the work of the provincial councils and apply this idea correctly helps maintain provincial councils and gives them the freedom to provide services and enable them to complete projects and increase levels of construction and reconstruction. The current study consists of two parts: Part I includes a theoretical study and Part II includes fieldwork. The field part of the research will conduct the field survey through preliminary interviews, questionnaires, interviewing experts, and building the proposed system for obtaining actual data on decentralization. The questionnaire was used as a means of gathering information and measuring the acceptability of decentralization in government institutions. The questionnaires were distributed to 100 persons. Eleven of the questionnaires were excluded due to technical errors and the adoption of 89 questionnaires. Analysis of the statistics by the statistical program,

2. Research Justification

The problem that seeks to address this message is to find solutions to obstacles that oppose the implementation of decentralization in Iraq, specifically, which is concerned with the financing and management of construction projects and the Iraqi government has not succeeded in decentralizing the required level and there is overlap in the roles between political and administrative decentralization and there is a sign of problems constrained by clear completion of projects during the implementation phase. The idea of decentralization in Iraqi institutions. Moreover, most of the research and studies on decentralization did not mention the importance of decentralization in the management of construction projects and that their application would address many of the project problems. To deal with decentralization in a good way, it is necessary to know the problem or obstacle to its application. The study aims to achieve several objectives, the most important of which are the following: -

- Assess the contribution of local management systems to improving community conditions. in the provinces to take their place and provide services directly to citizens.
- To know the extent of convergence between the systems of local administration in the provinces and how to build a system does not Centralized among them.
- Review the new concepts of administrative decentralization that enhance the capacity of the units, in line with technological progress and management values of partnership and transparency and accountability.
- Increasing the disclosure of the reality of administrative decentralization systems in ministries. The study will conduct a comparative analysis between the decentralized ministries and the work of their projects. the study is also trying to provide a futuristic vision for the decentralized systems of other ministries, through organizational structures, and building a model in local administration compatible with environmental factors and conditions facing local administration in all ministries.

3. Types or Forms of Decentralization

Three types of decentralization can be distinguished as mine [5].

3.1. Decentralization Policy

Political decentralization aims to give more elected authorities or elected representatives making decisions. [6] It assumes that decisions made through broad participation will be better and more connected to the diverse interests of society than those that are carried out by the Government The electoral system also allows citizens to know better about their representative's Politicians. Collins also allows elected members and employees to know better about the needs and desires of their constituencies [7,8].

3.2. Administrative Decentralization

Administrative decentralization refers to the distribution of administrative obligations between the central government (authority) in the capital, and the local bodies or autonomous bodies that have an independent legal personality exercising their jurisdiction under the supervision and control of the authority [9]. Decision-making to administrative bodies is not the work of the central authority, and these independent bodies must not follow the central authority as a general asset, but are subject to some sort of prevention and supervision by the central authority through the so-called administrative trusteeship [10]. The standard between administrative centralization and administrative decentralization is the unity or multiplicity of administrative authority. If there is one administrative authority, there will be a centralized system, but if the administrative powers are very diverse, there are decentralized authorities with a juristic personality and independence in exercising their powers under the supervision of the central authority [9,11].

3.3. Financial Decentralization

Financial responsibility is an essential element of decentralization, if local governments and private organizations are implemented Decentralized functions effectively, Perhaps the most significant criticism of the decentralized system is that its application in the administrative environment results in the phenomenon of wasting public expenditures. Recognition of local bodies and public utilities of all kinds will be followed by financial independence [12]. Contrary to a centralized system that is characterized by economy of expenditure by reducing the number of disbursements who are the representatives of the central authority as the independence of the region legally as well as the independence of the facility requires the recognition of financial autonomy from the state as previously stated, which ultimately transfers the authority to order disbursement a high degree pyramid to many other levels, such as (the state, municipality, institution) [13]. It is supposed to result from the multiplicity of the two-drainage negative phenomenon is the exaggeration or excessive exchange of public expenditures [14]. However, this phenomenon can be mitigated by triggering observation tools, whether exercised by the trusteeship authority or the proficient bodies of a financial nature [15].

4. Sample of Research

4.1. Sample survey

The purpose of it to verify the extent of the sample members' understanding of the articles and instructions of the scale they have according to the time taken to answer and identify the difficulties facing the respondent, the survey sample is randomly selected, which represented by distributing 30 forms to some ministries.

4.2. Sample Statistical Analysis

The sample of the statistical analysis was randomly selected in classes form with an equal distribution of 50 forms distributed according to the three governorates (Diwaniya, Najaf, Babel). This number is appropriate to represent the society study and the personal information of persons and experts is represented in the questionnaire by Gender, age, academic achievement, show Figs. 1 to 3. The ministry and governorate, department name-mail, mobile number, and position.

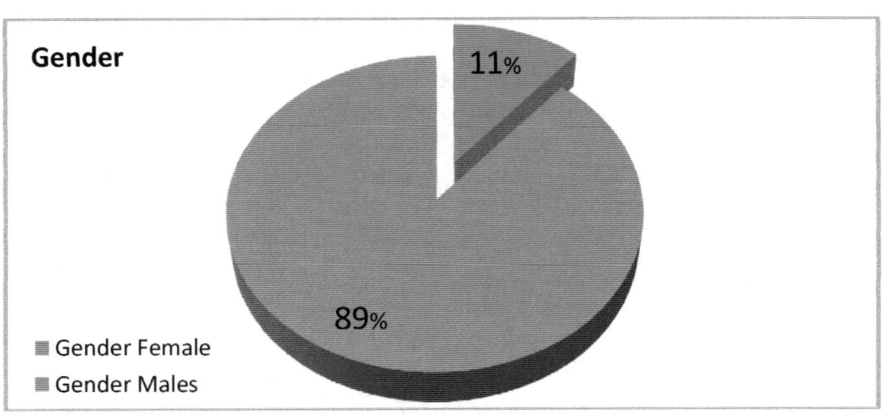

Figure 1. Percentage of study sample gender.

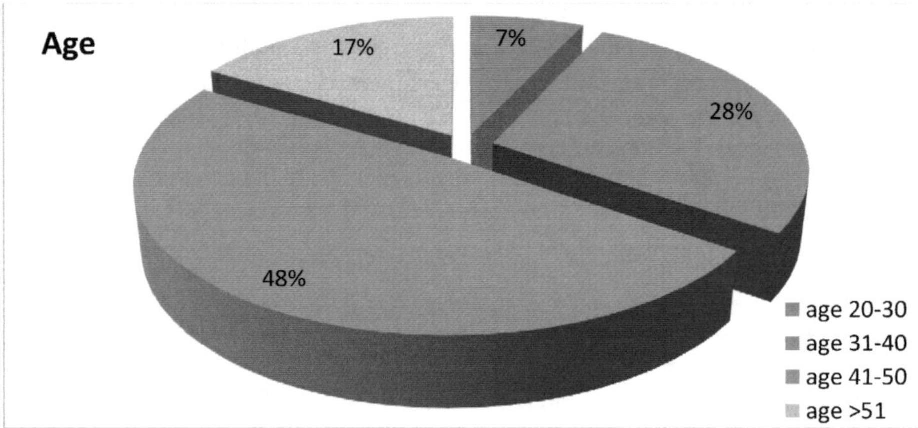

Figure 2. Percentage of study sample of age.

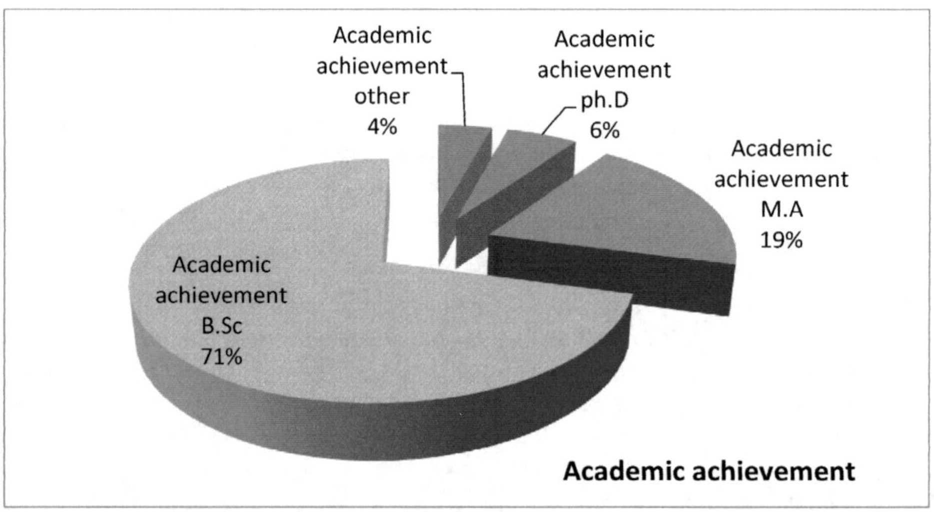

Figure 3. Percentage of study sample of academic achievement.

5. Analysis of The Variables of Questionnaire

The first question in Table 1 shows the existence of experience in the field of decentralization, whether or not in the selected staff. The answers have come most with the availability of experience and the answers to the second question have come with the experience that the department or section has for the rules to execute and implement the decentralization in Iraq, the remaining questions also have come with the large answer and most questionable for the questions third, fourth and fifth were assuring the experience in the field of decentralization, the sixth question also assured that its application is complex, but it reduces the routine, so the table included the balanced arithmetic averages ($\overline{x}$) the standard deviations (Si), and the Coefficients of variation (C.V) on the level of the research sample.

It is clear from the results of Table 1 that there is a rise in the general weighted arithmetic mean of the dimension (decentralization and administrative authorization), since reached (1.2), which means the prevalence of the knowledge and the idea of decentralization relatively to the level of the surveyed with a standard deviation of (0.33) and Coefficient of variation of (0.28%) expressed significant dispersion in the sample answers of the included in the research. As a result, the general weighted arithmetic mean of the experience in the decentralization field is greater than the satisfactory mean of (2). This means that after the (idea of decentralization) is clear specifically for a sample of respondents according to their point of view, and this because of visibility of the study sample individuals about the importance of adoption (decentralization in the Governorates).

Table 1. of the second part information of experience (Researcher).

Questions	Response measurement			Arithmetic mean, $\bar{x}$	Standard deviation Si	Coefficient of variant ion C.V.
	To a large extent	To a medium degree	To a small degree			
1	0	3	69	1.04	0.199	0.191
2	0	4	68	1.055	0.229	0.217
3	0	6	66	1.083	0.276	0.255
4	1	2	69	1.055	0.283	0.268
5	0	1	71	1.013	0.117	0.115
6	1	13	58	1.208	0.438	0.363
7	4	64	4	1.944	0.328	0.169
The general mean, the general standard deviation and the coefficient of variation				1.2	0.33	0.28

6. Summary of the Questionnaire

- Grant administrative and financial space to departments and institutions to take their decisions.
- To reduce the dominance of ministries and raise the level of efficiency of decisions taken where they are closer to reality.
- Create a flexible administrative and financial environment in the provinces and departments.
- Ease of conducting administrative procedures and the speed of execution of the work without the need for lengthy communications and restricted by the previous routine.
- Decentralization implementation will encourage the execution of projects that serve the need of the governorate as local administration is more aware of its projects and their needs that touch the lives of the citizen.
- Speed in the achievement of projects and provide services to citizens directly where there is freedom in the distribution of funds according to need and priority.
- Determine the aspects of exchange without reference to finance through the preparation of the local budget.
- Ease of achievement of transactions that serve the citizen.
- Simplifying contractual procedures for projects.
- Elimination of time waste between the governorates and the center, especially those far from them.
- Allow ministries to take their basic role and practice their basic works, for example, the Ministry of Education is concerned with matters of teaching and curriculum instead of building schools and educational institutions.

7. Conclusions

In its practical side, the research reached a set of conclusions on which a set of conclusions were based, as follows: The research leads to the failure to prove the first main hypothesis and its subsidiary hypotheses, i.e., there is no significant statistical relationship between the dimensions of decentralization and the exclusion of construction projects, which proves that it did not adopt the dimensions of decentralization in the studied construction projects and attention to its basic dimensions, Construction workers to adopt these dimensions in the future. Also, it proved that there is a significant effect of the indicators of decentralization in the indicators of construction projects, reflecting the state of compatibility between them is that decisions are rational as long as the construction projects studied the dimensions of decentralization in the work of art.

References

[1] C.R. Kothari, Research Methodology – Methods and Techniques, Second Revised Edition, New Age International (P) Limited, Publishers (2004).

[2] A.S. Hornby (2006). OXFORD Advanced Learner's Dictionary, New 7th Edition1- United Nations Development Program (UNDP), Program on Governance in the Arab Region (UNDP-POGAR), Geneva (2002)
[3] World Bank, Decentralization and State Economy, Electronic Reference Book on Decentralization and Local Governance. Program on Governance in the Arab Region 2005 UNDP, POGAR.
[4] K. Al-Munufi, Institutional Reform between Centralization and Decentralization, Cairo University, Center for Public Administration Studies and Consultations, March 2001.
[5] Department of Studies, The Concept of Decentralization in Educational Administration, Ramallah, Palestine, (2005).
[6] A.A. Al-Shaikhli, The relationship between the central government and local administration-a comparative study, Seminar on the relationship between central administration and local administration Arab Urban Development Institute, Beirut, September (2002).
[7] S. Al-Mubaidain, Centralism and Decentralization of the Local Administration Organization, Amman: Dar Al-Yazuri Publishing and Distribution, 2011.
[8] H.M. Hussein, Local Administration of the Approach, II, Algeria: Bureau of University Publications, 1982, p.6.
[9] F. Rababa, The Role of the Councils of Shared Services in Development, M.Sc. Thesis, University of Jordan, Amman, 1995.
[10] K. Al-Zoghbi, Administrative Law and its Applications in the Hashemite Kingdom of Jordan Amman: House of Culture, (1986).
[11] M.S. Al-Azawi, Views on Democracy, I (Amman: Dar Wael Publishing and Distribution, 2000), 77
[12] S.M. Abdel Wahab, Decentralization and Protected Governance II. Dar al-Jallal for publication and distribution Cairo. 2001.
[13] A.B. Abd Allah, Administrative Law, Alexandria Knowledge Establishment, 1993.
[14] N. Kanaan, Administrative Law, First Book, I1, Amman: House of Culture, 2002.
[15] M. Khalil, Contemporary Administration, Principles, Jobs 2005 Beirut.

Developing a Computer Program to Solve the Implementation Problems of Public Investment Projects in Iraq

Sanaa M. Almissary[1,a*] and Abbas M. Burhan[1,b]

[1]Civil Engineering Department, University of Baghdad, Baghdad, Iraq.
[a]mss.snaa2010@gmail.com, [b]abbas_alshemosi@yahoo.com

Abstract. The implementation of public investment projects in Iraq in terms of cost and time specified within the general investment budget and in terms of quality; requires analysis the obstacles facing the implementation of these projects and propose appropriate solutions to them through developing a proposed program to analyze and address the problems through project phases. The aim of research for developing a computer program comprises the identified problems as input to the program and the proposed treatments as output for it; to facilitate the work of the departments involved in projects in making decisions and to avoid repeat the previous mistakes. The researcher conducted personal interviews to identify (55) problems facing the public investment projects in Iraq grouped in (5) groups; legal and administrative, financial, technical, problems with documentary credits and other problems, conducting questionnaire survey to clarify the impact of these problems on the implementation process, using a statistical program (SPSS) to compute relative important index (RII) for them to identify the group that most impact on the implementation of public investment projects. A visual basic program with database (SQL) was utilized to analyze and treat the problems related to the subject of research. This program is equipped with the ability to withdraw reports on projects as (word, pdf, or excel). The results from research showed that there is (absences coordination between different service departments, delays in land acquisition procedures for projects, weakness of the possibility of most consultancy offices designed for projects, weakness of personals specialized in the preparation of technical and economic feasibility studies and inaccuracy of signed contracts with contractors…etc.).

Keywords: Investment projects, implementation, investment project, successful investment project, public investment project.

1. Introduction

The investment project can be well-defined as the optimum set of activities of investment based on sector and logical planning. It represented by a grouping of humans, material, and resources that lead to economic and social development [1]. Characteristic investment projects comprise the construction of new buildings, hospitals, roads, power plants, water tanks, and other infrastructure elements. Also it may include replacing the old facilities, rebuilding of existing facilities, acquirement of new facilities, or buying of equipment. Investment projects generally are large, non-repetitive expenses that comprise multi-year funding, have a useful life for more than five years, are based on a comprehensive requirements assessment, meet an important public purpose, and require public answerability for funding [2-5].

The expenses of the investment include: expenses to acquire the land, infrastructure expenditures, design and technical help expenditures (permits, approvals, designing, project review, technical support, and construction site supervisor), expenditures through the basic investment comprises (constructions and works of interference), costs with independent acquisitions, other expenditures such as the association of the construction site, comities, taxes, legal issues, financing expenses, others unexpected expenditures, operating expenditures and training the workers for utilization, recent technologies, court case, getting expertise [3-6]. Most of the public investment projects do not meet the anticipations of many clients. Some are delivered too late, at a greater cost, and do not meet approved quality standards. They might not be able to achieve the expected goals.

The government needs to do something because public resources are wasted [4]. For successful investment project, it is required a common concept to the problem and its explanation together with the clients, this concept include: Any investment project to be a success, it should be accurate, clear, and exact understanding of the constructing way, a complete analysis of the options in the problem to be addressed to solve or to meet the need [2].

2. Preliminary Interviews

To gather information about the problems of implementation weakness in the public investment projects of Iraq, personal interviews were conducted with a group of experts (eleven engineers), who have more than fifteen years, experience in investment projects. They are working in the field of design, planning, management, and execution of the investment projects. The results from these interviews identified (55) causative factors, which represent the origins of implementation weakness problems in the public investment projects under (five) groups as shown in table (1). The cause and effect diagram was helped the researcher in design the form of the questionnaire survey as shown in Fig. 1.

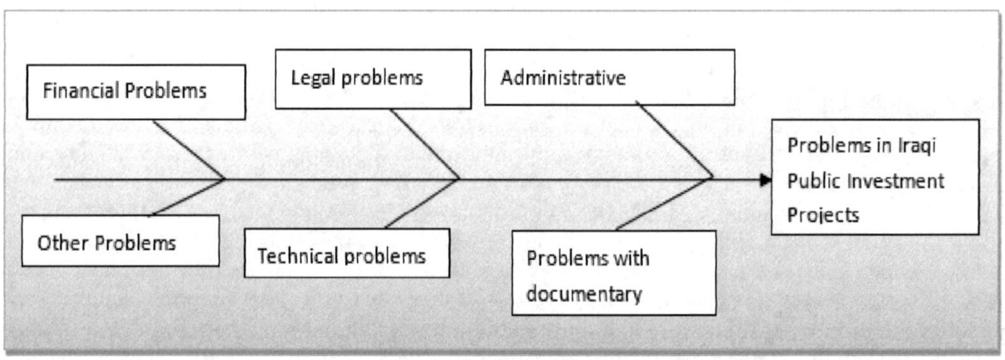

Figure 1. Cause and effect diagram for problems in Iraqi public investment projects.

3. Closed Questionnaire

The closed questionnaire is an important source of obtaining field information about weakness implementation problems in the public investment projects of Iraq. Sixty questionnaires were distributed to a group of experts in those departments and only Fifty-four were answered. The results of the questionnaire were analyzed using the statistical program SPSS and the following Tables 1 and 2 show the results.

Table 1. Relative important index (RII) for the implementation weakness causative factors.

Key Causative Factors	N	AM	S.D	α	Effect Level	RII
1-Administrative and legal problems						
1- Delaying ratification of the state budget	54	4.0377	0.96001	0.851	High	4.243
2- Lack of allocations allocated to projects	54	3.9811	0.72032	0.848	High	4.183
3- Lack of a specific mechanism for carrying out research projects	54	3.7358	0.73774	0.844	High	3.925
4- Exceeding the plots of land allocated to investment projects	54	3.8868	1.15460	0.873	Very high	4.084
5- Delays in land acquisition procedures for projects	54	3.8491	1.11625	0.845	High	4.044

Table 1. Continued.

Key Causative Factors	N	AM	S.D	α	Effect Level	RII
1-Administrative and legal problems						
6-The existence of disputes over the ownership of buildings and land between ministries and state departments	54	3.8679	1.07485	0.845	High	4.064
7-The difficulty of administrative procedures between the Ministries	54	3.9057	0.92537	0.843	High	4.104
8-Limited administrative powers granted to project managers.	54	3.7547	0.78215	0.843	High	3.945
9-Weak coordination between ministries when allocating or buying land plots	54	3.7358	0.76337	0.843	High	3.925
10 The lack of accuracy in the evaluation of companies executed	54	4.1509	0.74411	0.845	High	4.362
11-Delay in resolving disputes between the contractor and the employer	54	3.9434	0.74460	0.843	High	4.144
12-Inaccuracy of signed contracts with contractors	54	3.7170	0.94822	0.843	High	3.906
13-The large number of cases of forgery	54	3.4717	0.89020	0.842	High	3.648
14- The reluctance of some contractors to perform maintenance work	54	3.3208	0.80320	0.845	Medium	3.489
15 Lack of technical and financial efficiency of most companies	54	4.1698	0.75284	0.846	High	4.381
16- Re - announcement of tenders for more than one time	54	3.5660	0.84374	0.841	High	3.747
17- Weak knowledge of the people working on the implementation of projects with their laws and instructions	54	3.6415	0.68203	0.843	High	3.826
18- Lack of classified contractors in many governorates	54	3.3208	0.89386	0.841	Medium	3.489
19- The length of time in the tenders	54	3.6792	0.77889	0.843	Medium	3.866
20- Inadequate r project management in governorates	54	3.9623	0.64933	0.842	High	4.163
21- Poor implementation capacity of some implementing agencies	54	4.0566	0.84159	0.841	High	4.262
22- There is not existence coordination between different service departments.	54	3.9245	0.97762	0.840	High	4.124
23- Some contractors contracted with subcontractors.	54	4.0189	0.90915	0.840	High	4.223
24- Transfer more than one contract to one company	54	3.6792	0.99564	0.842	Medium	3.866
25- Executing companies ignore the period of delay penalties.	54	3.7925	0.84030	0.842	High	3.985
2- Financial problems						
26-The length of the procedures for the payment and receipt of payments by the ministries.	54	3.2075	0.71679	0.846	Medium	18.398
27-The inadequacy of the financial ceilings contained in the direct implementation instructions	54	3.3019	0.84546	0.845	Medium	18.939
28-The multiplicity of regulatory bodies (Financial Control, Integrity, Inspector General).	54	3.7170	0.71730	0.843	High	21.321
29-Lack of funds financed for executed projects.	54	3.5660	0.90955	0.845	High	20.454
30-Lack of proportion of the amounts allocated to the investment platform and the development program of regions of the state budget.	54	3.6415	0.87912	0.845	High	20.888
31-The failure of specialized agencies to open documentary credits	54	3.6038	0.81664	0.842	High	32.263

IOP Conf. Series: Materials Science and Engineering **901** (2020) 012029 doi:10.1088/1757-899X/901/1/012029

Table 1. Continued.

Key Causative Factors	N	AM	S.D	α	Effect Level	RII
3- Problems with documentary credits						
32-The length of procedures followed in opening documentary credits for the importation of materials, machinery, and equipment.	54	3.7925	0.81709	0.839	High	33.953
33-Slow shipping procedures, customs clearance, cargo insurance.	54	3.7736	0.84675	0.840	High	33.784
4- Technical problems						
34-The inability of some departments to prepare designs for their projects.	54	3.9811	0.77187	0.842	High	6.222
35-Lack of engineering staff with high experience in the implementation of projects.	54	4.1509	0.92811	0.839	High	6.488
36-The weakness of the possibility of most consultancy offices	54	3.8868	0.95395	0.842	High	6.075
37-Non-conformity of the structural materials to the specifications.	54	3.4717	0.97278	0.841	Very high	5.426
38- Old means of work and technology used by the local contractor.	54	3.9057	0.94593	0.840	High	6.104
39-Lack of number of modern construction laboratories.	54	3.1132	0.93357	0.844	Medium	4.866
40-The length of time required to examine the construction materials in laboratories.	54	4.000	5.53312	0.895	High	6.252
41-Lack of highly qualified companies (financial and technical)	54	4.0377	0.83118	0.845	High	6.311
42-Most of the projects lack technical and economic feasibility studies	54	4.2075	0.76858	0.842	Very high	6.576
43-Weakness of personals specialized in the preparation of technical and economic feasibility studies	54	4.0566	0.86414	0.843	High	6.34
44-Weak technical capabilities and practical experience of the personals working in the provinces.	54	3.9057	0.86077	0.840	High	6.105
45-The difficulty of guessing the prices of imported materials.	54	3.3962	1.02544	0.839	Medium	5.308
46-The majority of the materials involved in the work are not subject to the soil examination particulars, which leads to make an examination of the materials through experience, which leads to an error rate.	54	3.3774	0.92459	0.841	Medium	5.279
47-The delay in obtaining the results of the laboratory tests of the construction materials.	54	3.3019	0.86791	0.843	Medium	5.161
48-The lack of machines and specialized equipment	54	3.2264	0.93318	0.841	Medium	5.043
49-Weakness of procedures of the supervision committees on the projects.	54	3.8491	0.92811	0.840	High	6.016
50-Lack of accuracy in the preparation of statements and plans for projects	54	4.1132	0.77609	0.841	High	6.429
5- Other problems						
51-Tribal problems, especially in remote areas	54	3.8302	0.82592	0.842	High	21.573
52-The difficulty of lifting the excesses on the plots of land allocated for the projects	54	3.9057	0.86077	0.844	High	21.998

Table 1. Continued.

Key Causative Factors	N	AM	S.D	α	Effect Level	RII
5- Other problems						
53-The proliferation of security control points.	54	3.4906	0.93279	0.840	High	19.659
54-Water scarcity in general, so sometimes requires exceptional measures such as canceling projects or changing priorities of other projects	54	3.2075	1.00687	0.843	Medium	18.065
55- Security reasons.	54	3.3208	0.95623	0.843	Medium	18.703

Table 2. (RII) for the groups of causative factors.

Group of Factors	AM	RII
Administrative problems	95.1697	46.309
Technical problems	63.9811	31.133
Other problems	17.755	8.639
Financial problems	17.4339	8.483
Problems with documentary credits	11.1699	5.435

4. Building the Proposed System

In building the proposed system for improving implementation efficiency in the public investment projects, the three main phases were adopted, namely: the planning and design phase, the contracting phase, and the construction phase. For building the proposed system, the factors causing implementing weakness in public investment projects have been distributed to the main project phases and proposed appropriate procedures for these problems in each phase as shown in Fig. 2.

5. Developing the computer Program

The following are the most important specifications of the program: Visual basic (vb.net) program version 2012 was used, the database designed by (SQL) server 2012, this facilitates the preservation of project data and accessibility at a later date for audit purposes, the program is equipped with the ability to withdraw reports on projects, which facilitates the work of the departments involved in projects as (word, pdf or excel), the program's English language was used similar to the research language, and the program can be used by all parties to the project such as (client, contractor). When the user runs the program, the main window which represents the main page of the program will appear as shown in Fig. 3.

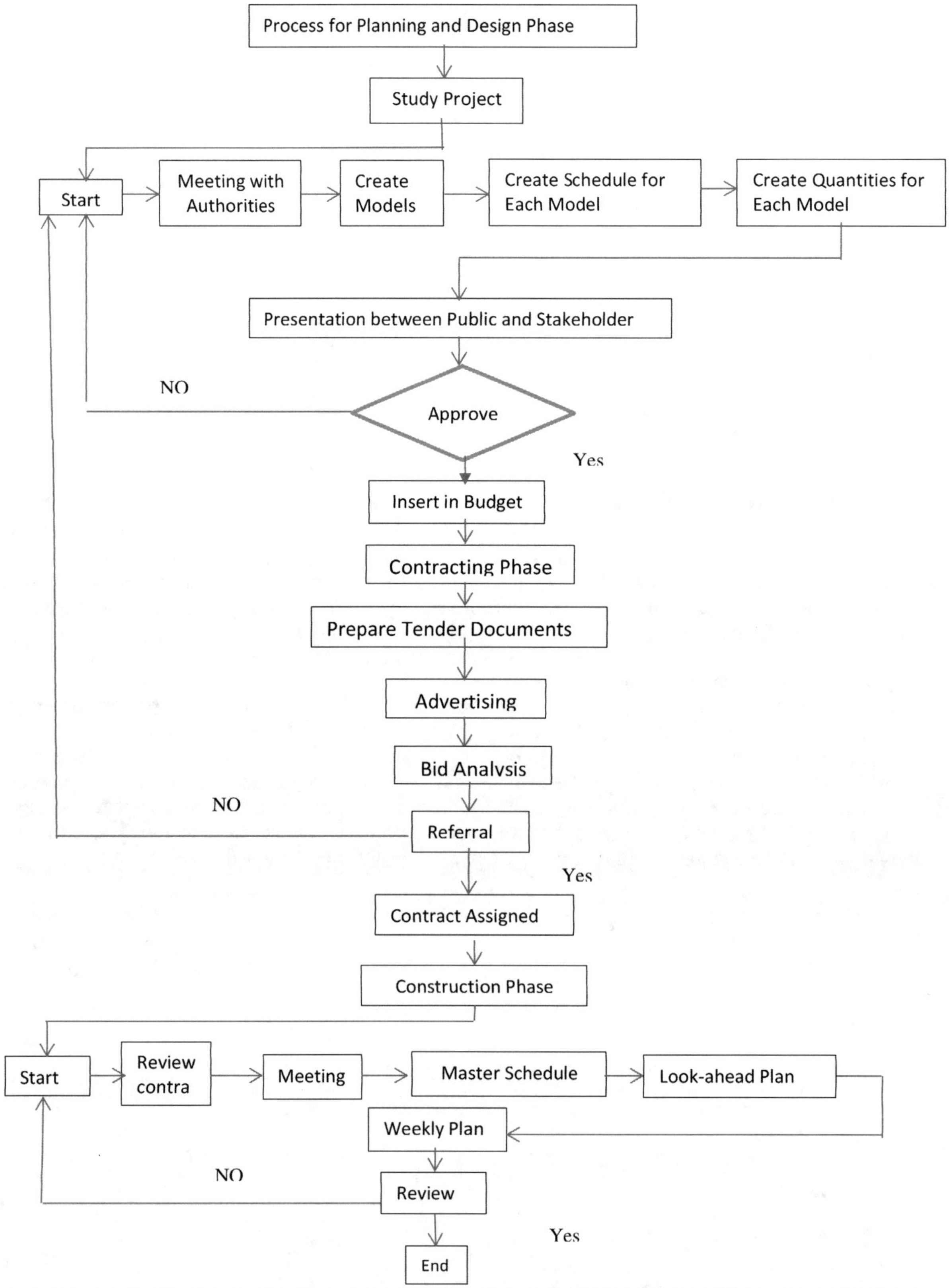

Figure 2. Flow chart of the proposed system.

Figure 3. The proposed system.

And when press log in appears the screen shown in Fig. 4.

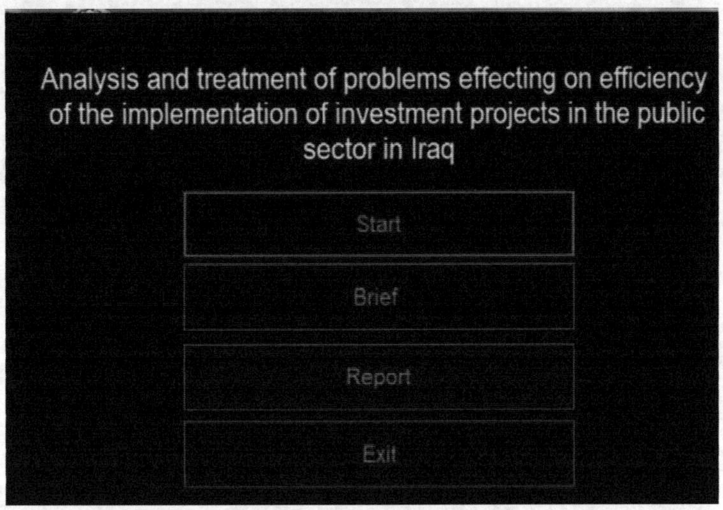

Figure 4. The beginning of the proposed.

When the brief option is pressed, the window appears as shown in Fig. 5.

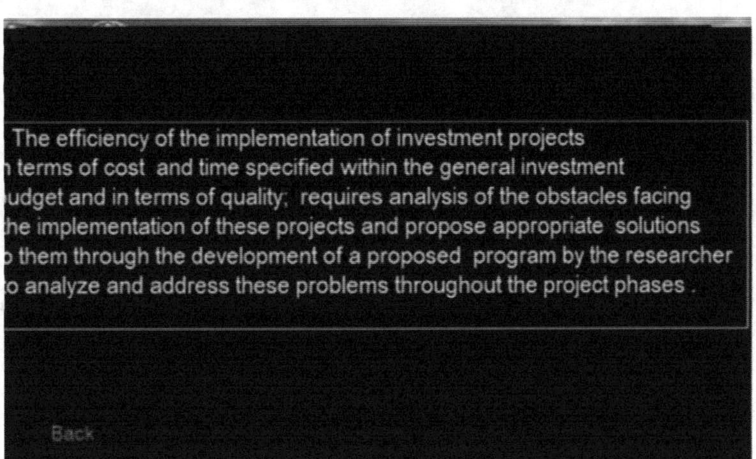

Figure 5. Brief of project.

When press on Back, return to previous figure and press on Start appears as shown in Fig. 6.

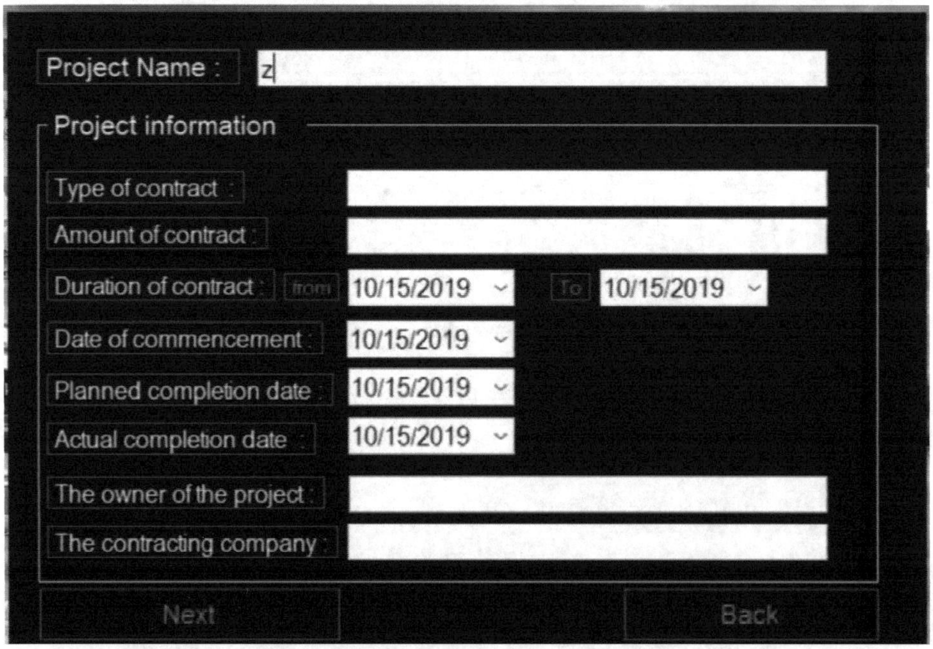

Figure 6. Input the name and Information of the current project.

When press on Next, the program asks about the requirements of any project as shown in Fig. 7.

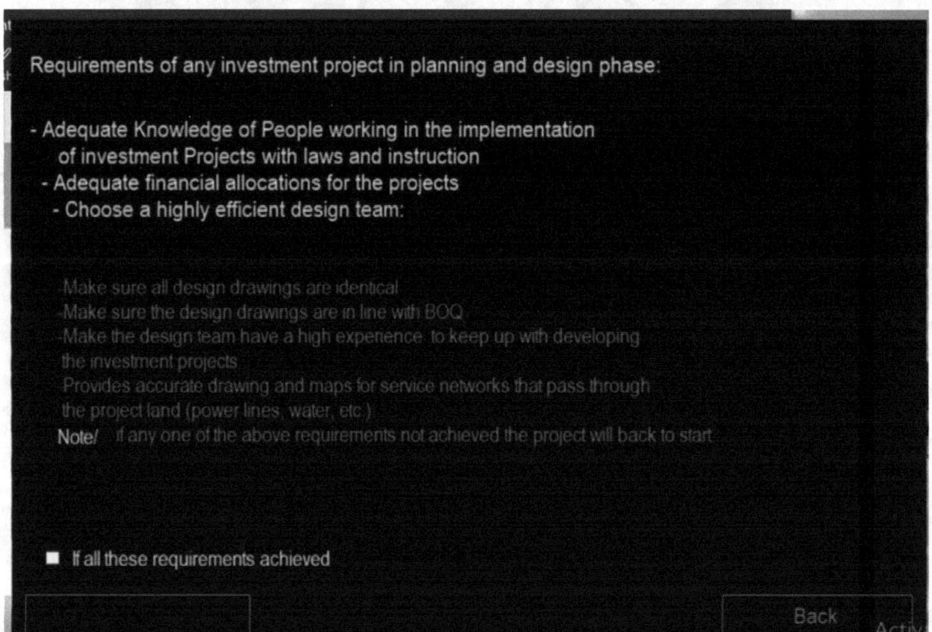

Figure 7. Check to achieve all requirements of any investment project.

When press (√) on (all these requirements achieved) option a list of project phases appears as shown in Fig. 8.

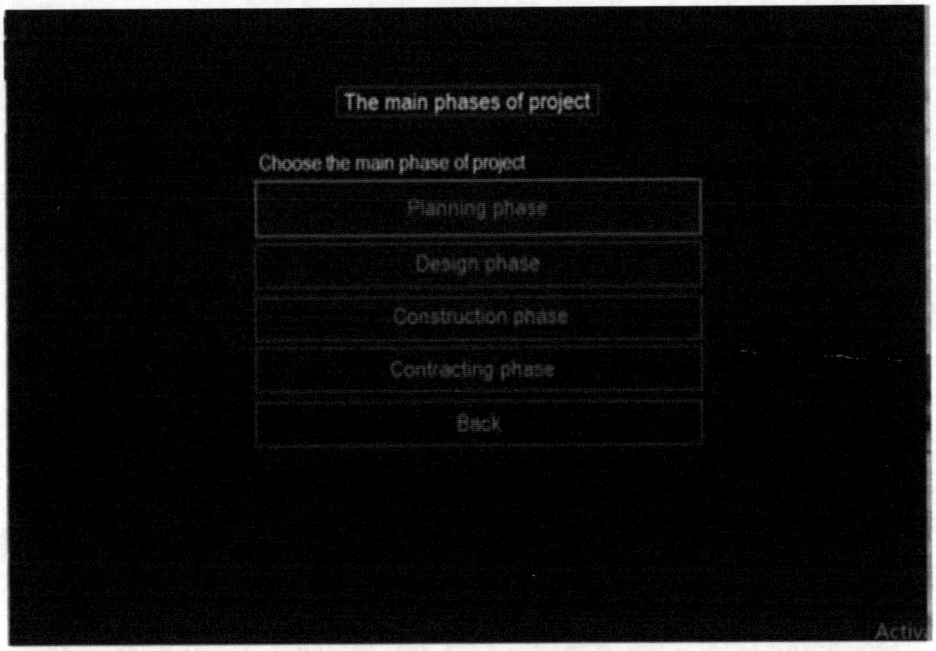

Figure 8. Choose the project phase.

When the project phases window appears, the program asks for the stage selection and when you click on a (back) option, you are returned to the previous page. When press on planning phase for example appears most important problems that may occur within this phase as shown in Fig. 9.

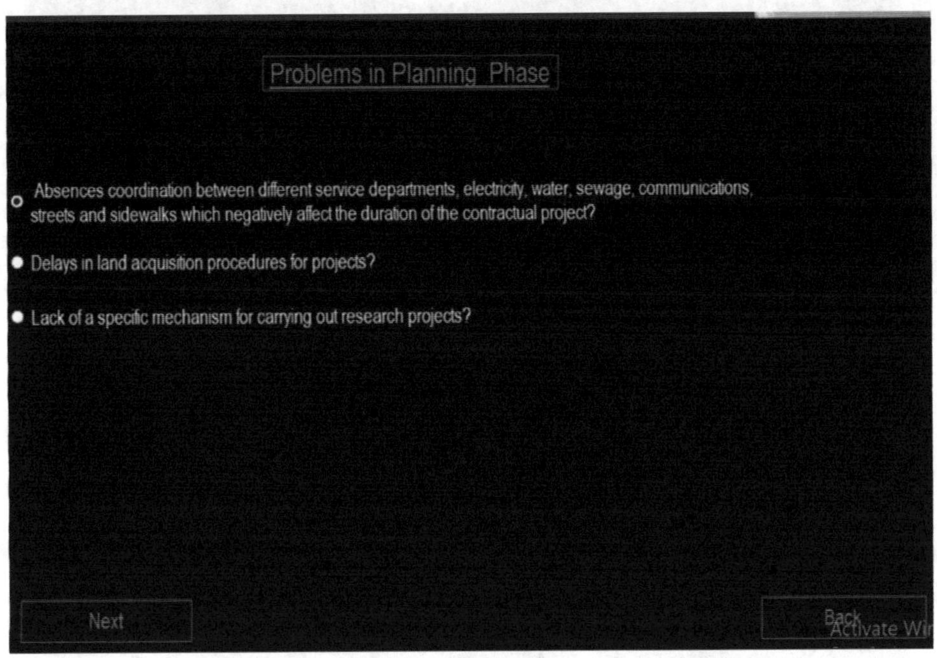

Figure 9. The problems in the planning phase.

When you press the option (absence coordination between departments) for example a window appears the impact of identified problem as shown in Fig. 10.

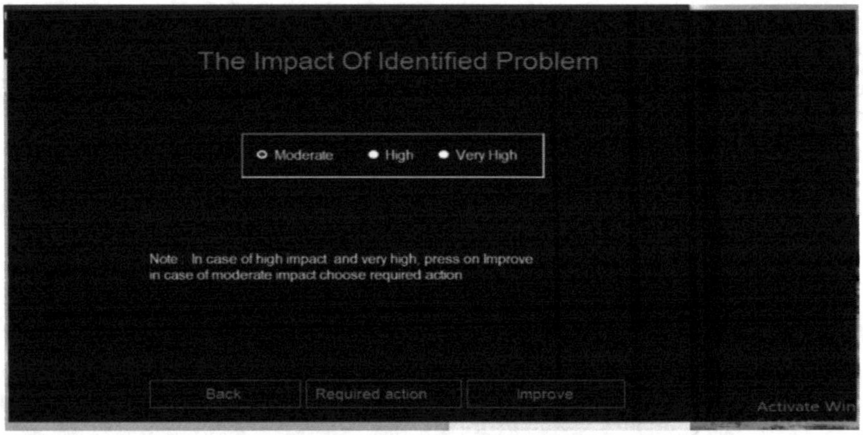

Figure 10. The impact of the identified problem.

In the case of high and very high impact choose (improve) action which shows the suggested treatments for the identified problem as shown in Fig. 11.

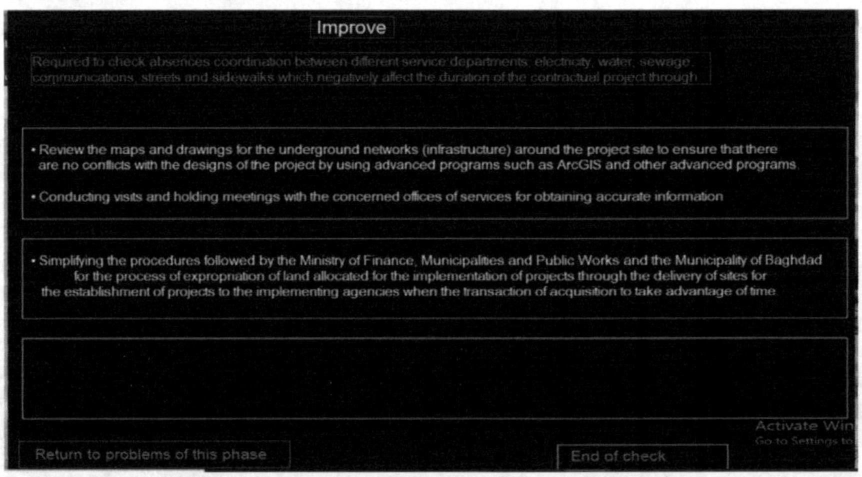

Figure 11. The proposed ways for improving.

In case of moderate impact choose (required action) option as shown in Fig. 12.

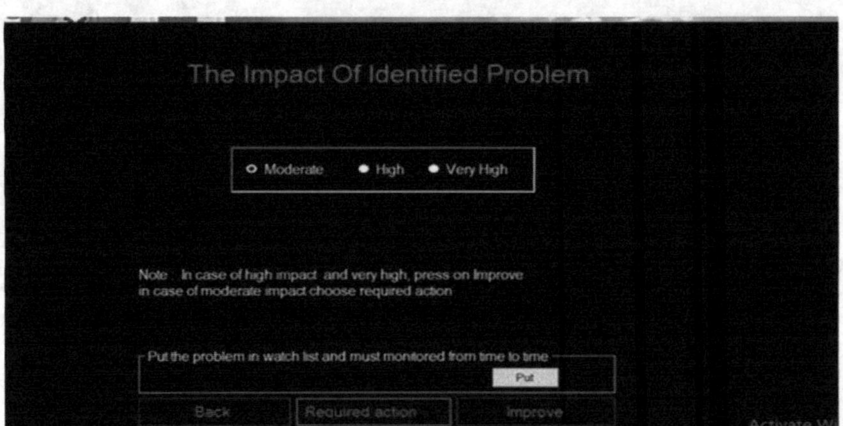

Figure 12. Required action.

Then, back to Fig. 11 and press on (return to problems of this phase) option to check all problems in Fig. 9, When press on (end of check) option in Fig. 11 and press (back) for returning to Fig. 8 and select another phase and When (end of check) for all phases of project return back to Fig. 2 and then press on (Report) option an interface appears that contains a report was discussed as shown in Fig. 13.

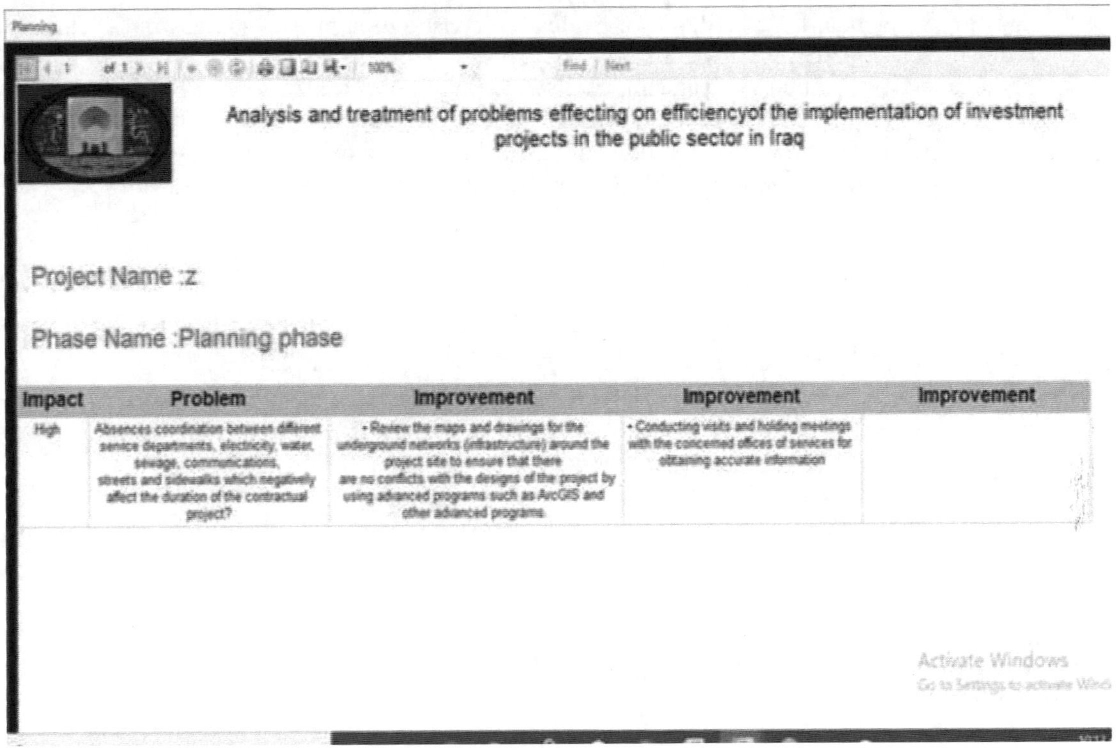

Figure 13. The final report.

6. Conclusion

The public investment projects in Iraq influenced by many factors. The results from research showed that (44) causative factors with high and very high impact on the implementation process, and the results from computing (RII) to all problems showed that the administrative problems in the first level and were (46.309%), the technical problems in the second level and were (31.133%), other problems in third level and were (8.639%), the financial problems in the fourth level and were (8.483%), finally the problems with documentary credits were (5.435%). The program will present a written report and presentation about definitions with the problems through the project phases and the ways to achieve the improvement process. It aims to satisfy the decision-makers to proceed with the project and to gain their consensus about the best way to proceed.

References

[1] L.C. Hurjui, Investment projects: general presentation, definition, classification, characteristics the stages, J. Sci. University of Economics and Public Administration, 2008, No.8.

[2] Ministry of public infrastructure, land transport, shipping, Investment project process manual, 2008.

[3] C. Burja1, V. Burja, The risk analysis for investments projects decision, J. Sci. University of Apulensis, 11(1) 2009 200.

[4] O.J. Klakegg, K. Samset, and O.M. Magnussen, Improving Success in Public Investment Projects. Lessons from a (2005).

[5] H. Schacter, A look at the Concept of Governmental Efficiency in Our Time Public Administration Review, 67(5) 2007 800-810.

[6] K.M. Immordino, Organizational Assessment and Improvement in the Public Sector. Routledge (2017).
[7] N. Gudienė, A. Banaitis, N. Banaitienė, and J. Lopes, Development of a conceptual critical success factors model for construction projects: a case of Lithuania, Procedia Engineering, 57 2013 392-397.
[8] C. Nguyen, A practical list of criteria for evaluation construction project success in developing countries, J. Sci. of ASEAN Engineering, 3(2) 2013 21.
[9] A.A. Frefer, M. Mahmoud, H. Haleem, and R. Almamlook, Overview Success Criteria and Critical Success Factors in Project Management, J. Sci., 7(1) 2018 1000244.
[10] H.K. Al Ageeli, A.S.J.A. Alzobaee, Critical Success Factors in Construction Projects (Governmental Projects as a case study), Journal of Engineering, 22(3) 2016.

IOP Conf. Series: Materials Science and Engineering **901** (2020) 012030 doi:10.1088/1757-899X/901/1/012030

Analyzing the Obstacles of Investment Process in Iraq Using (5 Whys) and Focus Group Discussion (FQD) Techniques

Asmaa M. Jaber[1,a] and Abbas M. Burhan[1,b]

[1]Civil Engineering Department University of Baghdad, Baghdad, Iraq
[a]AsmaaMutar2020@gmail.com, [b]abbas_alshemosi@yahoo.com

Abstract. The Iraqi government faces fluctuations in the investment process with oil revenues of the country and other revenues as these revenues cover only the operational budget and salaries of employees. The government seeks to overcome the financial crisis that hit the country through investment and privatization of some projects. The Iraqi plan for sustainable development (2030) Launched by the Ministry of Planning. But the implementation of this plan collides with a series of obstacles. This research aims to analyze the obstacles facing investment projects by using several approved techniques in managing construction engineering projects such as (5 whys) and (FGD) techniques. The open questionnaire was the theoretical source that the researcher adopted to arrive at to identify obstacles, which is conducting personal interviews and field visits and asking several questions to a group of experts in the Investment Authority through which (15) a obstacles was reached that caused delays in the implementation of construction investment projects. The most prominent of these obstacles was routine and slowed procedures Admin. The responses of the experts were arranged in a closed questionnaire form and distributed to the research sample for which a statistical analysis was conducted. These obstacles were arranged based on the importance of relativity in descending order. The researcher concluded that there is a benefit from using techniques that help analyze constraints such as (5 whys) through reaching the main roots and technique (FGD) is a panel discussion for several experts to arrive at through which corrective actions are possible.

Keywords: Investment, economic growth, productivity, job opportunities, investment opportunities, advanced technology.

1. Introduction

The need for investment, in general, arises from several reasons, including the existence of the savings gap resulting from the fact that the capitals required for investment are greater than the capacity of the national economy, as well as the inability to provide the hard currencies necessary for investment, and in this case, this gap can be filled through investment [1]. Supporting economic growth, where investment leads to an increase in the country's productive capacity, which is an effective element in changing the environment of the national economy as well as modifying its economic imbalances [2]. By increasing its size, the rate of economic growth increases by increasing the added value and productivity [3]. Creating additional job opportunities in the local labor market, and developing technical and administrative skills[4]. It encourages local investment by creating new investment opportunities for local companies and increases the country's exports, which affects the increase in local savings [5]. Transfer of advanced technology where transfer of technology by investment through the adoption of systems and methods of planning, organization, production, marketing, and technical knowledge, which contributes to the transfer of technological progress in the remaining sectors of the national economy [6].

The most important reasons and obstacles that lead to the failure of the investment projects were the reasons (administrative, Legal, economic, financial, documentary credits, technical, infrastructure, other constraints) [7]. The lack of investment projects to accurately define their objectives and the development of strategies, policies, plans, and time programs required for the process of logical implementation of these stages of the project all lead to the delay of the project and financial losses[7]. The process of administrative and engineering planning of the project rules, controls, and indicators must be followed to develop plans for each project according to the circumstances and needs and degree of importance[8]. After the literary reviews of these obstacles facing the investment process, the researcher decided to adopt administrative engineering techniques that analyze and fragment these obstacles to reach the root of the obstacles and monitor appropriate solutions to it as explained below: -

A. (5 whys) Technique. The researcher used two techniques (5whys) and (FGD), one of which complements the other, to analyze the obstacles and reach the root causes as shown in Fig. 1. To counter these constraints and attempt to resolve or mitigate their negative effects on the investment process, the researcher used (5 whys) technique. The five techniques reasons are to ask five questions about the causes of the main obstacles and try to reach through these questions to the root causes and develop until appropriate measures are taken to overcome them. There are three essential elements for effective use of 5 whys technique [8]: (1) Accurate and complete data for problems; (2) Full honesty in answering the questions; and (3) Determination to reach the root causes and main causes of problems and their solution [9].

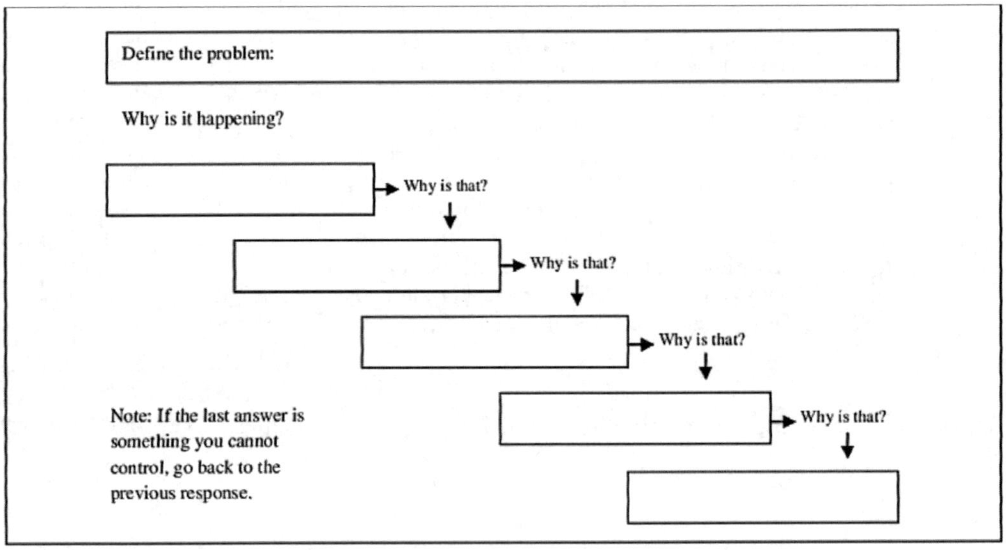

Figure 1. 5 Why Analysis.

B. Focus Group Discussion (FGD) Technique. Focus Group Discussion (FGD) is a qualitative research method and data collection technique in which a select group of people discusses a particular topic or issue in-depth, facilitated by a professional external supervisor. This method serves to solicit participants' attitudes, perceptions, knowledge and experiences, and Practices as shown in Fig. 2. This technique is based on the assumption that the group's operations are activated, this technique was adopted to subtract and present technical outputs (5 whys) and the root causes of this technique and presenting it to a group of experts and specialists in investment and investment project management to arrive at possible procedures [10].

Figure 2. How technique works (FGD).

2. Field Work

The closed questionnaire is the main data source that the researcher adopted to obtain the required information. After visiting approximately (Twentysix) an investment project in Al-Muthanna governorate, (A hundred) a questionnaire forms were distributed to the research sample to answer (Fifty-three) an obstacle who faces the implementation of construction projects. (Fifteen) obstacles were selected that represent the most important. The results of the analysis of the questionnaire were reached using the social sciences package program (Spss) as shown in Tables 1 and 2. The obstacles were arranged in descending order based on the relative importance of each obstacle.

Table 1. Level of importance according to the relative importance index.

RII values	Importance level (IL)	
0.8≤ RII ≤1	High	H
0.6≤ RII ≤0.8	High- medium	H-M
0.4≤ RII ≤0.6	Medium	M
0.2≤ RII ≤0.4	Medium- low	M-L
0 ≤ RII ≤0.6	low	L

Table 2. Statistical analysis of obstacles to investment projects in Iraq.

No.	Obstacles	N	Mean	RII	I.L	Category of obstacles
1	Routine, slow administrative procedures.	100	4.7	0.94	H	management
2	Lack of prior coordination with some ministries and provincial councils	100	4.5	0.90	H	legal
3	Tribal problems especially in remote areas.	100	4.3	0.86	H	others
4	Lack of development of support services such as transportation, insurance, and others	100	4.1	0.82	H	infrastructure
5	Difficulty in obtaining vacant locations suitable for investment projects.	100	4.05	0.81	H	legal

6	The existence of excesses on the plots of land allocated for investment projects.	100	4	0.80	H	legal
7	Insufficient financial ceilings in direct execution.	100	4	0.80	H	economical
8	Unstable financing policies.	100	4	0.80	H	economical
9	The length of the procedures for financing the departments of the ministries in the etc.)	100	3.8	0.76	M-H	financial
10	The multiplicity of supervisory bodies	100	3.7	0.74	M-H	legal
11	Weak mechanisms for the private sector.	100	3.6	0.72	M-H	legal
12	Poor services for projects located in remote areas.	100	3.55	0.71	M-H	technical
13	Inadequate sources of funding	100	3.5	0.70	M-H	financial
14	The existence of disputes over the returnees of land between ministries and state departments.	100	3.5	0.70	M-H	legal
15	Limited administrative powers granted to project managers.	100	3.45	0.69	M	management

Symbols represent the following: N= Number of samples, RII= Relative Importance Index, IL= Importance level.

3. Techniques (5 whys) and Focus Group Discussions (FGD) Features and Results

Depending on the results of the statistical analysis obtained, the obstacles were arranged in descending order according to their relative importance, so that the most important obstacles are (routine, bureaucracy and slowed government procedures), as they were used as a model that can be analyzed using techniques (5 whys) and (FGD), a path in which these obstacles go through a series of treatments passing through the root causes and reaching the procedures required to face these challenges as shown in Tables 3 and 4.

Table 3. Root Causes for constraints that were reached from the technique (5 whys).

Define obstacles:(Routine, bureaucracy, slow administrative and government procedures)	
Series whys	Root Causes
The authorities intervene between the provincial and governorate councils because the provincial councils are a supervisory body and not an executive authority (why_1?) Lack of prior coordination with some ministries and provincial councils that seek the work of some ministries and intersection in the implementation of projects in the provinces(why_2?) The existence of excesses on the plots of land allocated for investment projects (why_3) Delay in land acquisition procedures for projects (why_4 ?) Difficult to implement the project because of the difficulty of clearance procedures(why_5?)	1.The existence of disputes over the returnees of buildings and land between ministries and state departments. 2. The limited administrative powers granted to project managers that impede the progress of work. 3.Weak coordination and follow-up between the ministries and the competent departments when allocating or buying land plots. 4. Delay in resolving disputes between the contractor and the employer.

Table 4. Possible Actions that were Reached from the Technique (FGD) to Address Constraints.

No.	Action
1	Relevant authorities should simplify and facilitate procedures related to obtaining plots of land.
2	The Municipality of Baghdad and the municipal departments shall prepare the plots of land in Baghdad and the governorates for the implementation of priority projects, while supporting the prices of land allocated for this purpose.

3	The Legal Department at the Ministry of Planning should strictly grant the highest rating to the contracting companies and adopt the principle of similar works and technical and practical experience and not rely on the capital of the company only, as this measure gives some companies with high liquidity advanced classification degrees, but cannot fulfill its contractual obligations as it lacks experience and staff Technical and engineering.

4. Discussion and Conclusions

- The delay is part of investment projects in our country, Iraq. In this paper, the researcher wanted to determine the current situation in this field by identifying obstacles to investment in the construction sector by surveying various stakeholders in investment projects where the results show the reasons for the delay and slow implementation of the project, where the researcher found it useful to use some techniques in managing engineering projects that help to analyze the problem, including a technical (5 whys) because it helps the researcher in determining the root cause of the obstacles, and determines the relationship between the different root causes of the obstacle and is one of the simplest tools where it is easy to complete without statistical analysis.
- Many constraints are facing the construction industry and the construction sector and need to study, follow-up, and analysis. Therefore, project management has provided many administrative engineering tools and techniques that enable the researcher to analyze constraints without statistical tools including techniques (5 whys) and (FGD).
- (FGD) technique facilitates the transmission of thoughts and ideas among the members of the seminar, and by drawing on the results of the previous techniques, possible measures can be reached.
- Depending on the type of constraints facing the team, different types of discussion can be prepared for the focus group closely or distant, for example, one focus group, two-way focus group, dual supervisor focus group, responsive supervisors' group, mini focus group, and focus groups online.

References

[1] J. Earle, Post-Privatization Ownership and Productivity in Russian Industrial Enterprises, SITE Working Paper 127. Stockholm, Sweden: Stockholm Institute of Transition Economics (1998).
[2] A Journey to the Alt-A Zone: A brief primer on Alt-A mortgage loans. Nomura Fixed Income Research, June 3, 2003. http://www.securitization.net/pdf/nomura_journey_060303.pdf
[3] Project management Techniques. http://www.tifb.state.tx.us/Handbooks/Project_Management.htm.
[4] [W.J. Dixon and F.J. Massey, Introduction to Statistical Analysis, McGraw-Hill, New York (1969).
[5] G.J. Alexander, J.D. Jones, and P.J. Nigro, Mutual Fund Shareholders: Characteristics, Investor Knowledge, and Sources of Information, Financial Services Review, 7 1998 301-316.
[6] B. Arathy, A.N. Aswathy, S.P. Anju, and N.R. Pravitha, A Study on Factors Affecting Investment on Mutual Funds and its Preference of Retail Investors, International Journal of Scientific and Research Publications, 5(8) 2015 1-4.
[7] M.P. Spinner, Improving project management skills and techniques. Englewood Cliffs, NJ: Prentice-Hall (1989).
[8] M.P. Spinner, Project management: Principles & practices. Englewood Cliffs, NJ: Prentice-Hall (1997).
[9] H.N. Ahuja, S.P. Dozzi, and S.M. Abourizk, Project management: Techniques in planning and controlling construction projects. New York: Wiley (1994).
[10] C. Oglesby, H. Parker, and G. Howell, Productivity improvement in construction. New York: McGraw-Hill (1989).

Application of Root Causes Analysis Techniques in The Contractor Selection for Highway Projects

Mustafa I. Ahmed[1,a*] **and Sedqi Esmaeel Rezouki**[1,b]

[1]Civil Engineering Deptartment, University of Baghdad, Baghdad, Iraq
[a]mustafa.civil.eng270@gmail.com, [b]sedki_rizqi@gmail.com

Abstract. Highway projects are one of the most important projects in this decade in Iraq. Highway projects suffer from corruption factors in the contractor selection phase. To succeed in transparency and best value contractor selection for awarding procedure for highway projects, the researcher used root cause analysis techniques to specify the main factors that lead to bad selection for the highway projects in Iraq. Many tools techniques used for the root cause analysis techniques like Fishbone, Pareto chart, and 5-Why. This study conducts the causes of bad contractor selection and analysis by root cause analysis techniques, Pareto chart conducted that internal relationships and administrative corruption are the main causes that covered near than 80% of the problem, and 5-Why conducted that the acceptance of the lowest price is the main cause for bad selection.

Keywords: Highway projects, Contractor selection, Root cause analysis techniques, Fishbone diagram, Pareto diagram, 5-Why analysis.

1. Introduction

Difficult to imagine the awarding procedure for construction projects without problems or restrict or failed during implementation procedure due to poor management and poor monitoring or poor supervision. The highway project is implemented in a complex environment. There is no reliable data and trust information to evaluate project needs. The awarding procedure of highway projects is the start of project success or project failure. There are many problems during the selection procedure for the project criteria and their evaluation with a right way in Iraq:

1) The absence of using modern procedures for the selection of the best contractor.
2) Projects are awarded to poor contractors who have a poor experience and the true ability to implement these projects.
3) Corruption is the main reason for the selection process.
4) Depend on the lowest price.

The main issue during the contract awarding procedure is using traditional methods that not compare with our needs to implements highway projects in this decade. Even the highway projects implemented with traditional methods, we may get many execution problems like does not succeed in project age and high maintenance after project operation. Therefore, RCAT can deal with the awarding issue. In Iraq, the root cause analysis technique culture is absent so, it's useful in project management and with selecting procedure for the best contractor especially. Jimmie [1] investigated the root causes of construction accidents. One source of information on the causes associated with many serious injuries and fatalities is maintained by the Occupational Safety and Health Administration (OSHA). The study presents recommendations on how the OSHA reports could be made more meaningful. First, injuries had been coded into one of the 20 possible cause categories, rather than the traditional five groups of falls, struck-by, electric shock, caught-in/between, and others. Additional or secondary cause codes also were developed. The results could then be utilized to focus greater attention on those areas for which modifications in the regulations are warranted and it would be more helpful to the construction industry by emphasizing the major causes of serious accidents.

Tariq [2] investigates the root causes of construction accidents. Their study presents an accident root causes tracing model (ARCTM).ARCTM proposes that accidents occur due to three root causes: (1) failing to identify any unsafe condition that existed before an activity was started; (2) deciding to proceed with a work activity after the worker identifies an existing unsafe condition; and (3) deciding to act unsafe regardless of initial conditions of the work environment. Also, ARCTM emphasizes the need to determine how unsafe conditions exist before or develop after an activity is started and proposes that these unsafe conditions are due to four causes: (1) management actions/inactions; (2) unsafe acts of worker or co-worker; (3) non-human-related event(s); (4) an unsafe condition that is a natural part of the initial construction site conditions.

Tommy et al. [3] aimed first, at gathering the perceptions of civil construction practitioners on how significant are the causes of delay; and, second, investigating whether the suggestions as stated in the report of the Construction Industry Review Committee (the committee comprises members with good standing and knowledge in the construction and related fields as well as those from other professions who are responsible for examining the current state of the construction industry in terms of its output quantity, the quality of work, its environmental friendliness, site safety, its workforce and the system of supervision) apply to and effective at mitigating the corresponding delays concerning a ranking order established using the mean score method. The extent of the differences in perception among the different respondent groups on these two issues was also examined using the rank agreement factor (RAF), percentage agreement (PA), and percentage disagreement (PD). This study had been applied on six projects. A strong consensus was found between the client and consultant groups on the significance of the various causes of delay (PA = 74%) and the effectiveness of mitigation measures (PA = 67%) compared with the other pairs of groups.

Yehiel [4] examines cost overruns phenomenon impact on the construction industry as a worldwide problem, identifies its root causes, ranks them (on a local basis), and analyzes them. Firstly, the study had been gathered 146 potential causes from international professional literature as well as from prominent local experts. Finally, they were filtered and merged into merely 15 independent universal root causes. These were further investigated through a cross-sectional survey among 200 local practitioners who ranked the 15 universal root causes according to their conceived local importance and influence on cost overruns. The survey revealed that the three most important causes out of 15 are, cause number 1 premature tender documents; cause number 2 is too many changes in owners' requirements or definitions, and cause number 3 is tender-winning prices are unrealistically low (suicide tendering). Faiq and Reem [5] investigated the cost management status of the highway construction projects in Iraq. The survey included investigating and evaluating the reality of cost planning and control and determine the causes of construction cost deviation. Causes of cost deviation had been investigated and collected by conducting the personal interviews and questionnaires with the site's engineers and by reviewing the previous literature. Iraqi Road and Bridge Directorate was selected as a sample study. It is worth to mention that the statistical analysis has been conducted by using the SPSS program (version 20). Cost deviation reasons had classified into three groups by the researchers, which are planning reasons, designing reasons, execution reasons. Planning reasons classification had taken relative importance equal to 2.7%, designing reasons classification had taken relative importance equal to 3% and execution reasons classification had taken relative importance equal to 2.45%. The main objective of this paper is to classify and specify the causes that lead to choose poor contractors for the highway projects and give a reliable solution for the awarding procedure applied with project specifications and available needs.

2. Problem Statment

The only important question raised in this study is: "Is the root cause analysis techniques can be applied in highway projects management to control the bad selection for contractors" To answer this question, we must including the research problem. The research problem is summarized depend on the literature review and some of the interview with the specialist engineers regarding the following:

1) There is a poor check for selection procedure problems for the contractor for highway projects because the available techniques are poor. Besides that, there is a need to use root cause analysis technique, because of its accuracy for construction work and important to ensure good management for our projects.
2) The projects are awarding to contractors who do not have enough experience to implement these projects, despite the big need for these projects in this decade.
3) The importance of highways for Iraq as a developing country, because;
   a) No civilized society can be survived without highways.
   b) Highway has a great cultural value.
   c) Highway has the most important role to play in the defines of the country.
   d) Highway is the symbol of a country's progress.
   e) Highway development and transport have the highest employment potential of all economic activities.

3. Research Important, Limits and Methodology

Research important summarized as per the following:

- This research highlights the importance of the awarding procedure for highway projects for the best contractor. Because of this type of project have a big important between the other projects.
- It provides a true vision of the difficulties faced by project managers and planners, when applying the methodology of cost management in highway projects, to overcome these obstacles and reduce them.
- Encourage researchers, lecturers and academics to conduct further researchers and integrate with the results of the current study. The motivation of this study is a limited number of researches in the area of awarding contracts about highway projects, and the need for selecting the best contractor in the highway projects.
- The research limited to the following limits:
- Time limits: gathering data were implemented in 2019.
- Spatial limits: This study was conducted on highway projects in the Republic of Iraq.
- A three-step study methodology is conducted, to obtain the objective of the study:
- Literature review
- Questionnaire and Personal interviews with engineers of exact specialization in highway engineering or at least experience in highway projects, this step had explained in reference [5].
- Statistical analysis.

4. Root Cause Analysis Concept

Root cause analysis (RCA) is a useful process for understanding and solving a problem. RCA is a popular and often-used technique that helps people answer the question of why the problem occurred in the first place. It seeks to identify the origin of a problem using a specific set of steps, with associated tools, to find the primary cause of the problem, so that you can:

a) Determine what happened.
b) Determine why it happened.

Root cause analysis (RCA) is a valuable management tool that can be readily learned by managers as well as frontline personnel. These techniques can be of enormous value. They capture both the big-picture perspective and the details. They facilitate system evaluation, analysis of the need for corrective action, and tracking and trending. Regarding trending, managers will be able to determine how often a particular error occurs. This analysis is as useful and perhaps even more efficacious. The technique can be applied not only to engineering but also they apply to other disciplines [6]. It is a method that is used to address a problem or nonconformance to get to "the root cause" of the problem. It is used to eliminate or mitigate the cause and prevent the problem from recurring. RCA is simply the application of a series of well-known common-sense techniques that can produce a systematic, quantified, and documented approach to the identification, understanding, and resolution of underlying causes [3]. Root cause analysis is not merely an arbitrary expression; rather, it is a systematic, formal, well-structured methodology, used as part of the total quality–management approach [5].

5. Root Cause Analysis Techniques

There are several tools of root cause analysis techniques for creative thinking such as Fishbone diagrams, Mind mapping, Pareto analysis, causal tree, brainstorming, nominal group technique, metaphorical thinking, and 5-Why analysis. The researcher will focus on Fishbone diagrams, Pareto diagrams, and 5-Why technique.

5.1. Fishbone Diagrams

Fishbone diagram was devised by Professor Kaoru Ishikawa, a pioneer of quality management, in the 1960s. The technique was then published in his book in 1990, "Introduction to Quality Control.", the diagrams are known as Ishikawa diagrams, cause and effect analysis, or Fishbone diagram because a completed diagram can look like the skeleton of a fish [6]. This tool is used to come up with new ideas like a brainstorming, but in a more balanced way [7]. A cause and effect diagram offers a structured approach to the search for the possible causes(s) of a problem. This tool helps to organize problem-solving efforts by identifying categories of factors that might be causing problems. Often this tool is used after Pareto or brainstorming sessions to organize the ideas generated [8]. This type of diagram identifies all the potential factors that could contribute to a problem. Once all the desired information had been captured and represented by the fishbone diagram, this can help in finding resolutions [9]. Figure 1 is an example that represents this technique.

5.2. Pareto Diagram

Pareto analysis is a technique for focusing attention on the most important problem areas. The Pareto concept, named after the nineteenth century by Italian economist Vilfredo Pareto, is that relatively few factors generally account for a large percentage of the total causes (e.g., complaints, defects, and problems). The idea is to classify the causes, according to the degree of importance, and focus on resolving the most important, leaving the less important [10]. Pareto analysis is a formal technique for finding the changes that will give the biggest benefits [8].

5.3. 5-Why Technique

It is one of the many brainstorming methodologies of asking "why" five times repeatedly to help in identifying the root cause of a problem. If a problem is repeatedly questioned, each time an alternative solution comes out which is linked to the root cause. However, asking why may be continued till getting an agreeable solution. Five is an arbitrary figure. The theory is that after asking "why" five times one is likely to arrive at the root cause [11].

6. Comparison of Common Root Cause Analysis

Many researchers of root cause analysis techniques state the process is too complicated and we should use several of them for each problem or select them based on problem type. The following table used to compare between root causes analysis techniques.

Table 1. Comparison of selected RCA methods and tools [12].

Method/Tool	Type	Define problem	Define all causal relationship	Provide a causal path to root causes	Delineates evidence	Explain how solutions prevent recurrence	Easy to follow report	Score
Events & Causal Factors	Method	Yes	Limited	No	No	No	No	1.5
Change analysis	Tool	Yes	No	No	No	No	No	1
Barrier analysis	Tool	Yes	No	No	No	No	No	1
Tree-Diagram	Method	Yes	No	No	No	No	No	1
Why-Why chart	Method	Yes	No	Yes	No	No	No	2
Pareto	Tool	Yes	No	No	No	No	No	1
Storytelling	Method	Limited	Yes	No	No	No	No	0.5
Fault Tree	Method	Yes	No	Yes	No	No	No	4
FMEA	Tool	Yes	Yes	Limited	No	Limited	No	2
Reality Charting	Method	Yes	Yes	Yes	Yes	Yes	Yes	1.6

7. Application of Root Cause Analysis Techniques

In this study, the researcher relied on the fishbone diagram to classify the sub-causes and the main reasons leading to the poor selection of the contractor in the highway projects and relying on his practical experience in the highway projects and the work of the contracting management as well as relying on personal interviews.

7.1. Cause and Effect Diagram

The diagram in Fig. 1) is characterized by:

- Four main causes (Client, Contractor, work environment, and Legislations and laws) and 18 secondary causes.
- Absence of any reason leading to choose a bad contractor for project implementation.

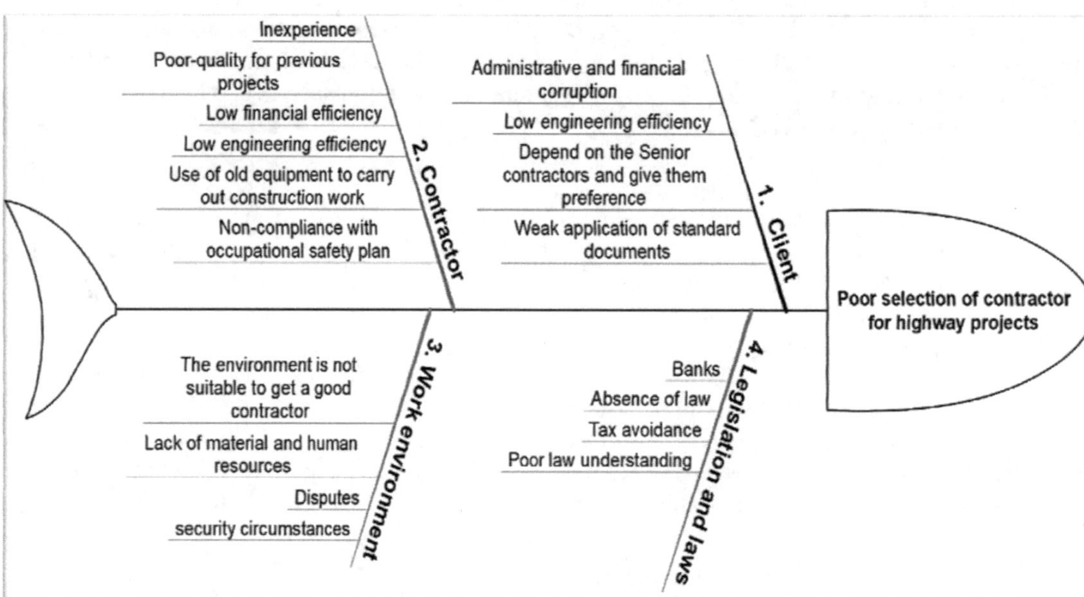

Figure 1. Cause and effect diagram.

7.2. Client

The Client is the first beneficiary of the special responsibility after the implementation of the construction work, so the decision of the client is a key decision to choose a good contractor, the employer's foundation must be rehabilitated to face any risk in the selection process, the following are the secondary reasons resulting from the Client:

7.2.1. *Administrative and financial corruption.* The phenomena of administrative and financial corruption is one of the riskiest events face the countries, where it took necroses in the body of their communities began to security and the subsequent shortage in the construction and economic process development, administrative and financial capacities, then the state's deficit to meet the defies of the age and reconstruction and build the infrastructure needs for growths. These corruptions are met care of many researchers and those who attentive agreed consensus on the need to develop and make an official framework designed to enclose the problem and salted through the serious, real steps in all its forms and manifestations in all areas of life to hurry the process of economic developments [13].

7.2.2. *Weak application of standard documents.* Due to the shortage of experience of the engineering staff and the lack of training courses necessary to prepare. standard documents need a stable security environment for application ideally and this does not help at the moment to achieve this.

7.2.3. *Low engineering efficiency.* The engineering staff is not qualified to run the expert work, due to the lack of training courses necessary to develop the experience and keep up with the modern methods, and the lack of sufficient opportunity for the youth to highlight their skills.

7.2.4. *It depend on senior contractors and give them preference.* The referral committee may play a major role in this regard. Recommendations may be made to the government contractor to evade the Integrity Commission or the reason for relying on the older contractor for financial corruption, leading to loss of the best contractor at the expense of the quality, time, and cost of the project.

7.3. Contractor

The contractor is a company or persons that perform works on a contract basis. the following are the secondary reasons resulting from the contractor:

7.3.1. Inexperience. This leads to improper construction work, such as relying on non-specialized staff and using old equipment to implement the required work, leading to loss of time, poor quality and waste of cost, and may lead to failure of the project and inability to continue to implement it.

7.3.2. Poor-quality for previous projects. Failure to achieve the required quality in previous similar projects is proof of the contractor's inability to carry out the required construction work, such as failure to achieve the project life span, failure to comply with the required specification and failure to deliver the project on time.

7.3.3. Use old equipment to carry out construction work. The use of modern equipment to carry out the construction work leads to faster implementation and minimizing the human resources required in the construction project. The non-use of modern equipment is due to the inability of the contractor to purchase equipment and the lack of operators to use modern equipment.

7.3.4. Low financial efficiency. This is one of the most serious risks of implementing the construction project. The most important problems are the inability to pay the financial dues, such as the wages of workers, machinery, and sub-contractors, which leads to a lack of production and interruption of intermittent work.

7.3.5. Low engineering efficiency. The lack of specialized engineers for the contractor leads to a high risk of wrong execution, lack of understanding of the conditions and specifications, in addition to the inability of the contractor to choose a specialized staff to carry out the required construction work.

7.3.6. Non-compliance with occupational safety plan. Lack of commitment and lack of attention to occupational safety plan leads to losses in materials and human resources, resulting in loss of time, loss of cost, injury of skilled workers, and disruption of important equipment, resulting in high maintenance costs.

7.4. Work Environment

The site where the construction project is carried out, which is an important factor in attracting the best contractors to carry out the work. the following are the secondary reasons resulting from the work environment:

7.4.1. The environment is not suitable to get a good contractor. This may be because the natural conditions at the worksite do not help to carry out construction work easily.

7.4.2. Lack of materials and human resources. The provision of human resources from a source far from the work site causes an increase in costs, such as housing, transport, and food. In addition to the fact that material resources are not available in the work environment, they will lead to the possibility that they will not reach the worksite.

7.4.3. Disputes. Maybe some disputes that prevent implementation, such as tribal conflicts, which prevent the contractor from using their material and human resources, and the use of resources available in the work environment, such as the use of unskilled workers, old equipment, and non-conforming raw materials. Sometimes the client cannot get a contractor from outside the work environment.

7.4.4. Security Circumstances. The absence of security in the work environment will lead to a serious risk to the lives of the workers. There is a great risk that the contractor will not be able to complete the work on time or not. The best contractors may not be able to compete on construction work in the environment has no security.

7.5. Legislation and Laws

Failure to understand this point leads to considerable disagreement between the contractor and the client directly affecting the quality of the construction work. the following are the secondary reasons resulting from the legislation and laws:

7.5.1. Banks. Such as giving a letter of guarantee at a low price, opening an account for the company without updating the company's information, and giving financial efficiency to the contractor without filing the required amounts as required.

7.5.2. Tax avoidance. Non-payment of taxes due to the State through the use of illegal methods. The reasons for non-payment of taxes is the imposition of non-strict sanctions, a sense of weakness in the imposition of state law, and administrative corruption. The choice of a tax evasion contractor is a major risk to the construction work.

7.5.3. Poor law understanding. Due to the absence of a specialized legal department in the institutions, there is a great challenge in legal problems and not to behave properly. To avoid or minimize bad selection for the best contractors for highway projects, the states must provide periodic trainings for all members, use modern technology for project management, give youth a chance to do their best, and commitment to international standards and specifications for project management.

Table 2. Results for online questionnaire

Defects	Frequency	Cumulative, %
Internal relationships	178	45.8
Administrative corruption	150	84.3
Lack of expertise for the Bid Analysis Committee	25	90.7
Accept the lowest prices	10	93.3
There are no correct criteria for choosing a contractor	8	95.4
Recommendations are made out of corruption by the Bid Analysis Committee	7	97.2
Selection of government contractor to evade responsibility	5	98.5
Non-compliance of contractors with quality standards and specifications	4	99.5
Depend on the old contractors and give them the right of preference	2	100.0

7.6. Pareto Diagram

The analysis of results using a Pareto diagram is shown in Fig. 2. The result of Pareto chart refers that the cause (internal relationship and administrative corruption) cause 84% of poor selection contractor for highway projects. Solving the causes (internal relationship and administrative corruption), we will correct 80% of the issue.

7.61. Internal Relationships. Internal relations are one of the most important factors of corruption that affect the selection of the contractor. This leads to a bias towards one party at the expense of the other, such as revealing the estimated cost of the project and disclosing the required specifications for the work. and accept one of the bidding without seeing competitors.

7.6.2. Administrative Corruption. Non-compliance with work, poor accuracy, and improper management, resulting from administrative corruption. Successful contractor selection is done in proper conditions free of corruption.

7.6.3. Lack of expertise for the Bid Analysis Committee. The analysis of bids requires a team with highly specialized expertise, in addition to training courses to develop skills in this field.

7.6.4. Accept the lowest prices. Acceptance of the lowest bids does not mean choosing the best. Neglecting other factors such as time and quality may negatively affect the implementation and achieve life span and demand maintenance for longer periods.

7.6.5. There are no correct criteria for choosing a contractor. To Choose the best, you must put the correct mechanism commensurate with the work required.

7.6.6. Recommendations are made out of corruption by the Bid Analysis Committee. Incorrect bid evaluation by the Bid Analysis Committee affects the project assignment committee and therefore there is an opportunity to choose a non-qualified contractor to carry out the work.

7.6.7. Selection of government contractors to evade responsibility. Relying on the government contractor and especially the self-financing companies is better than the private contractor even if the price is offered more by a private contractor. The reason for relying on the government contractor may also be to evade the Integrity Commission.

7.6.8. Non-compliance of contractors with quality standards and specifications. Non-compliance with the necessary specifications and quality of work, and consideration by the employer, and acceptance of tender in this cause means there is a risk to implement the construction work.

7.6.9. Depend on the old contractors and give them the right of preference. Dependence on the oldest contractor without the decision to choose new contractors is a loss to achieve better results than before, a new decision does not mean failure, but it means looking for the best.

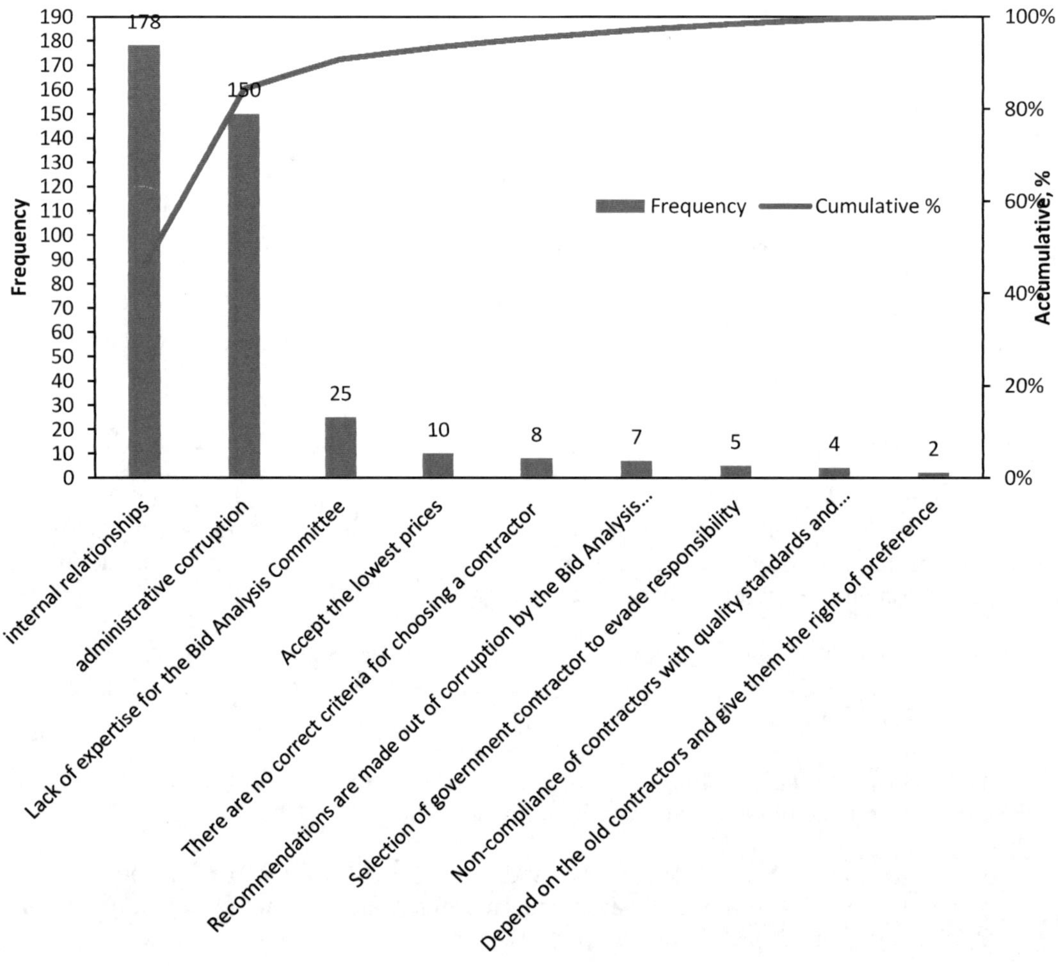

Figure 2. Pareto chart.

7.7. 5 - Why Technique

During the online survey, the most 4 factors affecting on the contractor selection are:

- Internal Relationships
- Administrative Corruption
- Lack of Expertise for The Bid Analysis Committee
- Accept the Lowest Prices

Therefore, the following figure illustrates the root cause factor for the bad selection.

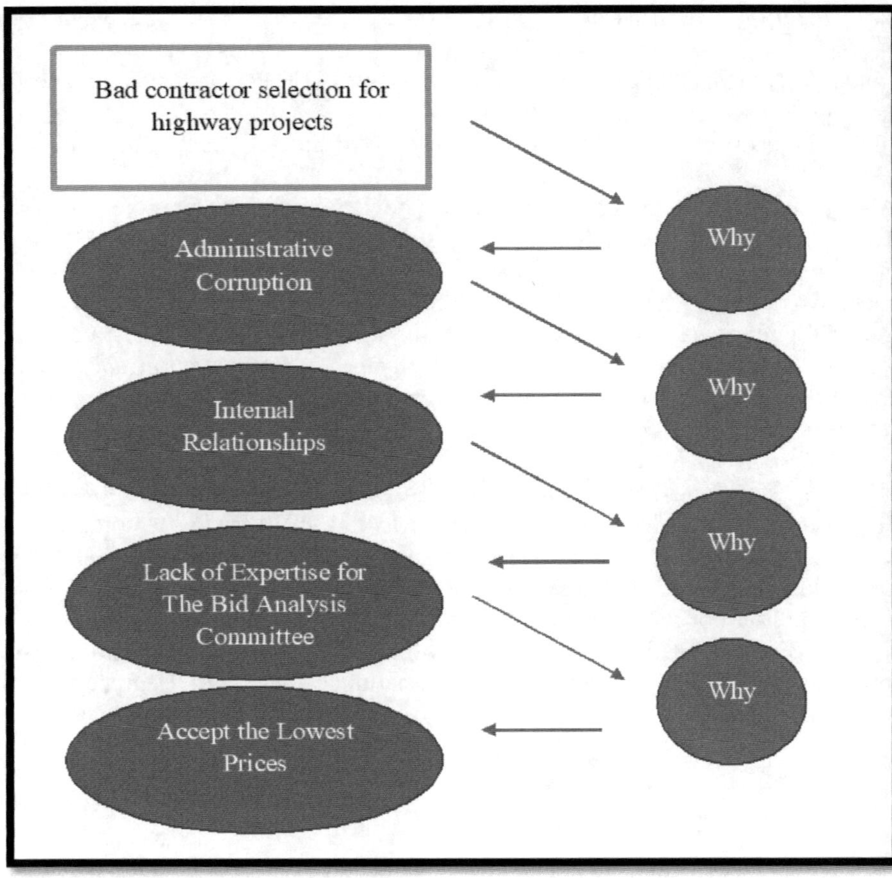

Figure 3. 5- why technique.

8. Conclusions

Major conclusions can be summarized as follows:

- In this research, root cause identification methodology has been adopted to diagnose the causes of contractor selection in highway projects. Root cause identification for project-related problems is a key and necessary step in the improvement and control of cost deviation. Fishbone diagram (FD), Pareto diagram (PD), and the 5-Why analysis used in this study to diagnose the causes of contractor selection in highway projects in the Republic of Iraqi.
- Eighteen causes concentrated mostly in the selection procedure explained by the Fishbone diagram.
- The result of Pareto chart refers that the cause (internal relationship and administrative corruption) cause 84% of poor selection contractor for highway projects, and Solving the causes (internal relationship and administrative corruption), we will correct 80% of the issue.
- Pareto diagram results had been filtered more by applying 5-Why analysis, this analysis concluded that the lowest bid price is a root cause for bad contractor selection for highway projects.

References

[1] H. Jimmie, Identifying root causes of construction injuries, J. Constr. Eng. Manag, 1(124) 1998 67–71.

[2] S.A. Tariq, Identifying root causes of construction accidents, J. Constr. Eng. Manag. 1(126) 2000 52–60.

[3] Y. Tommy, W. Ivan, and C. Karen, Construction delays in Hong Kong civil engineering projects, J. Const. Eng. Manag. 6(136) 2006 636–649.

[4] R. Yehiel, Root-cause analysis of construction-cost overruns, J. Constr. Eng. Manag, 1(140) 2004 04013039.

[5] F.M. Al-Zwainy and A. Reem, Investigation cost deviation of highway project, Res. J. Appl. Sci. Eng. Technol, 13 (11) 2016 843–855.

[6] D. Mahto and A. Kumar, Application of root cause analysis in improvement of product quality and productivity, J. Ind. Eng. Manag, 2(1) 2008 16–53.

[7] G. Nyren, Product development according to six sigma and the DAMIC improvement cycle, M.Sc. thesis, submitted to the Department of Business Administration and Social Sciences, Industrial Business Administration, Lulea University of Technology (2007).

[8] V. Geoff, Mini Guide to Root Cause Analysis, 1st ed. Quality Management & Training, United Kingdom (2008).

[9] M.W. Patricia, Techniques for root cause analysis, BUMC Conf. Proc. 14 2001 154–157.

[10] W.J. Stevenson, Operation Management, 8th ed. McGraw- Hill/Irwin, Boston (2005).

[11] G.D. Oberlender, Project Management for Engineers and Construction, 2nd and. McGraw-Hill Science/Engineering/Math, United States of America (2000)

[12] D.L. Gano, Apollo Root cause Analysis – A New Way of Thinking, 3rd edition. (2007).

[13] S. Ali Omer Burak, Abdalla Mohamed Almabrouk Alghandouri, Administrative & Financial Corruption in Developing Countries, The International Journal of Engineering and Science (IJES). (2018).

IOP Conf. Series: Materials Science and Engineering **901** (2020) 012032 doi:10.1088/1757-899X/901/1/012032

Quantifying and Modeling the Cumulative Impact of Changes Order for the Projects in Iraq

Soadad M. R. Hassan[1,a*] **and Sedqi Esmaeel Rezouki**[1,b]

[1]Civil Engineering Department, University of Baghdad, Baghdad, Iraq
[a]suadad1981@yahoo.com, [b]sedki_razqi@yahoo.com

Abstract. This research investigates the change orders projects of hospital construction in Iraq, for several reasons. Because these types of projects have a relatively long duration, which leads to a change in some requirements or the emergence of new developments with the owner or gets a new technical development in this field, also these projects contain a large number of documents, specifications, plans, bill of quantities, and a large number of engineers with various specializations participate in preparing them, which makes the probability of change possible due to the large work and poor coordination between those specializations or the development of new government regulations. This current research is based on a database of eleven projects for hospitals. The researcher used combined methodology: the qualitative method represented by interviews of many construction professionals, and the quantitative method represented by questionnaire surveys and gathering data through a survey the documents of change orders for the projects. In this study, we adopted the analytical statistical approach by used F-tests and then used a one - way ANOVA analysis. The researcher managed to reach mathematical models that can predict the added cost of change requests that occur through the planning or implementation phase for the project, to avoid or minimize project changes. A series of statistical model selection criteria were applied to carefully identify the best models, for ease of use of search results and to obtain a tool that can be used by the parties to the project.

Keywords: Modeling, causes of the change order, projects, and quantitative.

1. Introduction

The term "demand change" has always received a bad reputation among the builders of the construction industry. Changes in construction projects are very common and likely to occur from different sources, various causes, at any stage of a project, and may have considerable negative impacts on items such as costs and schedule delays. A critical change may cause consecutive delays in the project schedule, re-estimation of work statement, extra demands of equipment, materials, labor, and overtime, this leads to additional cost. Changes if not resolved through a formalized change management process, can become the major source of contract disputes, which is a severe risk contributing to project failure. In project management, a change order (variance order) is a component of the change management process, where changes in the scope of work or project activities by the owner, contractor, and consultant are common in most projects [1], especially in large and complex projects. After the scope is formed, contract requirements, the total price to be paid and the specific work to be completed, the owner may decide that the original plans do not represent the best definition of the final project [2].

Accordingly, the owner will propose an alternative method [3]. Researchers at Choy have expressed concern about the negative impacts of project change orders [4]. The change order is defined as "a written authorization by the Engineer or approval to change each of the original plans, specifications or other contract documents, as well as the change in cost", and with appropriate signatures, the change order becomes a legal document. "Change is defined as any an event resulting in an adjustment to the original scope, execution time, or cost of action" [5]. Any additions, deletions or other revisions to the project objectives and scope are considered changes, whether they increase or decrease the cost, schedule or quality of the project [6].

2. Types of Changes

Classifications of changes in general terms apply to changes in the construction domain. "Based on time, the change could be anticipated or emergent, proactive or reactive, or pre-fixity or post-fixity. Based on need, the change could be elective or required, discretionary or nondiscretionary, or preferential or regulatory. Based on effect, the change could be positive, neutral, or negative" [7]. Changes can be classified in many different ways depending on the basis and the purpose of classifications. In this review, the most common classifications will be presented: -

- According to the cause that forced them [1].
- Changes according to the net impact on the scope [8, 9].
- Changes according to the procedure used to introduce them [8, 9]. This classification is important in discussing the legal aspects of changes.

3. Ripple Impact or Cumulative Impact

Ripple impact is a term used in the construction industry to describe a cascade of changes [10]. High ripple projects are ones with a high number and monetary amount of successive changes because of a previous change, whereas low ripple projects are the ones with a low number and monetary amount of successive changes because of a previous change. For high ripple projects, the ripple effect can have a substantial impact on cost and schedule growth [11].

3.1. Data Collection

Previous researches have included many studies on residential, educational, commercial, and industrial projects. Health sector projects (hospitals) were selected to study because of the complexity of this type of project. The study community consists of 11 hospitals distributed in the provinces of Iraq and different clinical capacities as shown in Fig. 1. These projects included architectural and civil works, works sanitary, electrical works, mechanical works, and medical devices. The percentage of changes for civil and architectural works was 47% of the cost of all changes, sanitary works was 22% of the cost of all changes electrical works was 14% of the cost of all changes, mechanical works was 12% of the cost of all changes and medical devices was 5% of the cost of all changes as shown in Fig. 2. The cost of changes can be calculated by the following formula [11].

$$\text{Cost of changes (\%)} = \frac{\text{Final cost} - \text{Original budget}}{\text{Original budget}} \times 100 \tag{1}$$

The sample selection depends on a few selection criteria. The first selection criterion is that participants must have experience work in hospital buildings for 10 years minimum. The working profile of the sample to be selected for data collection must be the contractor, consultants, and owners. The reason is that these three features are the key in any construction projects and they are at the top of the pyramid in the construction activities. The researcher found that the most appropriate way to collect information initially is the questionnaire, so a questionnaire was designed that corresponds to the objectives of the study. Many owners, consultants, and contractors were asked about the main causes of change orders, and the impact of change orders in the construction of the hospitals to have a general preliminary study from their experiences and their point of view, which were used in formulating the questionnaire Closed-ended questions were chosen because they are easy and quick to answer. It only requires a short answer in an agreed format or not always, rarely, etc. [12].

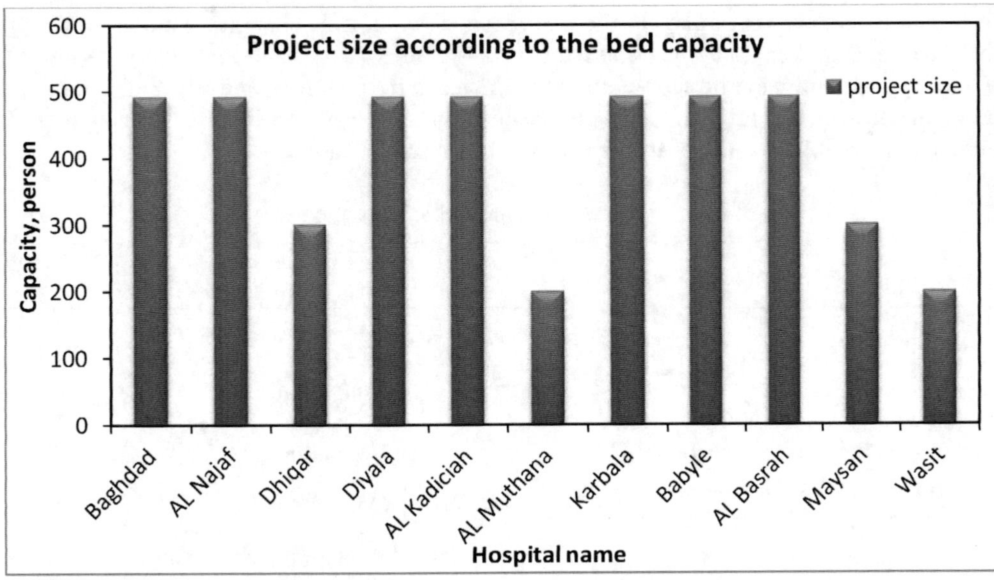

Figure 1. The bed capacity for each project.

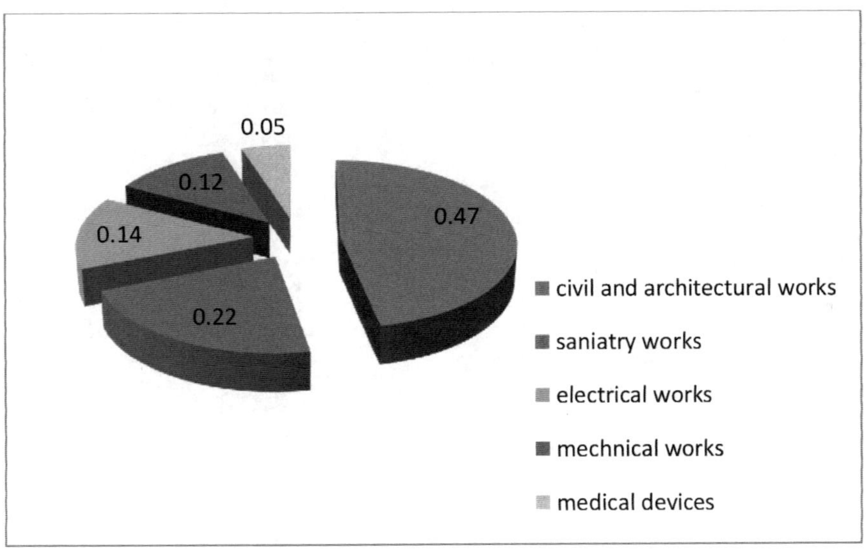

Figure 2. The percentage of changes in the cost for each part of the works from the total changes in the cost.

4. Analysis of the Data of the Questionnaire Form

Project managers were interviewed and contacted to collect data related to change orders. The researcher distributed questionnaires to a sample of 75 participants, 67 questionnaire forms were retrieved. Then deal with the results using the Microsoft excel package and statistical package for social sciences (SPSS) and give explanations for these results.

4.1. Possible Causes of Change Orders in Construction

Many literature reviews have indicated that the action of the project's stakeholders (e.g. owner, consultant, and contractor) may be the direct causes of the change order in construction [6]. Through a survey conducted by the researcher for the hospitals' projects documents of capacity 400, 300, and 200 beds, the main causes that led to occur changes in these projects, were documented and a questionnaire was conducted for the participants in these projects. Table 1 shows the main reasons that led to changes in the work. These reasons were linked to the main responsible for the changes. Number 1 was given to the owner, 2 to the consultant, 3 to the contractor, and 4 to other parties and then analyzed using (SPSS V. 22). The documents of each project included an increase in quantities and the addition of works to the scope of the project due to changes in the description of the project, due to poor planning of the owner and lack of feasibility study of the project as shown in the Table 2.

Table 1. The main responsible for changes.

No.	Reasons for Changes	First Responsible	Frequency	Percent %
1	Financial problems	Owner	46	68.7
2	Change in plans or scope of work	Owner	46	68.7
3	Replacement of materials or work procedures	Consultant	46	68.7
4	Change original design	Consultant	45	67.2
5	Detailed designs are insufficient	Consultant	43	64.2
6	Detailed designs do not conform to the functional requirements of the project	Consultant	44	65.7
7	Errors and omissions in design	Consultant	45	67.2
8	There is ambiguity in the design	Consultant	45	67.2
9	No one of the parties to the contract is involved in the design	Owner	56	83.2
10	Project activities are insufficient	Owner	46	68.7
11	Lack of experience.	Contractor	55	82.1
12	Contradictions of contract documents	Consultant	46	68.2
13	Unlimited scope of work	consultant	46	68.7
`14	Value Engineering.	Consultant	46	68.7
15	Change technology	Consultant	46	68.7
16	Different location conditions	Contractor	46`	68.7
17	Lack of knowledge of local conditions	Contractor	46	68.7
18	Required work skills are not available	Contractor	46	68.2
19	Materials do not meet specifications	Contractor	45	67.2
20	Desire to improve financial conditions	Contractor	46	68.2
21	New government regulations	Others	46	68.2

Table 2. Types of changes (Prepared by the researcher depending on documents of change orders of the hospitals' projects issued by the Ministry of Health.

No.	Type of change orders	The engineers' opinions (%)					Arithmetic Mean
		Always	Frequently	Occasionally	Rarely	Never	
1	Delete any part of the works.	0 0	0 0	46 68.7	10 14.9	11 16.4	2.52
2	Change the description, quality, or class of the works.	46 68.7	21 31.3	0 0	0 0	0 0	4.52
3	Increase the quantity of any work included in the contract.	46 68.7	10 14.9	11 16.4	0 0	0 0	4.52
4	Decrease the quantity of any work included in the contract.	0 0	0 0	12 17.9	44 65.7	11 14.6	2.01

IOP Conf. Series: Materials Science and Engineering **901** (2020) 012032 doi:10.1088/1757-899X/901/1/012032

4.2. Impact of Change Orders on the Construction Projects

Several studies have been conducted to demonstrate the impact of change orders and to identify the main cause of changes in construction and their effects [3,4,8,9]. The impacts of change orders on the construction industry were ranked by using a 5 points scale as the following: never = 1; rarely = 2; occasionally = 3; frequently =4, and always =5. Cost overrun came in the first position with a mean score (4.69), followed by a decrease in productivity which was given a mean score (4.67). Whereas delay in the schedule was ranked the third with a mean score (4.66). Moreover, many of the change orders' adverse impacts were ranked according to their mean scores as shown in Table 3.

Table 3. Impact of change orders.

No.	Impact of change order	The engineers' opinions					Arithmetic Mean	Standard deviation
		Always	Frequently	Occasionally	Rarely	Never		
		No. (%)						
1	Delay in the schedule	44 65.7	23 34.3	0 0	0 0	0 0	4.66	0.478
2	Increase the cost of the Project	46 68.7	21 31.3	0 0	0 0	0 0	4.69	0.467
3	Decrease in productivity	45 67.2	22 32.8	0 0	0 0	0 0	4.67	0.473
4	Dispute between the owner and the contractor	22 32.8	45 67.2	0 0	0 0	0 0	4.33	0.473
5	Low in quality	24 35.8	43 64.2	0 0	0 0	0 0	4.36	0.483
6	Access to materials and equipment	21 31.3	46 68.7	0 0	0 0	0 0	4.31	0.467

4.3. Causes of Change and Its Impact on the Cost Factor of Project

The results of the questionnaire reviewed the causes of changes and their impact on the project. According to these results, the researcher selected the impact of the reasons mentioned previously on the cost element, included an increase in quantities and a change in specifications, which had a significant impact on the cost contracted, and it exceeded by up to 25 %, which is a very high percentage of what is allowed in the law of contracts and standard documents in Iraq. So previous studies have been interested in trying to control this factor, and not overtaken to a certain extent.

4.4. One-Way ANOVA

A one - way ANOVA analysis is used to verify the existence of statistically significant differences between more than two groups in one dependent variable. This is done by comparing between the averages of the groups, the available data about three sets of hospitals, 200, 300, and 400 beds, and to conduct this analysis some conditions must be achieving:

- The samples are random and independent.
- The communities of these samples have a normal distribution.
- The homogeneity of the variance of the community from which random sampling was taken.

4.5. Tests of normality

Table 4 shows that the value of each Kolmogorov-Smirnov Shapiro-Wilk is greater than 0.05.

Table 4. Tests of normality.

Bed capacity	Kolmogorov-Smirnova			Shapiro-Wilk		
	Statistic	df	Sig.	Statistic	df	Sig.
200 bed	0.260	2	.			
300 bed	0.260	2	.			
492 bed	0.206	7	0.200*	0.901	7	0.343

4.6. Homogeneity Test Between Groups

A test was performed homogeneity between three groups of hospitals 200, 300, and 400 beds. Table 5 shows that significance level at 0.362 larger than 0.05 which means achieved the homogeneity.

Table 5. Test of Homogeneity of Variances for Percentage of Change.

Levine Statistic	df1	df2	Sig.
1.175	2	8	0.362

4.7. One - way ANOVA analysis

After fulfilling the conditions of one-way ANOVA analysis were achieved, the analysis was performed using (SPSS V.22) software. This analysis is based on the F-test attributed to a popular probability distribution F-distribution, which has many applications in hypothesis tests. The researcher tests the existence of differences between the averages of the three groups of hospitals or not. F calculated > F tabulated as shown in Table 6

1. Null hypothesis (Ho): $u1 = u2 = u3$
2. The alternative hypothesis (H1): $u1 \neq u2 \neq u3$

Table 6. One-way ANOVA analysis.

Relation	Sum of Squares	df	Mean Square	F	Sig.
Between Groups	0.004	2	0.002	34.371	0.000
Within Groups	0.001	8	0.000		
Total	0.005	10			

F tabulated 4.46

5. Model to Quantifying the Cost of the Change of Hospitals Projects Iraq

Many researchers use administrative and engineering programs and use technological development to find solutions to changes that will difficult to predict or control. The researcher used the data of contract cost and data of change orders cost for hospitals 400 beds, to find a relationship between the two variables according to the reasons given for the previously mentioned change orders. These changes were carried out in an indiscriminate and un-studied manner during the implementation of the projects, which negatively reflected on the increased schedule work and the low quality and productivity of the project, which impact on the increased contractual cost of the project. The researcher managed to reach a mathematical model that can predict the added cost of changes that occur through the planning or implementation phase for the project. The mathematical model was selected from among several models each of which represents an equation: (a linear first-degree-second-the third-equation exponential-logarithmic, etc.), by testing the correlation coefficient for it, and select the model has the largest coefficient, which indicates the greater representation of the sample studied.

6. Model for Calculating the Cost of Changing for All Works of the Contracting Cost

After testing the mathematical relationships in the statistical analysis program (SPSS V .22), show that the relationship is the equation of the third degree:

$$Y = b0 + b1\,X + b2\,X^2 + b3\,X^3$$

Where Y: represents the cost of project changes, X: contractual cost of the project, and provided that the value of X >0. Through the compensation of those constants is the final form of the equation. The equation that represents this model is:

$$Y = 1.35*10^{11} - 2.72\,X - 2.11*10\text{-}^{11}X^{2} - 4.87*10^{-23}\,X^{3} \quad (2)$$

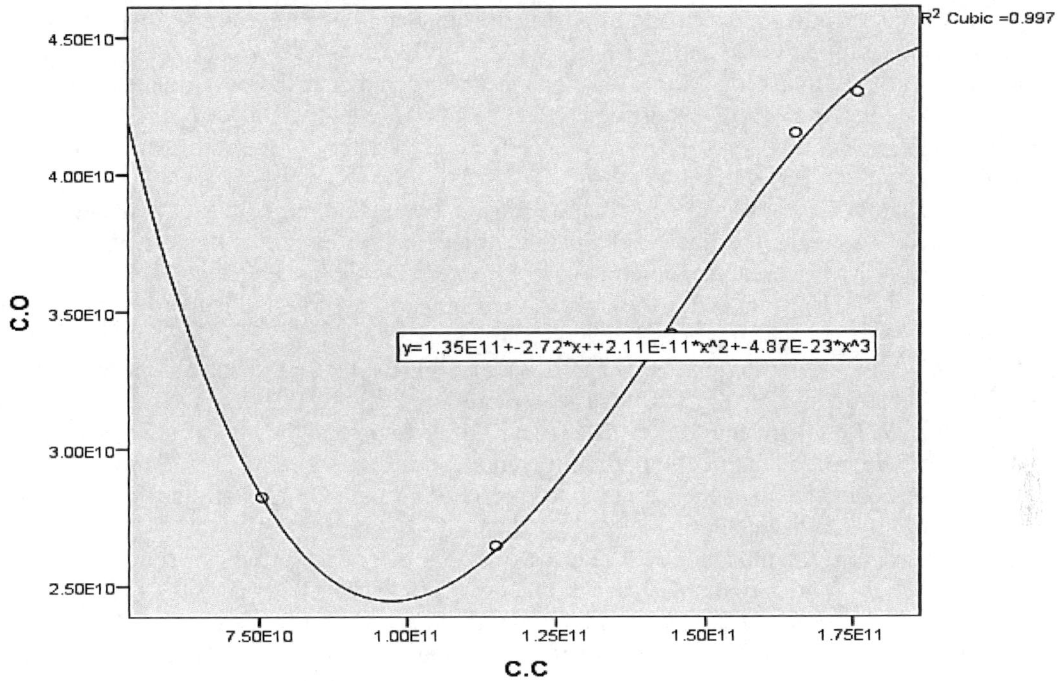

Figure 3. Model for calculating the cost of changing for all works of the contracting cost.

7. Conclusions

Based on the field survey conducted and the results presented the following can be concluded:

- Lack of feasibility study for the project and choose the appropriate type of contract to contract in this type of projects.
- Lack of financial and human resources to manage and follow up on the project.
- Non-contracting with a consultant specialized in this type of project, preparing the initial plans and contracting documents, supervising the contracting procedures, and then participating with the contractor in preparing detailed plans, supervising the worksite, preparing reports, and quantity survey for the site.
- Lack of field experience in this type of projects and lack of specialized staff on the part of the contractor to design, manage and implement the project.

Recommendations

Based on the results of this research, recommendations are made:

- Preferably contracting for this type of projects turnkey.
- Hire experienced workers, engineers, and construction managers to avoid work repetition.
- It is recommended to educate contractors on the negative impacts of change orders.
- Contractors should consider the direct and indirect impact of changes to check feasibility.
- Use updated lists of materials to avoid erroneous material specifications.

References

[1] N.Z. Muhammad, A. Keyvanfar, M. Zaimi, M.Z. Abd Majid, A. Shafaghat, A.M. Magana, and N.S. Dankaka, An evaluation of the causes and effects of variation on building project execution in Nigeria, Jurnal Teknologi, 77(6) January 2015.

[2] F.M. Arain, L.S. Pheng, and S.A. Assaf, Contractors' views of the potential causes of inconsistencies between design and construction in Saudi Arabia. Journal of performance of constructed facilities, 20(1) 2006 74-83.

[3] M.F. Al-Hams, Simulation Model of Change Orders and their Impact on Building Projects Performance in Gaza Strip, Master theses, The Islamic University, Gaza, 2010.

[4] F.M. Arain, and L.S. Pheng, The potential effects of variation orders on institutional building projects, Facilities, 23(11/12) 2005 496-510.

[5] E.R. Fisk and W.D. Reynolds, Construction project administration, 10th Ed., USA, 2014.

[6] M.M. Staiti, Assessing the Impact of Change Orders in Construction Engineering Sector in the West Bank, Master theses, An-Najah National University, Nablus, Palestine, 2015.

[7] C.W. Ibbs, L. D. Nguyen, and S. Lee, Quantified Impacts of Project Change, J. Prof. Issue Eng. Educ. Pract., 133(January) 2007 45–52.

[8] CII, RR282-11 — Improving Project Performance by Estimating, Controlling, and Managing Indirect Construction Costs. The Construction Industry Institute, 2014a.

[9] A. Hanna, W. Lotfallah, and M. Lee, Statistical-Fuzzy Approach to Quantify Cumulative Impact of Change Orders, Journal of computing in civil engineering, 16(4) 2002.

[10] K.A.S. Iskandar, Modeling the Cumulative Impact of Change Orders, Master theses, University of Wisconsin – Madison, 2016.

[11] I.A. Motawa, C.J. Anumba, S. Lee, F. Peña-Mora, An integrated system for change management in construction, Automation in Construction, 16(3) 2007 368-377.

[12] S.G. Naoum, Dissertation Research and Writing for Construction Students, Third Edition, 2012.

[13] J.K. Ssegawa, K.M. Mfolwe, B. Makuke, and B. Kutua, Construction variations: a scourge or a necessity, Proceedings of the 1st CIB-W107 International Conference on Creating a Sustainable Construction Industry in Developing Countries, Cape Town, South Africa, 2002.

IOP Conf. Series: Materials Science and Engineering **901** (2020) 012033 doi:10.1088/1757-899X/901/1/012033

Achieving the Benefits and Requirements of Integrated Project Delivery Method Using BIM

Mahdi S. Salim[1,a] * and Ahmed Mohammed Raoof Mahjoob[1,b]

[1]Civil Engineering Department, University of Baghdad, Baghdad, Iraq.
[a]m.salim1001@coeng.uobaghdad.edu.iq, [b]ahmed.mahjoob@coeng.uobaghdad.edu.iq

Abstract. Through the last decade, Integrated Project Delivery (IPD) methodology considers one of the new contractual relations that are also on the way to further integrate the process of combining design and instruction. On the other hand, Building Information Modeling (BIM) made significant advancements in coordinating the planning and construction processes. It is being used more often in conjunction with traditional delivery methods. In this paper, the researcher will present the achievement of IPD methodology by using BIM through applying on the design of the financial commission building in Mayssan Oil Company in Iraq. The building has not been constructed yet and it was designed by using 2D CAD drawings. The researcher will redesign the project by using BIM to eliciting the abilities that are consistent with the requirements and benefits of integrated project delivery (IPD). The results show that BIM is an appropriate technology that can be used with the IPD method because it has achieved most of the requirements and benefits of IPD.

Keywords: Building information modeling (BIM), integrated project delivery (IPD).

1. Introduction

Integrated Project Delivery (IPD) is a new approach, its main requirements is the involvement of the major project participants from initial phases of the project, and that can be achieved by using BIM. BIM is a tool that recently adopted by architecture, engineering, and construction industries in the world. BIM technology provides a precise digital virtual model of building without conflicts or clashes. It promotes design during all its phases, enabling better control and analysis when the model completed, it become a curate digital model contains all data that support the construction and other phases of the project. Using the IPD method with BIM relatively is a new approach to the design and construction phases of the project [1].

2. IPD Definitions and Characteristics

Many literatures define what IPD is about. Although they have some differences in some parts of the definition, however, there are more similarities. The most common definition is presented by American Institute of Architect California Council, which is "IPD as a project delivery approach that integrates people, organizations, business structures and exercises into a process that collaboratively harnesses the talents and insights of all project participants to optimize the results, increase value to the owner, reduce waste, and maximize efficiency through all stages of design, manufacturing and building"[2]. This detention point to the development of an integrated team which guided by the principles that shown in Fig. 1.

These principles of IPD can be a chance to be used under a variety of contractual exercises to accomplish in cooperation. This will initiate starting from the early phase of design and continue through to deliver the project [4]. The main emphasis in delivering the integrated project is the final value that has been formed for the structure and the owner, instead of focusing each participant exclusively on their share of the building without considering the implications of the entire project process. This IPD system will bring the whole participants together early with cooperative incentives to maximize the value of the project to the customer. This integrated system will allow participants to make decisions in early project designing so they can prepare most of the costs. Close teamwork can get clear of a large waste trade in the scheme, and permit data to be shared simultaneously between the planning and construction team, which removes a large obstacles to increasing project productivity [4].

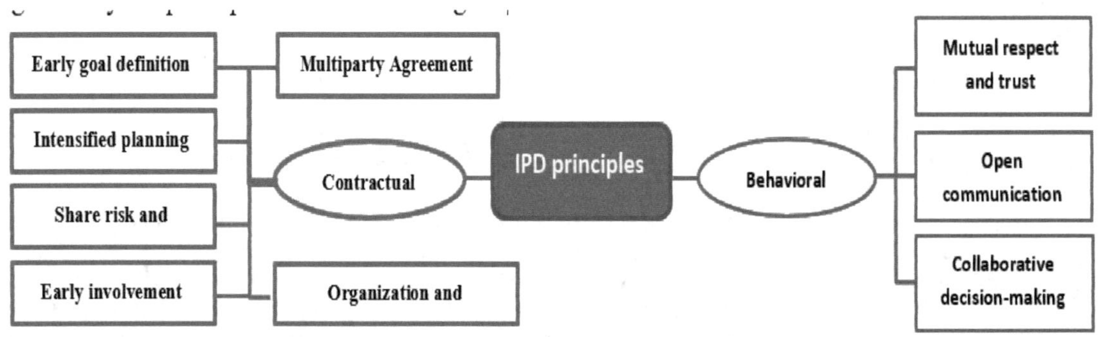

Figure 1. Principals of IPD [2, 3].

3. Building Information Modeling (BIM)

The US National Building Information Model Project committee has a great starting point in illustrating BIM. Although there are many definitions, they defined BIM as " Building Information Modeling (BIM) is a digital representation of physical and functional characteristics of a facility. BIM is a shared knowledge resource for information about a facility forming a reliable basis for decisions during its life-cycle; defined as existing from earliest conception to demolition. Another definition describes BIM as "A basic premise of BIM is collaboration by different stakeholders at different phases of the life cycle of a facility to insert, extract, update or modify information in the BIM to support and reflect the roles of that stakeholder" [6].

As a practical substance, the committee also notes that the BIM represents the management of information, this means data added to and shared by all project members. The right data to the right individual at the right time. Also, BIM represents a project delivery interoperable process - determining how individual teams work and how many teams work together to design-build and operate the facility. As well as BIM represents integrated technology solutions for design, encourages creativity, provides more feedback, and team empowerment. So BIM is a software tool that the project team uses to virtually design and build the building over the Internet by integrating model simulations, such as service constructability and service configurations that can be simulated during construction [7].

4. Relationships Between BIM and IPD

AIA and AIA California Council Guide reflect the interrelationship between BIM and IPD as "It is understood that integrated project delivery and building information modeling (BIM) are different concepts – the first is a process and the second a tool. Certainly, integrated projects are done without BIM and BIM is used in non-integrated processes. However, the full potential benefits of IPD and BIM are achieved only when they are used together" [2]. Moreover, IPD and BIM match each other by improving project management by increasing data exchange and collaboration between stakeholders, this leading to less risk of defects and rectification; less materials waste and less disputes during construction

IPD is the catalyst that allows parties to easily exchange ideas, information, and intellectual property, and this along with BIM, creates efficiencies throughout the project life. This methodology also removes the " us vs them " attitude because participants must share protected intellectual property in another way to achieve project outcomes.

5. Application of IPD-BIM Integration

To examine the achievement of IPD-BIM integration, a construction project will be used as a case study. The project locates in the campus of Mayssan Oil Company in the center of Al-Amarah city / Mayssan province in Iraq. The area of the project 2240 m^2. The building includes three stories, the ground story consist of sixteen office rooms, two holes, two rooms for manager of commission, fortified room to store the money, and a room for electricity adapters. The first floor also consists of the same described facilities except one hole was divided into two rooms and set an office room instead of the protected room, while the second floor was the same with the ground floor. The building scheme also consisted of one elevator, two water cycles for each story, and two emergency exits.

The building design involves all the items in the structural design in which the concrete was used for foundation, columns, ceiling, and walls that make up the protected room, as well as the brick for the walls of the other rooms of the building. While the architectural interfaces were designed in a way that to be compatible with nearby buildings interfaces. Fig. 2 shows the architectural drawings for the ground floor.

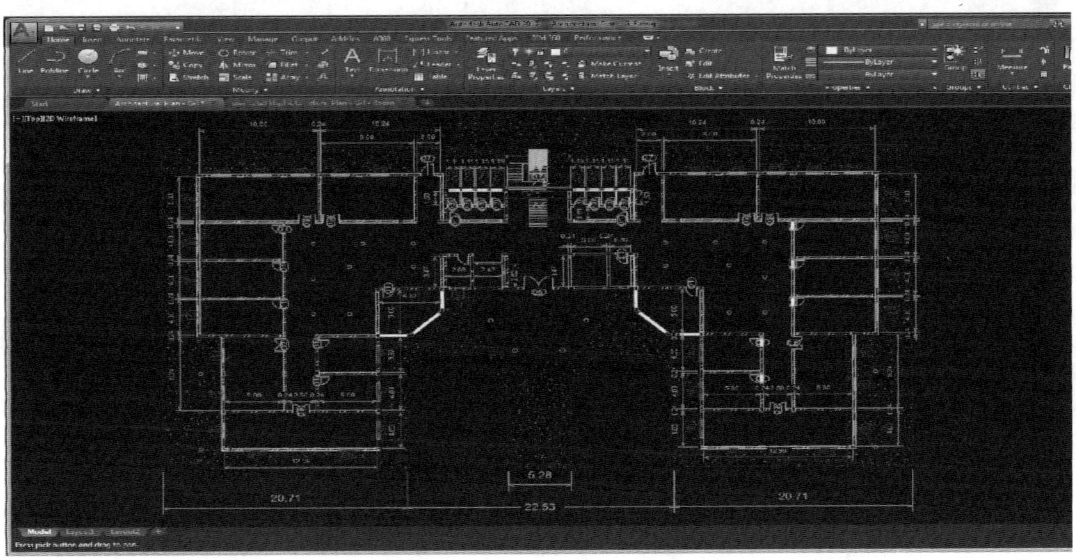

Figure 2. Architectural drawing for the ground floor of the financial commission building.

6. Current Design Workflow

In this case study, the workflow of design consists of several steps that usually adopted in the Iraqi government firms when implementing the work using traditional ways such as direct implementation or Design-Bid-Building. After obtaining the approval of the establishment of the project, the design department begins to form the schematic design using 2D CAD according to the building requirements stated by the requesting commission for the building, and then presented this conceptual design to the same body to make notes on the plan. After amendment on the plan according to the notes, the design department begins to prepare a detailed design for the architectural, structural, electrical, and mechanical designs separately using 2D. They use to description of the plumbing works illustrated in the bill of quantities only without being represented in the drawings, depending on the implementation on the site by the engineers at the construction phase. The bill of quantities were performed by each division according to the specialist, and then combined in one bill and estimate the cost for the project and obtain the approval for the budget to prepare the contract documents for bidding. Fig. 3 summarized the current workflow of the designing phase in the department of design.

In general, there is no single model that combines all the components of the building design to detect conflicts that cause delays in the construction phase. Also, the steps followed in the design stage suffer from fragmentation and were not characterized by cooperation where each division design their part and prepare the related bill of quantities without reference to other components of the building design. This may cause many inconsistencies and errors during the construction phase that result in a delay in the implementation time and increasing the construction cost of the project due to rework. Also, the majority of these activities in the design phase were made by correspondence between departments and between designers while the face-to-face communication may be nonexistent.

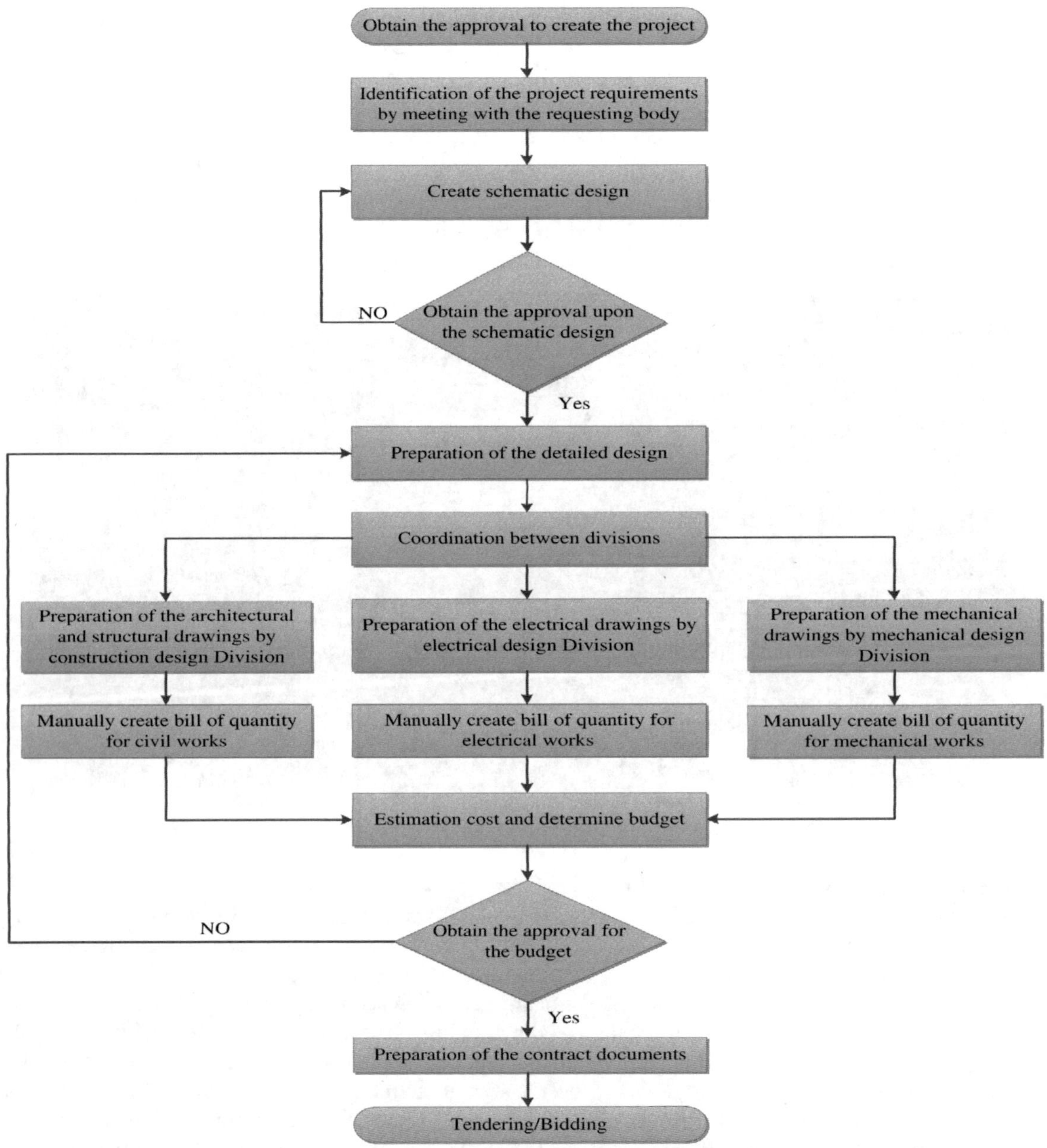

Figure 3. the current design flow chart for the case study during the design phase.

IOP Conf. Series: Materials Science and Engineering **901** (2020) 012033 doi:10.1088/1757-899X/901/1/012033

7. Application of BIM

7.1. Design Requirements

The Design of the financial commission building was performed according to the requirements that were stated during the meeting between the Director of the financial commission, the Director of the projects commission, the department's Directors of the financial commitment and the designers, which were as following:

- A building that accommodates a staff of six departments with all divisions.
- Design the floors and rooms according to the relation between the departments so that each two departments have more communication will locate on one floor.
- Specify the ground floor to the departments related to the payment of salaries of employees and financial dues to facilitate the delivery of dues through the external windows of the building.
- Adopting the halls system for the employees to facilitate the communication between them and allocate one hall for training.
- Placing emergency exits with fire alarm system in the building for safety requirements.
- Installing external roofs around the building on windows where salaries are delivered to prevent the arrival of sunlight as well as the rains when salaries are delivered to employees.
- Identify the money storage room on the ground floor as it is intended to pay salaries and dues.

The main goal was to fulfill these requirements so that they construct one building collecting all the departments of the financial commitment to facilitate the communication between them and decrease the time consumed to perform the transaction. Meeting with most of the stakeholders of the project to identify the requirements of the building before starting the design process, means involve the participation of all parties concerned from the early stages of the project. This step will achieve one of the most important principles of the IPD method (Early Involvement of Key Participants), which enhances the ability to define the project goals (Early Goal Definition). Thus, creating a long list of requirements are achieved, and then shorten this list to the requirements only to being creating a schematic design and produce it to the owner or the requesting body for approval. Then starting the 2D architectural plan. To redesign the building using BIM methodology, the architectural model will be allocated after conversion from 2D to 3D as a central file to be imported to create local files to complete the other components of the building (Structural and MEP design). The design in each local file will be synchronized to the central file to show all the updates in the central file and show the work of each other local files in this local file and vice versa. The researcher will apply the flowchart diagram that illustrated in Fig. 4 to redesign the building using BIM methodology.

IOP Conf. Series: Materials Science and Engineering **901** (2020) 012033 doi:10.1088/1757-899X/901/1/012033

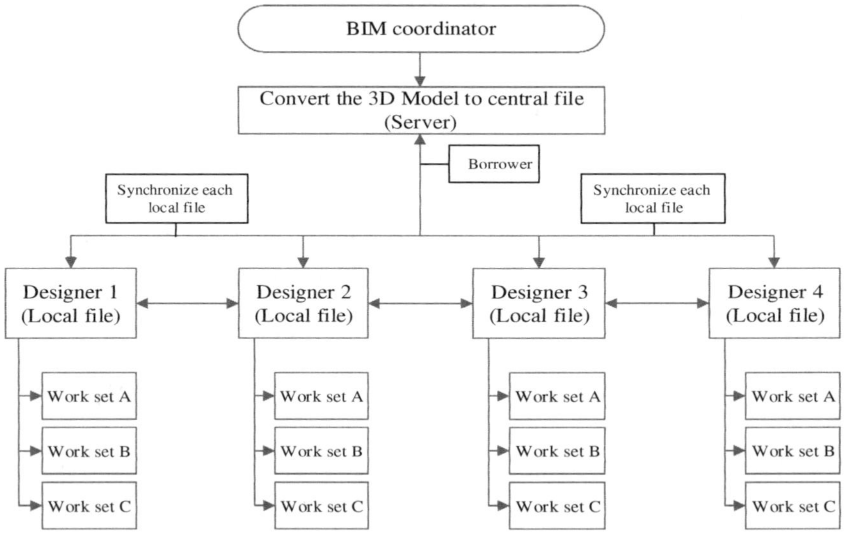

Figure 4. the flowchart diagram of BIM implementation.

7.2. Architectural model
The researcher used Revit 2019 to create a 3D architectural model and integrate all other components of the case study. The start was by import the 2D plan for each story of the building from cad to Revit. Based on the description of some items such as interface's items, the finishing of walls, the location of columns in the 2D drawings, location, and dimensions of the windows and doors. The architectural model was built to become the original model for all other discipline designers, which later will be imported to complete the design details of other components of the building. Fig. 5 shows the 3D architectural model.

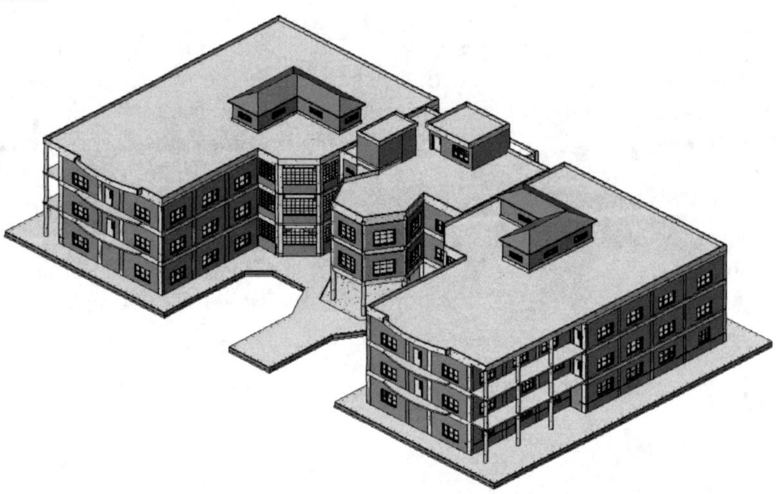

Figure 5. The 3D architectural model.

7.3. Structural Form
After the architectural design was completed, the model was imported to create a local file related to the structural design. The structural design began by applying the live and dead loads using the structural analysis programs to determine the dimensions of the foundations, columns, beams, and the slab based on carry out the required calculations, as well as calculating the required reinforcements. Then establishing the elements cross-sections of building to cover all structural details that will be needed in the construction phase of the project. The structural model was synchronized to the central

file to merge the structural and architectural design in one model. Fig. 6 shows the structural model for the building.

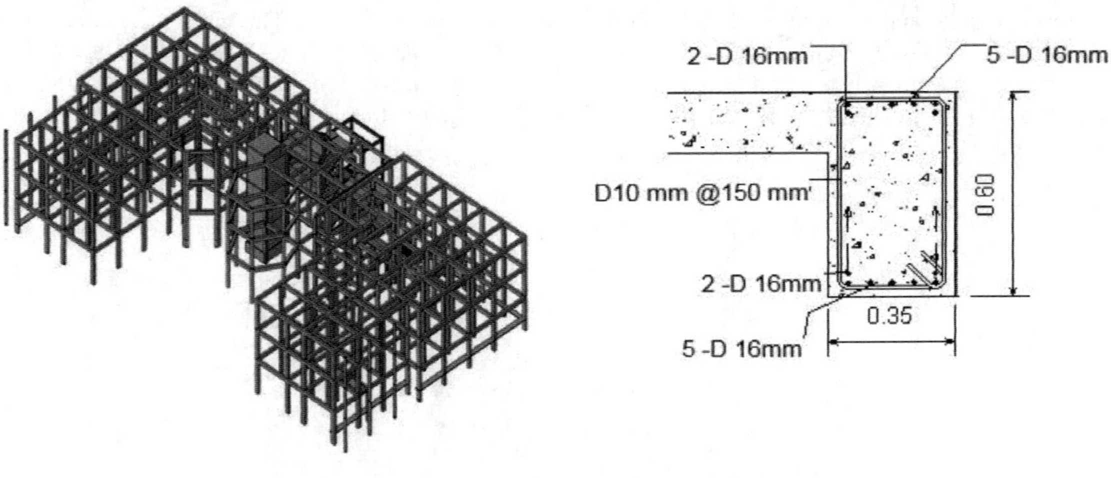

Figure 6. the structural model and section of beam B4.

7.4. Mechanical, Electrical & Plumbing (MEP) Design

The case study does not contain any mechanical works, therefore the researcher will focusing on the electrical and plumping designs in this domain.

7.5. Electrical Design

A local file was created related to electrical design and subdivided into four work sets represent the lighting, power, telephone, and fire alarm system to be synchronized to the central file. All the electrical drawings were presented in a 3D model based on 2D CAD and the description of electrical items in the bill of quantities. The interior ceiling light fixture and wall lighting were represented in the 3D model for each floor. All these lighting fixtures and other electrical fixtures such as fans and power sockets 13A, 32A, 45A, were linked to the electrical secondary panel for lighting and power. Fig. 7 shows the distribution of ceiling light and its wiring system.

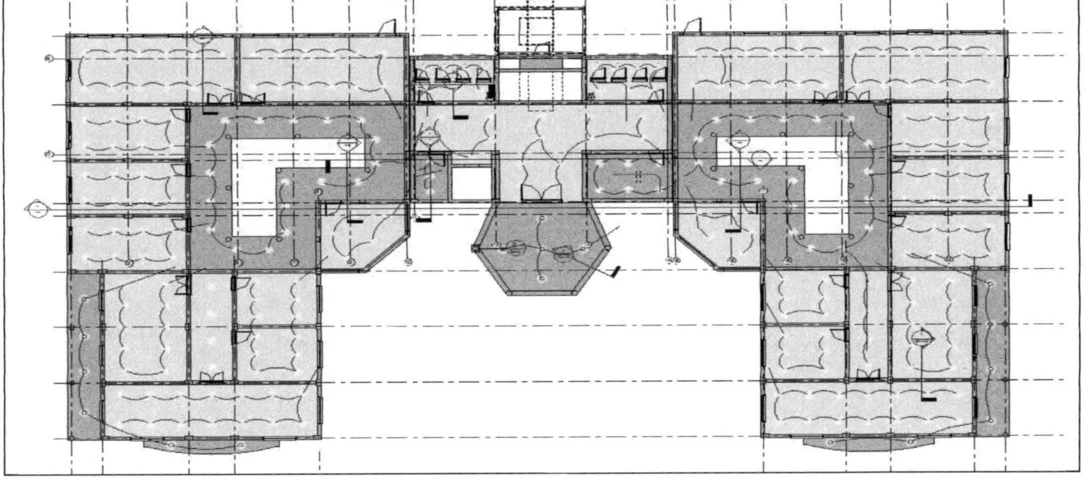

Figure 7. The ceiling light fixture and wiring system.

The fire alarm system was also represented in the 3D model based on the 2D CAD and the items description in the bill of quantity. Smoke detectors have represented in the ceiling for each story and linking them to the 4-zone panel of fire alarm. The electrical and telephone cables will be represented by draw conduits and marking each conduit to represent a specific cable. The cables included in the building consist of the cables from the main boards to the sub-panels on each floor, in addition to the interconnecting cables between the mainboards, as well as the interconnecting cables between the main boards and the main feeders of the building. The specification of each cable was fixed in the properties as shown in Fig. 8.

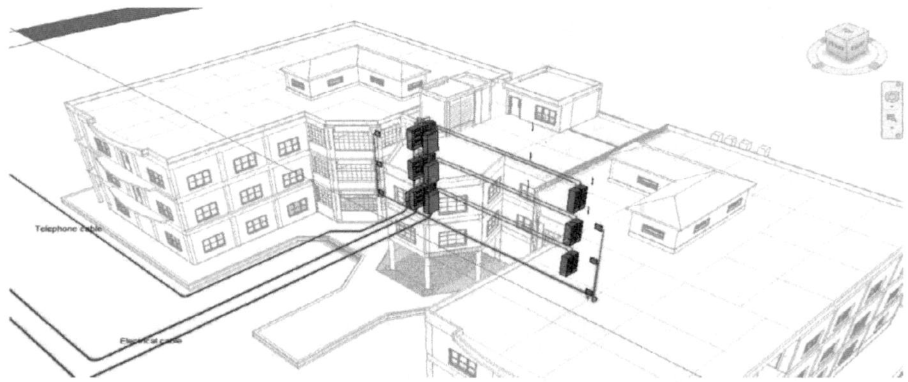

Figure 8. Telephone and Electrical cables and panels.

7.6. Plumping Design

In the same manner, a local file will be created from the center file to complete the plumping design. The plumping components are not included in the 2D CAD drawings because they used to perform the plumping works based on the description of the item in the bill of quantities only, while the electrical drawings made in 2D CAD as in architectural and structural aspects. Therefore, the researcher represented the plumping items in the drawings based on the requirements of the requesting body and the description of this item in the bill of quantity. The roots of the sewage drains, cold and hot water pipes were identified in all stories of the building, as well as the plumbing fixtures, which consist of western and eastern toilets, heaters, water tanks in above the building, pipefittings and manholes. Then after complete the plumping design, the model was synchronized to the central file to show all the architectural, structural, plumping design in one model. Fig. 9 shows the design of the plumping works in the 3D model.

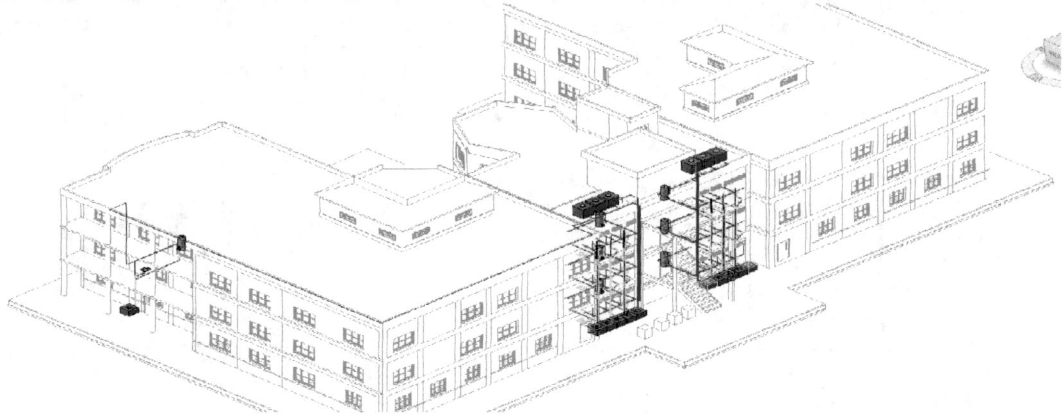

Figure 9. The plumping works.

This teamwork in the design phase was done through the work-sharing feature in Revit 2019 that allowed all team members to communicate and collaborate by exchanging information and synchronizing each work to the central model. Therefore, the communication was open between all parties and mutual trust in addition to mutual understanding and a high degree of coordination between them that led to the completion of the design of the building in all its domains. This means that several IPD requirements have been achieved, which are (Collaboration during all phases of the project, Mutual understanding concerning each other's objectives, Open communication among parties, Mutual trust among the parties).

7.7. Clash Detection

One of the most important features of the BIM methodology is identified of all clashes in the design phase, where a conflict is identified between any two items in the design phase that can lead to project delay or increases the cost during the construction. Clash detection leads to preventing rework during construction, which saves large amounts of money in addition to reducing the schedule and preventing many problems in the construction phase. Revit provides this feature but the researcher has chosen the (Autodesk Navisworks Manage 2019) to implement the clash detective as it has better features than Revit in the detection of clashes, it provides a detailed photo-enhanced report for each conflict.

After the coordination of all domains of the building in one 3D model, the researcher starts to export the model to Autodesk Navisworks through (Add-Ins) tab and save the file in the computer as (nwc) then open Autodesk Naviswork and open the saved file. The clashes were initially identified between the plumping and beams, as well as the clashes between the conduits and structural framing, and the electrical work with architectural and structural design, so that there were no clashes between them as shown in Fig. 10. Regarding the clashes between the pipes with walls and between the structural framing with walls, the results showed that there are 45 and 29 clashes, respectively. The most of the pipes and walls clashes are considered acceptable, as most pipes works pass through walls, and this is what is implemented in projects, so all the clashes were approved except three of them that were resolved because the pipes were outside the boundaries of the wall as shown in Fig. 11. While all the clashes between the structural framing and walls were approved because they overlap simply so as not to cause delay or rework, where most of them overlap with upturned bridges.

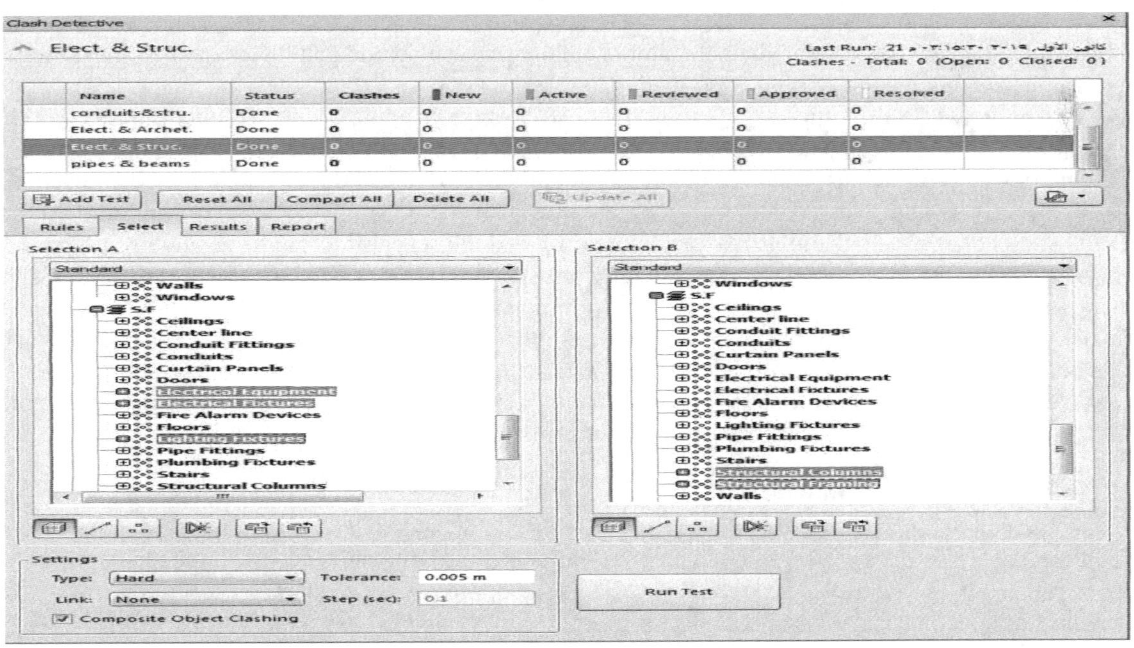

Figure 10. Results of clash detection for some items.

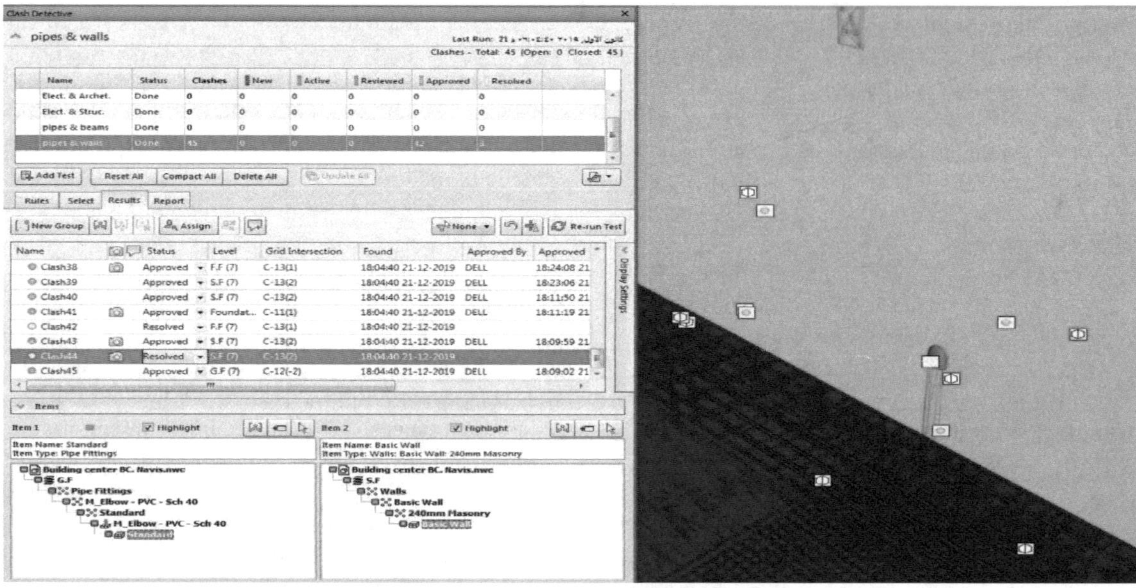

Figure 11. The view of a clash for pipes and walls clashes detection.

Using BIM methodology in the design phase has immediate and clear benefits. This methodology requires the collaboration to obtain the benefits of this technique for all participants, which enhances mutual trust, exchanging of information and open communication. Therefore, most of the requirements and principles of IPD method have been achieved, so BIM technology can be considered as the appropriate technology because it catalyzes the processes of the IPD method, and thus fulfills one of the principles and requirements of IPD method (appropriate technology).

7.8. Achievement of IPD Benefits

BIM is the only tool that can facilitate the work of more than one participant in the same project model and can make all members of the project teamwork on the same model. This helps to enhance cooperation between all participants, enable them to arrange meetings based on this model, prevent duplication of information, and complete the work faster as well as prevent any collision in the building design. This means obtaining high-quality building, which leads to reducing project costs and waste, as well as improving contract documents to exclude the use of the requests for information (RFIs) and change orders then the cost of the building and the scheduling can be accurately estimated. These gained benefits from using BIM are fully agreed with the benefits of the IPD method that is (Arrange collaborative meetings based on a shared model and a virtual reality application, get better quality of design, with less waste, shorter timelines, decrease the construction costs, interference and collision detection, improved contract documentation to eliminate using RFIs and Change Orders, improved preconstruction estimating, procurement, expediting and scheduling, faster and more effective processes through sharing and reusing information).

8. Conclusions and Recommendations

The construction industry needs to adopt new technology and delivery systems that differ from traditional methods if they want to overcome all obstacles in traditional systems to reduce costs, time, and waste and thus increase productivity. IPD method provides a collaborative environment not provided by the traditional approaches of delivery by involving all the main parties from the early stages of the project and cooperation during all stages and open communication and mutual trust between the parties in addition to effective coordination. BIM is a suitable technical tool that needs a collaborative environment to obtain the highest benefits. Therefore, it has achieved many requirements and benefits of IPD when used together. Also, the benefits of IPD have been achieved when using BIM methodology so BIM is the appropriate technology to be used with the IPD method to improve the performance of construction projects. Adopting the use of IPD method in the public sector (as a philosophy), which means adopting this method through its principles because of the difficulty of creating a contract among all parties, will help in providing better and faster building in the future. As well as encourage using the IPD method with BIM to obtain the greatest benefit for BIM and helps to providing faster and excellent buildings.

References

[1] J. Neil, The use of an Integrated Project Delivery Approach in conjunction with BIM, M.Sc. MSc Building Information Modelling and Integrated Design, University of Salford, 2015.

[2] A.I.A. Guide, Integrated project delivery: A guide. American Institute of Architects, California (2007).

[3] B.M. Perlberg, Contracting for Integrated Project Delivery: Consensus Docs, The 48th Annual Meeting of Invited Attorneys, Victor O. Schinnerer & Company, Inc., 2009.

[4] R. Anderson, An Introduction to the IPD Workflow for Vectorworks BIM Users, Nemetschek, Vectorworks, 2010.

[5] S.A. Rahim, M.N. Mohd Nawi, and F.A.A. Nifa, Integrated project delivery (IPD): A collaborative approach to improve the construction industry, Adv. Sci. Lett., 22(5–6) 2016 1331–1335.

[6] B. McNamara, BIM: What Can It Do For Me?/Bentley Building Solutions, presented to the CADD/GIS Center Symposium. https://tsc.wes.army.mil/symposium/2006/Thursday%5CMcNamara.pdf," 2006.

[7] A. Mahjoob and M.M. Abed, Assessing BIM Integration with Sustainable Requirement for Buildings Construction, International Journal of Science and Research (IJSR), 6 2017 2258-2265.

IOP Conf. Series: Materials Science and Engineering **901** (2020) 012034 doi:10.1088/1757-899X/901/1/012034

Evaluating Implementation of Electric Power Generation Projects in Iraq

Ameer A. Nadhum[1, a *], **Kadhim R. Erzaij**[1, b]

[1]Civil Engineering Department, University of Baghdad, Baghdad- Iraq
[a]a.nathum1001@coeng.uobaghdad.edu.iq, [b]kadhim1969@yahoo.com

Abstract. Iraq has witnessed development in many fields and the electricity production sector is one of the most important of these fields. This is attributed to the State's philosophy in supporting this sector through the development of economic development projects to serve and raise the level of the Iraqi individual. Electricity production in Iraq is affected by many variables such as load growth, economic and financial conditions, planning, etc. All these led to a disparity in the production of electric power to cover the demand for energy. The paper deals with the electricity generation sector overview in Iraq, Analysis of the effects of variables on the implementation of power generation projects in Iraq, and Creating a mathematical model. The results enable decision-makers to prioritize the implementation of power generation projects in Iraq.

Keywords: Electrical power planning, Implementation of power generation projects, Multi-linear regression.

1. Introduction

The electricity sector is a vital sector that plays an important role in the economic development of many countries that have faced wars such as Iraq [1]. The implementation of power plant projects in Iraq is one of the important elements in improving the reality of electric power, which suffers from a deficit in the provision of electric power over the past decades for several political, security and economic reasons. The authors provide an assessment of the factors of implementation of power generation projects in Iraq and consider how to draw policy can determine the priority of project implementation according to these factors. They discuss the requirements of the management of the implementation of these projects, in which the objectives of this policy can be achieved to implementation of projects better.

Electrical power planning has been determined in varying degrees of inclusiveness. Some scientists have given relatively narrow definitions. There are several definitions in this area where energy planning is known, Hiremath et al. (2007) states that "energy planning is limited to having sources and switching devices in groups that ideally meet the energy requirements of various tasks. This can occur at the central or decentralized level" [2]. A process aimed at finding a solution, often at the lowest cost, for a particular centralized or decentralized demand. Later, however, the scope of energy planning was greatly expanded.

Both Rojas-Zerpa and Yusta (2014) and Loken (2007) have debated that energy planning problems are limited to multiple criteria for decision-making. so that a simple and globally optimal planning does not exist [3,4]. However, both views reflect[2]. Energy planning, often by focusing on finding many suitable options for power supply. There are comprehensive definitions of energy planning by other researchers. Summarized under the term integrated energy planning. Bhattacharya (2012), in his paper on off-grid electricity planning, mentioned that "successful implementation... requires careful planning of all phases associated, which goes beyond stripping technology selection or selection decision of component" [5]. Along the value series and from other stages that he considers important for energy planning for obtaining on best estimates of demand, energy delivery, and appropriate implementation mechanisms Bhattacharyya remarkably highlighted the importance of energy planning analysis to draw valuable lessons for future projects.

Mirakyan and De Guio (2013) display a broader definition in their review of integrated energy planning. It again features the multi-criteria nature of energy planning, its applicability at different units of analysis, and their diffusion across different stages of the value chain for different power stages including generation, transmission, distribution [6]. Söderholm (1999) notes that lifetime costing approaches tend to miss considerations of maximum energy demand, rather than focusing on actual energy provided for a longer period. Most technologies may be economically viable in these cases, despite their high lifetime cost to prevent temporary power outages. In general, he points out that economic standards are of global importance for finding more cost-effective solutions for supplying electricity to developing [7].

Technological factors generally include functionality, reliability, efficiency, safety, and energy quality measured in exergy content. For example, Okello et al. (2014) found several technological advantages to the use of bioenergy technology including reliability, safety, and functionality [8]. Researchers P. A. Trotter, M. C. McManus, and R. Maconachie (2017), a systematic review of the planning and implementation of electricity in sub-Saharan Africa, analyzed decision factors and criteria that ensure economic, social, environmental, technological and political standards [9]. The objective of this study is to identify the factors that most influence the implementation of electric power generation projects in Iraq.

2. Methodology

One of the most important methods used to provide engineering solutions to problems. Many data that includes many variables can be used by analyzing it, and results can be used and entered in many applications. In this paper, a statistical method for prediction will be used by analyzing the factors statistically and creating a mathematical model using the SPSS statistical analysis program through which a strategic plan is developed to implementation of electric power generation projects in Iraq and determine the priority of implementation of these projects.

3. Electricity generation sector overview in Iraq

Most of the existing stations in Iraq were built between 1955 and 1986. Many were damaged during the first Gulf War in 1991 and did not receive adequate rehabilitation or maintenance due to lack of spare parts and the shortage of skills imposed by the economic sanctions for more than 12 years following the Gulf War, with a few small gas stations that operated in 2003. Iraq's other generating capacity weaknesses include low generating availability, high technical losses, and load factors resulting in an inability to cope with peak daily demand. Most of the existing electric power plants are thermal electric plants that use crude oil backed by gas and fire plants [10].

Types of electrical generation plants in Iraq are thermal power plants, gas power plants, diesel power plants, and hydropower plants. Within the formations of the Ministry of Electricity, there are four public companies for the production of electricity distributed by geographical regions of Iraq, namely: North Production Company, South Production Company, Middle Production Company, and Middle Euphrates Production Company. The total power generation according to the installed capacity of the four companies is more than 27182 MW and the total power generation for the operating units is 22592 MW and the actual production rate is 9758 MW[11]. These stations are composed of 375 generating units, of which 277 units are in operation with a generating capacity ranging from 1.7 to 610 MW Alone. These generating units are distributed according to their eight hydroelectric power plants with a combined design capacity of 1674 MW, eight thermal power plants with a total capacity of 5,985 MW, 33 gas turbine power plants with a total capacity of about 13,414 MW and 12 diesel engine stations with a total capacity of 1,594 MW. The country also imports about 914 MW[11] from neighboring countries and local investment stations. Fig. 1 shows an overview of electricity generation, which relies mostly on fossil fuels, which accounts for more than 86.3 percent of the main electricity sources in the grid [10].

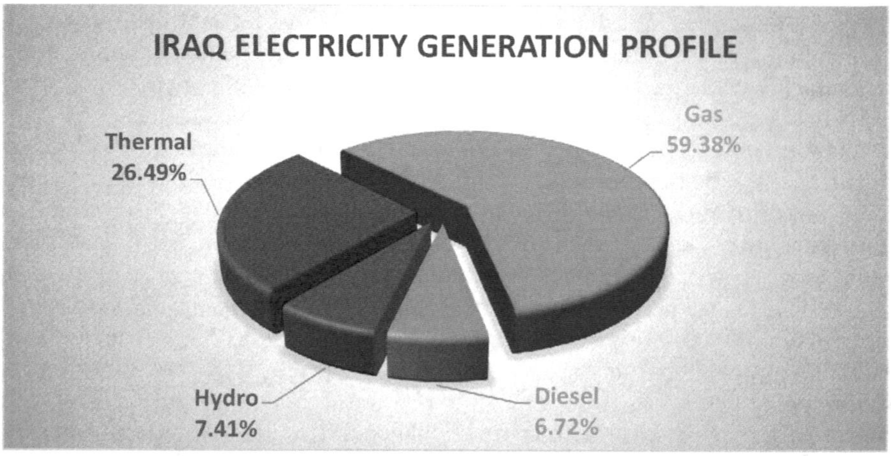

Figure 1. Iraq electricity generation profile (Source: Iraqi Ministry of Electricity).

4. Database

The factors that interfere in this context are based on the data collected through the work of a questionnaire in cooperation with decision-makers, experts and specialists in the energy field in the Iraqi Ministry of Electricity as well as university professors after investigating the interviews and making the required adjustments to create the questionnaire and distribute it to specialists and experts and determine the factors affecting the implementation of generation projects Electrical energy in Iraq.
These factors are classified by the researcher to implementation of generation projects electrical energy in Iraq that include major factors financial and economic aspect, availability of infrastructure, and planning aspect, and sub-criteria as shown in Table 1.

Table 1. Main criteria affecting the implementation of power plant projects and their sub-criteria.

Code	Main criteria	Code	Sub-criteria
C1	Financial and economic aspect	C1-F1	Capital cost of project
		C1-F2	Operating and Maintenance Cost
		C1-F3	Project duration
C2	Infrastructure availability	C2-F1	Near the location of the transmission lines
		C2-F2	Near the site from the discharge source loads
		C2-F3	Near the site of the source of water to cool the station
C3	Planning aspect	C3-F1	Capacity of the station
		C3-F2	Project area

5. Linear Regression

When there is a linear relationship between the variable factors it is called linear regression. There are two types of linear regression, the first is simply one variable and the second is multiple variables [12]. If more than one independent variable is used, the multivariate linear regression (MLR) is used. Eq. 1 shows the linear model of MLR formula.

$$Y = a + b1 * X1 + b2*X2 + b3* X3 + \ldots + bn*Xn + Error \quad (1)$$

Where
a: is the intercept
b1, b2, b3 and bn: the slopes.
Y: is the dependent variable
X1, X2, X3 and Xn: the independent variables [13] .

5.1. Create the mathematical model

In this paper, the factors that influence the implementation of electric power generation projects in Iraq will be analyzed to create a mathematical model financial and economic aspect, availability of infrastructure, planning aspect.

5.2. Financial and economic aspect

To determine the financial and economic aspects of energy resources, "Levelized-cost" is generally used. This cost takes into account all the costs of the plant from the capital, operation, and maintenance. The initial cost of construction and his modification cost can be classified at the capital cost. This cost is covered and is amortized over by the energy output expected to be generated during plant life. As for the daily costs associated with operating the electric power plant, it is classified under maintenance and operation costs [14]. Dealing with the variables financial and economic aspect that has the most important impact on the factors of implementation of projects of electric power production, which are appropriate with the formula is as shown in Eq. 2

$$FEA = C + (F_1 \times I_1) + (F_2 \times I_2) + (F_3 \times I_3) \quad (2)$$

Where:
FEA: Financial and economic aspect
C: Constant
I_1: Plant Capital Cost
I_2: Operation and Maintenance Cost
I_3: Project Duration

Through analyzing the questionnaire using the SPSS statistical analysis program and analyzing the parameters, it is shown to us in Table 2. From the analysis of parameters, the equation for the financial and economic aspects as shown in Eq. 3.

$$FEA = 0.14 + 0.579\, I_1 + 0.286\, I_2 + 0.125\, I_3 \quad (3)$$

Table 2. Parameter Estimates of variables for financial and economic aspects.

Parameter	Estimate	Std. Error	95% Confidence Interval	
			Lower Bound	Upper Bound
Constant (C)	0.14	0.346	-0.558	0.839
F_1	0.579	0.098	0.381	0.776
F_2	0.286	0.100	0.084	0.487
F_3	0.125	0.110	-0.097	0.346

This equation Eq. 3 is used to calculate the effect of the financial and economic factor, to use into the calculation of the main equation. Table 3. shows a summary of the mathematical model shown in Eq. 3 and shows the reliability of this equation for its sub-factors with the main factor.

Table 3. Model Summary of dependent variables for financial and economic aspects.

Model	R	R Square	Adjusted R Square	Std. Error of the Estimate
1	0.882[a]	0.778	0.762	0.490

a. Predictors: (Constant), Project duration (Project schedule until the station is operational and connected to the network), Operating and Maintenance Cost, the cost of setting up the project
b. Dependent Variable: Financial and economic aspect

6. Availability of Infrastructure

Generation power plant projects are experiencing problems of locating more than other different projects due to the difficulty of linking these plants with power transmission lines because they usually face the problems of land possession with many different owners. These problems may lead to opposition from the local people and may require lengthy procedures to address and eliminate these difficulties and obstacles, which may lead to time delays. As such, it proposes locating the nearest power transmission networks as well as their proximity to the loads discharge centers [15]. The independent variables of the availability of infrastructure that have the most important influence on the factors of implementation of electric energy production projects, which are appropriate with the formula is as shown in Eq. 4.

$$AI = C + (A_1 \times I_1) + (A_2 \times I_2) + (A_3 \times I_3) \quad (4)$$

Where:
AI: Availability of infrastructure
C: Constant
I_1: Distance of Project Site from The Transmission Lines
I_2: Distance of Project Site from The Source of Discharge Loads
I_3: Distance of project site from the water source (cool plant)

Through analyzing the questionnaire using the SPSS statistical analysis program and analyzing the parameters, it is shown to us in Table 4.

Table 4. Parameter Estimates of variables for Availability of infrastructure.

Parameter	Estimate	Std. Error	95% Confidence Interval	
			Lower Bound	Upper Bound
Constant (C)	0.234	0.252	-0.274	0.742
A_1	0.289	0.167	-0.048	0.627
A_2	0.415	0.197	0.017	0.813
A_3	0.207	0.173	-0.143	0.558

From the analysis of parameters, the equation for availability of infrastructure as shown Eq. 5

$$AI = 0.234 + 0.289\ I_1 + 0.415\ I_2 + 0.207\ I_3 \quad (5)$$

This equation Eq. 5 is used to calculate the effect of Availability of infrastructure, to use into the calculation of the main equation. Table 5 shows a summary of the mathematical model shown in Eq. 5 and shows the reliability of this equation for its sub-factors with the main factor.

Table 5. Model Summary of dependent variables for financial and economic aspects.

Model	R	R Square	Adjusted R Square	Std. Error of the Estimate
1	0.914[a]	0.836	0.824	0.404

a. Predictors: (Constant), Near the site of the source of water for the cooling of plant, Near the location of the transmission lines (the presence of a transport network close to the center of the load to connect the plant to the plant's link to the national grid transmitting energy), Near the site from the discharge source loads
b. Dependent Variable: Availability of infrastructure

7. Planning aspect

The planning aspect includes many sub-factors and tasks that fall within the planning process. It is not possible to deal with all aspects and problems of planning in one way. Therefore, studies that combine several factors, stages and different ways to achieve the goals of all aspects of planning are important

[6]. Dealing with the independent variables for Planning aspects that have the most important impact on the factors of implementation of projects of electric power production, which are appropriate with the formula is as shown in Eq. 6.

$$PA = C + (P_1 \times I_1) + (P_2 \times I_2) \tag{6}$$

Where:
PA: Planning aspect
C: Constant
I_1: Plant Capacity
I_2: Project Area

Through analyzing the questionnaire using the SPSS statistical analysis program and analyzing the parameters, it is shown to us in Table 6.

Table 6. Parameter Estimates of variables for Planning aspect.

Parameter	Estimate	Std. Error	95% Confidence Interval	
			Lower Bound	Upper Bound
Constant (C)	1.21	0.287	0.632	1.789
P_1	0.392	0.082	0.226	0.557
P_2	0.344	0.081	0.180	0.508

From the analysis of parameters, the equation for the planning aspect as shown in Eq. 7.

$$PA = 1.21 + 0.392\ I_1 + 0.344\ I_2 \tag{7}$$

This equation Eq. 7 is used to calculate the effect of the planning aspect, to use into the calculation of the main equation. Table 7 shows a summary of the mathematical model shown in Eq. 7 and shows the reliability of this equation for its sub-factors with the main factor.

Table 7. Model Summary of dependent variables for the planning aspect.

Model	R	R Square	Adjusted R Square	Std. Error of the Estimate
1	0.824[a]	0.679	0.664	0.428

a. Predictors: (Constant), Project area, Capacity of plant (the contribution of plant to meet the demand for energy)
b. Dependent Variable: Planning aspect

8. Main Factors Equation

From the results of sub-criteria, were dealing with the independent variables for Factors affecting the implementation of power plant projects that have the most important impact on the factors of implementation of projects of electric power production, which are appropriate with the formula is as shown in Eq. 8.

$$FAIPPP = (W_1 \times FEA) + (W_2 \times AI) + (W_3 \times PA) \tag{8}$$

Where:
FAIPPP: Factors affecting the implementation of power plants projects
FEA: Financial and economic aspect
AI: Availability of infrastructure
PA: Planning aspect

Through the use of the SPSS statistical analysis program for the questionnaire, finding the arithmetic mean, and obtaining the weight of each of the main equation factors so that weights (W) appeared as shown in Table 8.

Table 8. Parameter weight of variables for implementation of power plant projects.

Parameter	W_1	W_2	W_3
Weight	0.339	0.326	0.335

From the analysis of parameters, the equation for variables for implementation of power plants projects as shown in Eq. 9.

$$FAIPPP = 0.339\ FEA + 0.326\ AI + 0.335\ PA \tag{9}$$

This equation Eq. 9. using to calculate the priority of implementing electric power generation projects in Iraq by entering the calculation of the effect of the factors of implementing power generation projects and their weights. This equation is used on the projects proposed to be implemented, showing us the order of priority of implementing these projects according to their importance. Table 9 shows data for a series of electrical power generation projects that were previously implemented according to the Iraqi Ministry of Electricity.

Table 9. Electrical power generation projects that were previously implemented in Iraq.

Generating station	Plant Capital Cost $/MW	Operation and Maintenance Cost $/MW	Project Duration Day	DPS from the Transmission Lines km	DPS from The Source of Discharge Loads km	DPS from the water source km	Capacity MW	Project Area m^2
A	870253	3600	540	2.27	6.34	4.63	500	632192
B	766856	5700	720	11.23	14	1.42	1250	485530
C	915660	5700	540	0.63	8	0.18	500	10000
D	787656	3500	510	1	3.22	2.26	292	118786
E	767490	5700	540	15	21.15	10	500	63364
	CLV	CLV	CLV	CSV	CSV	CSV	CLV	CLV

CSV: compared with the smallest value.
CLV: Compared with the largest value.

Through the use of Eq. 9, the implementation of electric power generation projects will be arranged according to their importance to factors, and this equation will be applied to several projects previously implemented by the Iraqi Ministry of Electricity as follows in Table 10.

Table 10. Prioritizing the implementation of electric power generation projects in Iraq.

Station	W_1	Eq.3: FEA %		W_2	Eq. 5: AI %		W_3	Eq.7: PA %		Eq. 9: FAIPPP %	Ranking
A	0.339 ×	82.60693	+	0.326 ×	30.13674	+	0.335 ×	51.29	=	55.01048	2
B	0.339 ×	89.73067	+	0.326 ×	14.02423	+	0.335 ×	57.1294713	=	54.12897	3
C	0.339 ×	96.015	+	0.326 ×	66.53775	+	0.335 ×	43.3095561	=	68.74909	1
D	0.339 ×	76.36148	+	0.326 ×	61.58967	+	0.335 ×	10.9112585	=	49.62005	4
E	0.339 ×	86.64575	+	0.326 ×	8.138603	+	0.335 ×	23.3536035	=	39.84955	5

The results show that Project (C) ranked first with a score of points 68.74909%, Project (A) ranked second with a score of points 55.01048%, while Project (B) ranked third with a total score 54.12897%, Project (D) ranked fourth with a score of points 49.62005%, finally Project (E) ranked fifth and last among the projects 39.84955% that were Applied the equation to its data, and in this formula, we get a sequence of projects that must be implemented according to their importance and by the sources of funding and government budgets.

9. Conclusion

After obtaining the results to successfully apply the equations and prioritize the stations for each project, the conclusion can be summarized as follows:

- The MLR multivariate linear regression model provided good results by validating the training data series.
- Through analyzing the questionnaire using the SPSS statistical analysis program and obtaining weights of the main factors (W), we realize that the financial and economic factor is the most influential on the implementation of electricity generation projects in Iraq.
- The mathematical model assists decision-makers in setting a priority implementation of future electric power generation projects in Iraq and strategic plans.
- The mathematical model equation can be developed by applying it to renewable energy plant projects by creating other factors that affect the priority of implementing renewable power plant projects such as solar energy, wind energy and other energies.

References

[1] C. H. Liu, S. J. Lin, and C. Lewis, "Evaluation of thermal power plant operational performance in Taiwan by data envelopment analysis," *Energy Policy*, 2010, vol. 38, no. 2, pp. 1049–1058.
[2] R. B. Hiremath, S. Shikha, and N. H. Ravindranath, "Decentralized energy planning; modeling and application-a review," *Renew. Sustain. Energy Rev.*, 2007 vol. 11, no. 5, pp. 729–752.
[3] J. C. Rojas-Zerpa and J. M. Yusta, "Methodologies, technologies and applications for electric supply planning in rural remote areas," *Energy Sustain. Dev.* , 2014, vol. 20, no. 1, pp. 66–76.
[4] E. Løken, "Use of multicriteria decision analysis methods for energy planning problems," *Renew. Sustain. Energy Rev.*, 2007, vol. 11, no. 7, pp. 1584–1595.
[5] S. C. Bhattacharyya, "Review of alternative methodologies for analysing off-grid electricity supply," *Renew. Sustain. Energy Rev.*, 2012, vol. 16, no. 1, pp. 677–694.
[6] A. Mirakyan and R. De Guio, "Integrated energy planning in cities and territories: A review of methods and tools," *Renew. Sustain. Energy Rev.*, 2013, vol. 22, pp. 289–297.
[7] P. Söderholm, "The political economy of power generation in Zimbabwe since 1980," no. Natural Resources Forum, (1999)
[8] C. Okello, S. Pindozzi, S. Faugno, and L. Boccia, "Appraising bioenergy alternatives in Uganda using strengths, weaknesses, opportunities and threats (SWOT)-analytical hierarchy process (AHP) and a desirability functions approach," *Energies*, 2014, vol. 7, no. 3, pp. 1171–1192.
[9] P. A. Trotter, M. C. McManus, and R. Maconachie, "Electricity planning and implementation in sub-Saharan Africa: A systematic review," *Renew. Sustain. Energy Rev.*, 2017, vol. 74, no. June 2016, pp. 1189–1209.
[10] Harry H Istepanian, "Iraq ' s Electricity Crisis," *Elsevier*, (2014)
[11] Information on https://www.moelc.gov.iq/uploads/attachments/1_2017التقرير.pdf.
[12] S. H. Rasool and F. M. S. Al-Zwainy, "Estimating Productivity of Brickwork item using Logistic and Multiple Regression Approaches," *Sholrs J. Eng. Technol.* , 2016, vol. 4, no. 5, pp. 234–243.
[13] F. M. S. Al-Zwainy, M. H. Abdulmajeed, and H. S. M. Aljumaily, "Using Multivariable Linear Regression Technique for Modeling Productivity Construction in Iraq," *Open J. Civ. Eng.* , 2013, vol. 03, no. 03, pp. 127–135.
[14] Y. I. Topcu and F. Ulengin, "Energy for the future: An integrated decision aid for the case of Turkey," *Energy*, 2004, vol. 29, no. 1, pp. 137–154.
[15] B. K. Sovacool, A. Gilbert, and D. Nugent, "An international comparative assessment of construction cost overruns for electricity infrastructure," *Energy Res. Soc. Sci.* , 2014, vol. 3, no. C, pp. 152–160.

Isolation and Identification of Local Bactria Produced from Soil-Borne Urease

N. A. Ali[1,a], M. O. Karkush[1,b*] and H. H. Al Haideri[2,c]

[1]Department of Civil Engineering, University of Baghdad, Baghdad, Iraq
[2]Department of Biology, University of Baghdad, Baghdad, Iraq
[a]Nawarakil_87@yahoo.com, [b]mahdi_karkush@coeng.uobaghdad.edu.iq,
[c]hhmalhaideri@gmail.com

Abstract—The hydrolysis of urea by the enzyme urease is significant for increasing the irroles in human pathogenicity, biocementation, soil fertilizer, and subsequently in soil improvement. This study devoted to the isolation of urease from urea-rich soil samples collected from seven different locations. Isolation of the various bacterial species was conducted using nutrient agar. The identity of isolated urease was based on morphological characteristics and standard microbiological and biochemical procedures. The urease producing strains of bacteria were obtained using the urease hydrolysis test. The bacterial isolates produced from soil samples collected from different environments and treated by different morphological processes helped in precipitation of large calcium carbonate ($CaCO_3$) crystal aggregates precipitated within bacterial colonies grown on agar. The different microbial species and functional attributes produced striking differences in the morphology of precipitated crystals. The phylogenetic sequencing of 16S ribosomal RNA genes produced several isolates that are mostly related to the Bacillus group. One strain of promising results was selected and the environmental and nutritional conditions were characterized. The growth curve of the selected strain with an optimized condition was investigated.

Keywords: local bacteria; isolation; calcium carbonate; hydrolysis; microbial analysis; soil improvement.

1. Introduction

Microorganisms including bacteria live in the soil and all over the earth's surface. Microorganisms play important roles in nutritional chains that are an important part of this biological balance in the life on our planet, and are necessary to complete nutrient and geochemical cycles such as the carbon, nitrogen, sulfur, and phosphorous cycle. Bacteria play an important role in soil without which the soil will not be fertile and organic matter such as leaves and straw will accumulate within a short period [1–3]. Microorganisms are widely Spread in the world. Urease (urea amidohydrolase, EC 3.5.1.15) is an enzyme that catalyzes for hydrolysis urea substrate to form carbon dioxide and ammonia [4]. urease has been well studied from a clinical perspective as it can indicate increased virulence properties in pathogenic bacteria [5–7]. In recent years, the study on the application of using the production of microbial ureases in civil and geotechnical engineering has increased because of the abilities of these microorganisms to induce calcite precipitation in the presence of urea and calcium ions [4]. This process is termed microbial induced calcite precipitation (MICP), an innovative technique that harnesses the activity of bacteria to improve the physical properties of soils as a sustained and environmentally responsible method for soil strengthening and other engineering applications [8, 9].

Microorganisms that can hydrolyze urea are abundant in natural soils, thus, making the process of urea hydrolysis common to soils worldwide [10]. Many soil bacteria can crush urease in the presence of ammonia and other nitrogen compounds. a few bacteria are known to be able to produce urease in the presence of urea [6]. A well-documented application of ureolytic microorganism has been the improvement of the mechanical properties of soil for construction and environmental purposes, known as biocementation. Biocementation enhances the strength and stiffness properties of soil through microbial activities and products [11]. Various studies have reported the isolation of ureolytic bacteria using enrichment culture from a variety of soil samples, groundwater samples, and cement samples [12–16]. Most Bacillus species can hydrolyze urea using urease enzymes [13]. The previous studies have shown that there is an increasing number of microorganisms being screened from extreme environments with the capability to produce essential enzymes useful for various industrial applications.

2. Materials and Methods

2.1.Soil sample collection and preparation

Ten soil samples were collected from different regions of soil enriched with urea. A total of 10 samples were aseptically collected randomly from various locations. During collection, a clean stick was used to clear the debris away and a 10 cm hole was dug. The soil sample approximately 5 g was collected using a sterile spatula and put in a clean, dry, and sterile polythene bag, then stored in an icebox.

2.2.Enrichment culture of samples

To enrich the samples for urease producing bacteria, 1 g of soil sample was added to 10 mL of nutrient broth (HiMedia, Laboratories Pvt. Ltd) containing 20 g urea (100 mL shaking flasks) and incubated under aerobic batch conditions at 37°C for 120 hrs. under shaking condition at 100 rpm.

2.3.Isolation of urea hydrolyzing bacteria

For bacterial isolation, 1 mL of soil culture was serially diluted (five-fold) and inoculated on urea nutrient agar. A loopful culture from serially diluted enrichment samples was inoculated onto Petri plates containing urea nutrient agar were then spread using a loopful spreader until the fluid was evenly distributed. Then the plates were incubated under aerobic conditions at 37°C for 48 hrs. When isolates of 20 g urea degradation were grown in Petri dishes containing nutrient agar, a subsequent sub-culture was performed until single colonies of bacteria were obtained.

2.4.Screening for ureolytic bacteria

Urea agar base (Oxoid, Thermo Scientific Microbiology Sdn Bhd) was used as a qualitative urease assay and to rapidly test for bacteria producing urease. Preparation of the medium was performed following the manufacturer's instructions. To test the production of urease, heavy inoculation was performed on the surface of the (UAB) medium in the test Petri plates and incubated under aerobic conditions at 37°C for 48 hrs. The urease test was studied through visual observation for color changes. The isolates that changed the medium from yellow to a pink color were selected for further studies based on their ability to rapidly produce urease qualitatively.

3. Identification and Characterization of Isolates

All isolated were further identified using biochemical tests such as Gram stain reaction, motility test, citrate test, catalase test, sugar fermentation test.

3.1. Isolation of genomic DNA

Genomic DNA was isolated using the Wizard Genomic DNA purification Kit (Promega).

3.2. Extraction protocol of DNA

According to the kit's protocol, the following procedure was implemented to extract genomic DNA from isolated bacterial strains:

3.2.1. For gram-positive bacteria

- One mL from an overnight broth culture of isolated bacterial was added to the Eppendorf tube and pelleted by centrifuge at 14000 rpm for 2 min, then remove supernatant.
- re-suspend cell pellet in (50 mm) EDTA to final volume 480 μL.
- a lytic enzyme (120 μL) (lysozomandor or lysostaphin) was added to the tube and mix again by vortex
- after that incubate at 37°C to 1 h., centrifuge for 2 min at 14000 rpm, finally, remove the supernatant.

3.2.2. For gram-negative bacteria

- Re-suspended cell pellet in Nuclei Lysis Solution to a final volume of 600 μL, and incubate at 80°C for 10 min.
- Five μL of RNase solution was added to the mixture, mix with a pipet for 2-5 times, and incubate at 37°C for 1 h.
- After RNase treatment, 200 μL of Protein Precipitation Solution was added to the tube, and mix the mixture by vortex at high speed for 20 s.
- After that incubate the tube on ice for 5 min, and centrifuge at 14000 rpm for 3 min.
- Transfer the supernatant to a clean tube contains 600 μL of isopropanol, mix gently and centrifuge at 14000 rpm for 2 min.
- Absolute ethanol of 600 μL (70%) was added to the tube and mixed again by vortex at high speed for 15 s.
- Drain the tube to absorbent paper and leave it to dray for 10-15 min.
- Consequently, about 50 μL of DNA rehydration solution was added to the Eppendorf tube and incubate at 65°C for 1 h, then the sample was centrifuged at 6000 rpm for 1 min.
- Finally, the extracted DNA was stored at -20°C freezer in the appropriate sample box until PCR amplification.

3.3. Measuring the concentration and purity of DNA

The concentration of DNA and purity were measured by the Nano-drop UV spectrophotometer (Quawell Technology Inc, USA). The elution buffer that used as the blank was dsH_2O, then 1 μL of DNA was measured at two wavelengths (260 and 280 nm) to conclude DNA concentration and purity respectively. The calculated ratio was gained from the operating software and displayed by the Nano-drop spectrophotometer (157 UV/Vis single wavelength detector).

4. PCR Amplification

4.1. Oligonucleotides primer pairs

The primer of the bacillus gene used in this study was purchased from Alpha DNA (Canada) in lyophilized, shown in Table 1, the bacillus sequence of isolated bacterial strains. The first twenty nucleotides represent the forward primer and the complementary primer represents the last seventeen nucleotides. Primarily, the stock solution was prepared by dissolving the lyophilized powder in nuclease-free water buffer according to the manufacturer's protocol to give a total concentration of 100 pmol μL^{-1} for each one. Then the working solution was prepared where 10 μL from Stock solution was added to 90 μL of nuclease-free water to give a total concentration of 10 pmol μL^{-1} for each reverse/forward primer.

Table 1. Primers used in this study (PCR)

Primer name	Primer sequence	Reference
27F	5`-AGAGTTTGATCTTGGCTCAG-3	Designed in this study
142R	5`-TACGGTTACCTTGTTACGACTT-3`	

4.2. Amplification kit

The PCR amplification mixture which used to amplified bacillus gene includes GoTaq® Green Master Mix, 2X (25 µL), 2.5 µL of 25 ng DNA template, 5 µL (1 mM) of each forwarded and reversed primers and then 22.5 µL of nuclease-free water was added to complete the amplification mixture to 50 µL (Table 2).

Table 2. PCR mixture components for the bacillus gene.

Component	Volume	Final concentration
GoTaq®Green Master Mix, 2X	10 µl	1X
Forward primer	1 µl	1µM
Reverse primer	1 µl	1µM
DNA template	3 µl	30 ng
Nuclease free water	5 µl	
Final volume	20 µl	

4.3. Agarose gel electrophoresis

Extracted DNA and PCR products were tested by (1% [w/v]) agarose gel electrophoresis with a 1X TAE buffer. Firstly, 1 g of agarose was dissolved in 100 mL TAE working buffer (1X) in a flask, then the mixture was heated on a hot plate for 5-10 min until the agarose was completely softened, the solution was left to be warm for 5 min, and 1 µL (10 mg/mL) of ethidium bromide was added. After that, the solution was poured in a gel tray with a good comb and left to solidify at 50-60°C for 20-25 min. For loading samples, DNA samples were mixed with 1X DNA loading dye, then 5 µL of PCR product and DNA samples along with DNA ladder was loaded into the gel wells. The running buffer is 1X TAE, and the gel was run at 100v/mAmp for 1 h and it visualized by Ultraviolet light (UV) device (158), see results in Table 3.

Table 3. PCR optimization program for the bacillus gene.

Optimization	Temperature, °C	m:s	Cycle number
Initial denaturation	95	05:00	1
Denaturation	95	00:30	
Annealing	60	00:45	30
Extension	72	01:00	
Final extension	72	07:00	1
Hold	10	10:00	

4.4. Sequencing

PCR product was sent to the Macrogen Company in Korea for sequencing. Unpurified PCR product at a volume of 50 µL, and forward primers at a concentration of (10 pmol μL^{-1}) were sent to the company for purification to remove all the unwanted elements, then Sanger sequencing was done by using AB13730XL, automated DNA sequences. The result analyzed by the BLAST website on NCBI, Mega, and Bioedite.

5. Measurement of Urease Activity

Urease activity was measured immediately after incubating an overnight bacterial culture. urease activity was determined by [17] In measuring the activity of the enzyme on the conversion of one micromol of urea to ammonia. The reagent used in this study are:

5.1. HEPES (N-2Hydroxy Ethyl Piperazine-N-2-Ethan)

HEPES solution was prepared by dissolving (50 mM of sulfuric acid, (1 mM) of EDTA (Ethylene diamine tetraacetic acid), (1 mM) of 2-mercaptoethanol) in 100 mL of distilled water and the initial pH was adjusted to 7.5 using 0.1 M of NaOH [18].This buffer is used to dilute the ammonium chloride storage solution of the standard curve as well as to dilute the samples to estimate the effectiveness of urease.

5.2. Indophenol assay reagents
The indophenol reagents are:

5.2.1. Phenol nitroprusside (A Reagent). Prepare dissolved (2.5 g) of (NaOH) in 500 mL of distilled water added (4.2 mL) of sodium hypochlorite (6%), shake well, and kept in a dark bottle in the refrigerator for a period not to exceed one month.

5.2.2. Alkaline hypochlorite (B Reagent). Prepare (2.5 g) of sodium hydroxide (NaOH) in 500 mL of distilled water and 4.2 mL of sodium hypochlorite (6%), shake well and keep in a dark bottle in the refrigerator for Do not exceed a month [19].

5.2.3. Urea stored solution (500 mM). Prepared by dissolving 300 mg of urea in 10 mL of distilled water.

6. Procedure of Testing
The procedure of testing is conducted as follows:

- Putting 215 μL of HEPES buffer solution and 25 μL of stored urea solution in test tubes and left in a water bath at (37°C) for three minutes.
- Adding 10 μL of the enzyme solution to the reaction components to become the final volume of the reaction solution 250 μL and the final concentration of urea is 50 mM. The tubes were incubated in a water bath at a temperature of 37°C for 15 min (reaction time by which urea was hydrolysis in the presence of urease).
- Adding 5 mL of reagent (A) and 5 mL of reagent (B) to the tubes with rapid shaking, and left in a water bath at 37°C temperature for 20 min for the blue color to appear.
- The light absorbance of the resulting color of the reaction was measured at a wavelength of 625 nm.
- The amount of ammonia released in micromol (μmol) was measured according to the standard curve of the ammonia solution as shown in Fig. 1. The enzymatic activity (unit/mL) was estimated according to the following equation:

$$\text{Activity (unit/mL)} = \frac{\text{Amount of ammonia } (\mu\text{mol})}{0.01 \times 15 \times 2} \quad (1)$$

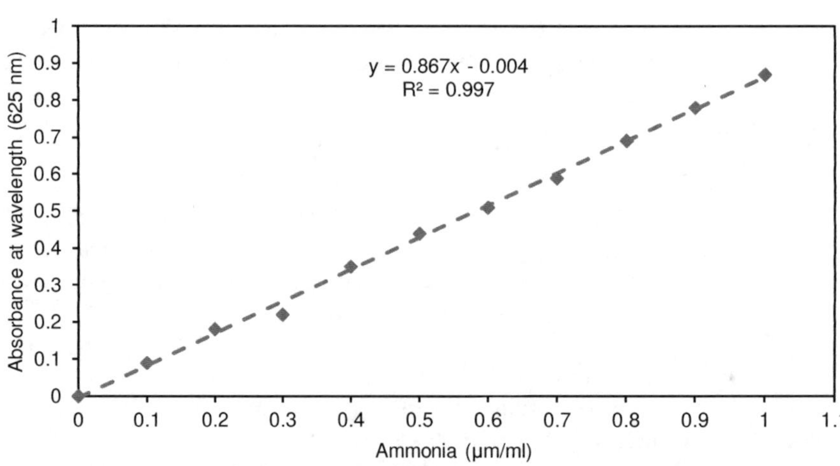

Figure 1. Variation of ammonia concentration with absorbance at a wavelength of 625 nm.

7. Results and Discussion

7.1. Selection of ureolytic bacteria

Among all six isolated bacteria, six bacterial isolates designated as C11, D11, H11, A11, B12, and D12 were selected based on their quick urease production when tested on urea agar base medium. Urea agar base media is used for the diagnosis of a variety of microorganisms, especially based on urease production [20]. Bacterial isolates that can produce urease turned the UAB medium in the Petri plate from yellow to pink as shown in Fig. 2, while the isolates which were unable to produce urease turned the media yellow. This medium contains urea and a pH indicator, phenol red. When urea is hydrolyzed, ammonia accumulates in the media which increases the pH of the environment [21]. This increase in pH causes the pH indicator to change from pale orange to pink, confirming a positive result for urea hydrolysis. Several studies have reported using urea agar base media as a preferred qualitative urease assay for isolation and differentiation of ureolytic microorganisms [13–16, 21–23].

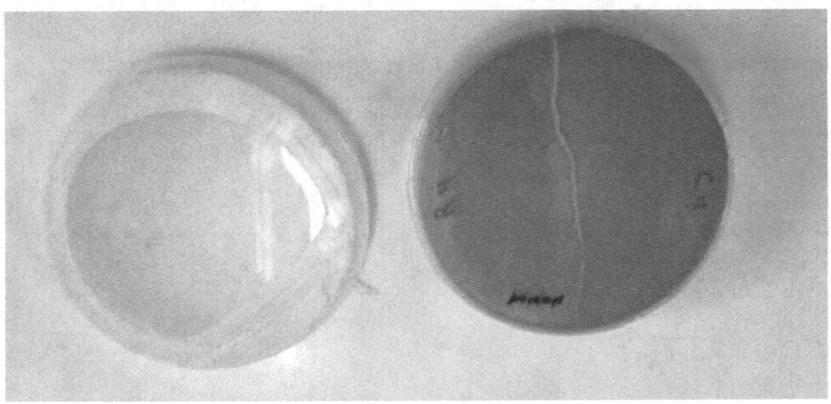

Figure 2. Urease producing test UAB medium.

7.2. Preliminary bacterial identification

The isolates C11, D11, H11, A11, B12, and D12 were characterized using standard methods. All six isolates were Gram-positive rod-shaped and smooth surfaces (Table 4). There were notable morphological differences among bacterial isolates. Bacterial community diversity in the soil environment due to its alkaline state, only organisms that can grow in these conditions can survive in such an environment [22]. The dominant species that may occur during the period of fertilization cultivation where bacteria are usually chosen in isolation and cultivation methods [24]. The phenotypic and biochemical properties of the bacteria isolates have a similar resemblance to those Bacillus species reported previously by [22].

Table 4. Biochemical characteristic of bacterial isolated

Characteristic	C11	D11	F11	A12	B12	C12
Gram stain	+ve rod	+ve rod	+ve rod	+ve rod	+ve rod	+ve rod
End spore staining	Spore forming	Spore forming	Spore forming	Spore forming	Spore forming	Spore forming
Catalase	+	+	+	+	+	+
Motility	+	+	+	+	+	-
Citrate	+	+	+	+	+	+
Anaerobic growth	+	+	+	+	+	-
NaCl 7%	+	+	+	+	+	+
NaCl 10%	-	-	-	+	_	+

IOP Conf. Series: Materials Science and Engineering **901** (2020) 012035 doi:10.1088/1757-899X/901/1/012035

7.3. Bacterial identification using 16S rRNA sequence analysis

The six selected urease producing bacteria isolated from soil were identified and characterized by the sequencing of 16S rRNA. These sequences were BLAST searched against the GenBank database using the BLASTN program. The nucleotide BLAST analysis of the 16S rRNA region showed a reasonable degree of correlation with the physiological characterization especially the morphological classification schemes of species within the genus. The BLAST analysis revealed that C11 had 99% similarity (100% coverage) with Bacillus sonorensis, D11 had 99% similarity (100% coverage) with Bacillus sonorensis, H11 had 97% similarity (100% coverage) with Bacillus sonorensis, A11 had 98% similarity (99% coverage) with Bacillus licheniformis, B12 had 92% (99% coverage) with Bacillus licheniformis, D12 had 95% similarity (99% coverage) with Bacillus sonorensis. Based on sequence data of the 16S of the rRNA region, all bacterial isolates from soil showed a high degree of similarity (91-100%) to their closest species. The relationship between the locally isolated ureolytic bacteria and their closest species are shown in the phylogenetic tree see Fig. 3.

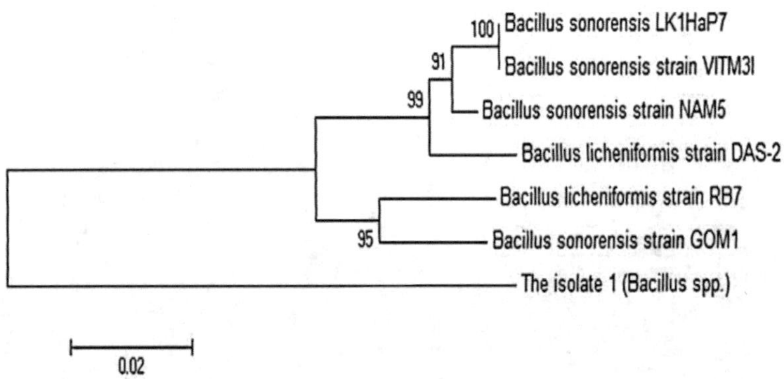

Figure 3. Phylogenetic tree based on the 16sRNA gene sequence of isolated bacteria in the current study along with sequences available in the GenBank database.

7.4. Measurement of urease activity

Among all the bacterial cultures, C11 had the highest urease activity (10.25 mM urea hydrolyzed/min/OD) when compared to other local cultures, while C12 showed the lowest specific urease activity (1.43 mM urea hydrolyzed/min/OD) as shown in Fig. 4. Numerous studies have reported that the usage of reported studies on locally isolated strains of Bacillus bacteria had urease activity 3.3 to 8.8 mM urea hydrolyzed/min/OD [25, 26].

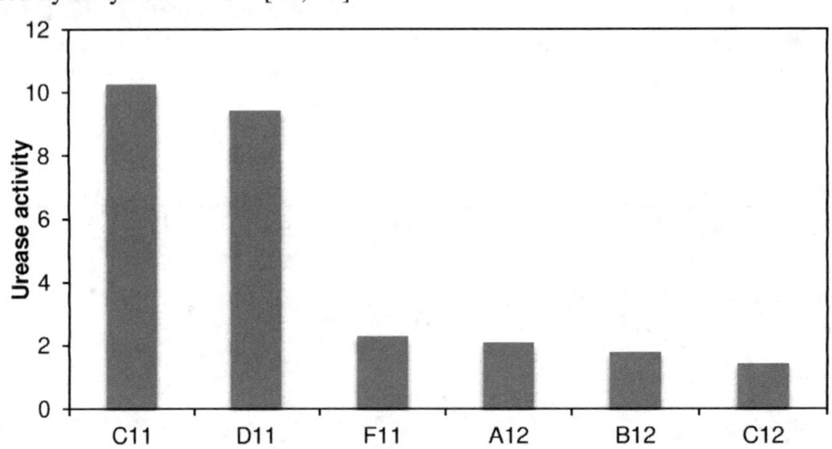

Figure 4. Comparison of urease activity among the isolated urease producing strains. Two bacterial isolates, C11 and D11 show high urease activity, almost comparable to control strain.

8. Conclusions

The BLAST analysis of the 16S rRNA gene sequence from the locally isolated ureolytic bacteria identified them as Bacillus sonorensis and Bacillus licheniformis. The specific urease activity of the measured bacterial strains indicates that these ureolytic strains serve as a good urea hydrolytic agent, potentially useful for biocementation applications.

Funding

The work was supported by the authors and the Department of Civil Engineering and the Department of Biology at the University of Baghdad.

References

[1] Kummerer, K. (2004) Resistance in the environment, Journal of Antimicrobial Chemotherapy, 45, 311-320.

[2] Sylvia, D. M., Fuhrmann, J. J., Hartel, P. G., and Zuberer, D. A. (2005) Principles and applications of soil microbiology, No. QR111 S674 2005, Pearson.

[3] Anyadoh-Nwadike, S. O., Ezeanyika, C., and Okorondu, S. I. (2017) Isolation and characteristic of soil-borne urease producing bacteria, European Journal of pharmaceutical and medical Research, 4, 30-38.

[4] Cheng, L. and Cord-Ruwisch, R. (2013) Selective enrichment and production of highly urease active bacteria by non-sterile (open) chemostat culture, Journal of industrial microbiology & biotechnology, 40, 1095-1104.

[5] Collins, C. M. and D'Orazio, S. E. (1993) Bacterial ureases: structure, regulation of expression and role in pathogenesis, Molecular microbiology, 9, 907-913.

[6] Mobley, H. L., Island, M. D., and Hausinger, R. P. (1995.) Molecular biology of microbial ureases, Microbiol. Mol. Biol. Rev., 59, 451-480.

[7] Lee S. G. and Calhoun, D. H. (1997) Urease from a potentially pathogenic coccoid isolate: Purification, characterization, and comparison to other microbial ureases, Infection and Immunity, 65, 3991-3996.

[8] DeJong, J. T., Soga, K., Banwart, S. A., Whalley, W. R., Ginn, T. R., Nelson, D. C., Mortensen, B. M., Martinez, B. C. and Barkouki, T. (2011) Soil engineering in vivo: harnessing natural biogeochemical systems for sustainable, multi-functional engineering solutions, Journal of the Royal Society Interface, 8,1-15.

[9] Gat, D., Tsesarsky, M., Shamir, D., and Ronen, Z. (2014) Accelerated microbial-induced $CaCO_3$ precipitation in a defined coculture of ureolytic and non-ureolytic bacteria, Biogeosciences, 11, 2561-2569.

[10] Mobley H. L. and Hausinger, R. P. (1989) Microbial ureases: Significance, regulation, and molecular characterization, Microbiological Reviews, 53, 85-108.

[11] Ivanov, V. and Chu, J. (2008) Applications of microorganisms to geotechnical engineering for bioclogging and biocementation of soil in situ, Reviews in Environmental Science and Bio/Technology, 7, 139-153.

[12] Rivadeneyra, M. A., Delgado, R., Delgado, G., Moral, A. D., Ferrer, M. R., and Ramos-Cormenzana, A. (1993) Precipitation of carbonates by Bacillus sp. isolated from saline soils, Geomicrobiology Journal, 11, 175-184.

[13] Hammes, F., Boon, N., de Villiers, J., Verstraete, W., and Siciliano, S. D. (2003) Strain-specific ureolytic microbial calcium carbonate precipitation, Appl. Environ. Microbiol., 69, 4901-4909.

[14] Achal V. and Pan, X. (2011) Characterization of urease and carbonic anhydrase producing bacteria and their role in calcite precipitation, Current Microbiology, 62, 894-902.

[15] Burbank, M. B., Weaver, T. J., Williams, B. C., and Crawford, R. L. (2012) Urease activity of ureolytic bacteria isolated from six soils in which calcite was precipitated by indigenous bacteria, Geomicrobiology Journal, 29, 389-395.

[16] Elmanama, A. A. and Alhour, M. T. (2013) Isolation, characterization and application of calcite producing bacteria from urea rich soils, Journal of Advanced Science and Engineering Research, 3, 377-399.

[17] Achakzai, A. K. K., Kayani, S. A., Yaqoob, M., and Nabi, A. (2003) Effect of fertilization and inoculation on lipase and urease activity of mature soybean cv. Williams-82 seeds, Asian Journal of Plant Sciences, 9, 692-695
[18] Todd, M. J. and Hausinger, R. P. (1991) Reactivity of the essential thiol of Klebsiella aerogenes urease. Effect of pH and ligands on thiol modification, Journal of Biological Chemistry, 266, 10260-10267.
[19] Weatherburn, M. W. (1967) Phenol-hypochlorite reaction for determination of ammonia., Analytical Chemistry, 39, 971-974.
[20] Atlas, R. M. (2010) Handbook of microbiological media, CRC press.
[21] Hammad, I. A., Talkhan, F. N., and Zoheir, A. E. (2013) Urease activity and induction of calcium carbonate precipitation by Sporosarcina pasteurii NCIMB 8841, Journal of Applied Sciences Research, 9, 1525-1533.
[22] Achal, V. (2010) Characterization of two urease-producing and calcifying Bacillus spp. Isolated from cement, Journal of Microbiology and Biotechnology, 20, 1571-1576.
[23] Dhami, N. K. (2013) Biomineralization of calcium carbonate polymorphs by the bacterial strains isolated from calcareous sites, Journal of Microbiology and Biotechnology, 23, 707-714.
[24] Stocks-Fischer, S., Galinat, J. K., and Bang, S. S. (1999) Microbiological precipitation of $CaCO_3$, Soil Biology and Biochemistry, 31, 1563-1571.
[25] Stabnikov, V., Jian, C., Ivanov, V. and Li, Y. (2013) Halotolerant, alkaliphilic urease-producing bacteria from different climate zones and their application for biocementation of sand, World Journal of Microbiology and Biotechnology, 29, 1453-1460.
[26] Al-Thawadiand, S. and Cord-Ruwisch, R. (2012) Calcium carbonate crystals formation by ureolytic bacteria isolated from Australian soil and sludge, Journal of Advanced Science and Engineering Research, 2, 12-16.

Modeling the Response of Wrap-faced Reinforced Lime Treated Soil Retaining Walls under Static Condition

Bisrat Gissila Gidday[1,a *] and Satyendra Mittal[1,b]

[1]Civil Engineering Department, Indian Institute of Technology Roorkee, 247 667, Uttarakhand, India
[a]bisratgissila@gmail.com, [b]satyendramittal@gmail.com

Abstract. This paper presents model studies conducted on geotextile wrap faced reinforced soil walls under static conditions in a rectangular model box. Untreated and lime treated clay was used as backfill soils. Numerical modeling has been done by PLAXIS software. The surcharge pressure, angle of the slopes, and different number of reinforcement layers were varied in model tests. The influence of these various parameters at different elevations of the retaining wall, settlement at the crest, and horizontal face displacement are presented in this paper. The tests were conducted at $45^0$and 60^0 slopes. It was seen that vertical crest settlement and horizontal displacement of reinforced soil wall decrease with an increase in reinforcing layers.

Keywords: Geosynthetics, Reinforced soils, Wrap-faced retaining walls, Lime, Numerical modeling.

1. Introduction

Reinforced soil technology has acquired broad recognition due to its constructional, functional, and cost-effective benefits for constructing walls even in less space [1]. The usage of this technology for soil retaining structures has increased enormously in infrastructural projects across the world. Retaining walls are amongst the earliest geotechnical structure since civilization. Depending upon the purpose of the project, these walls can be permanent or temporary [2]. Various researchers have done work on laboratory models of reinforced soil walls under different conditions [3, 4], and at the same time, various field studies have been done by different investigators [5, 6, 7]. Some work [8] has been done to examine the behavior of the reinforced soil wall subjected to overburden pressure and seismic loads with backfill as cement treated. The facing element in this wall was geotextile. One of the significant outcomes of this work was that the cement-treated backfill soil can provide more stability to such walls, even during earthquakes.

Some of the researches have also been done on reinforced soil structure based on numerical modeling [9, 10, 11] to study the behavior of such walls under static and dynamic conditions. The wraparound walls are built by folding an extended reinforcing component (geogrid and geotextile) through 180° [12] to form the facing and fixing it back into the fill (Fig.1). The present paper is based on the ongoing research work by authors, in which studies are made to investigate the bearing capacity improvement, horizontal face displacement, settlement reduction of a wrap faced reinforced soil retaining wall with untreated, and lime treated clay soil backfill

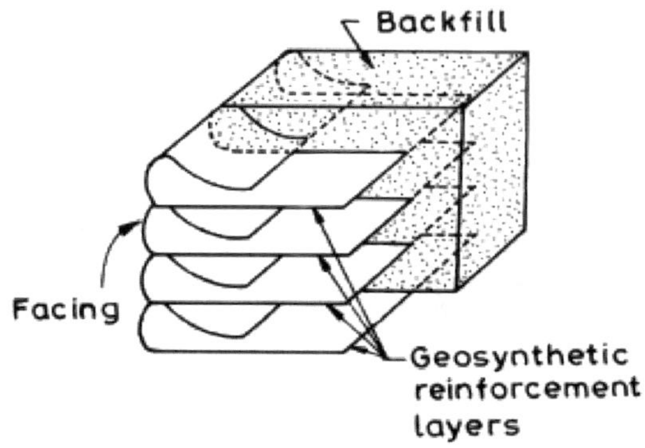

Figure 1. Schematic diagram of a geosynthetic-reinforced soil retaining wall [12].

2. The Material Used in The Study

2.1. Laboratory Model Tests

Figure 2 depicts a graphical observation of the experimental model device applied in this study. The test box is a rectangular-shaped steel box, having inner dimensions of 1000 mm x 1000 mm and 800 mm (L×W×D). The walls of the tank have a thickness of 6 mm. The tank walls were supported from the outer surface using the horizontal steel beam fixed at the half-depth of the tank. The inner wall of the tank is made to decrease friction with the soil to a minimum by applying a galvanized coat on the inside wall. The loading system consists of a hand-operated mechanical jack and pre-calibrated proving ring to apply the load manually to the footing soil system.

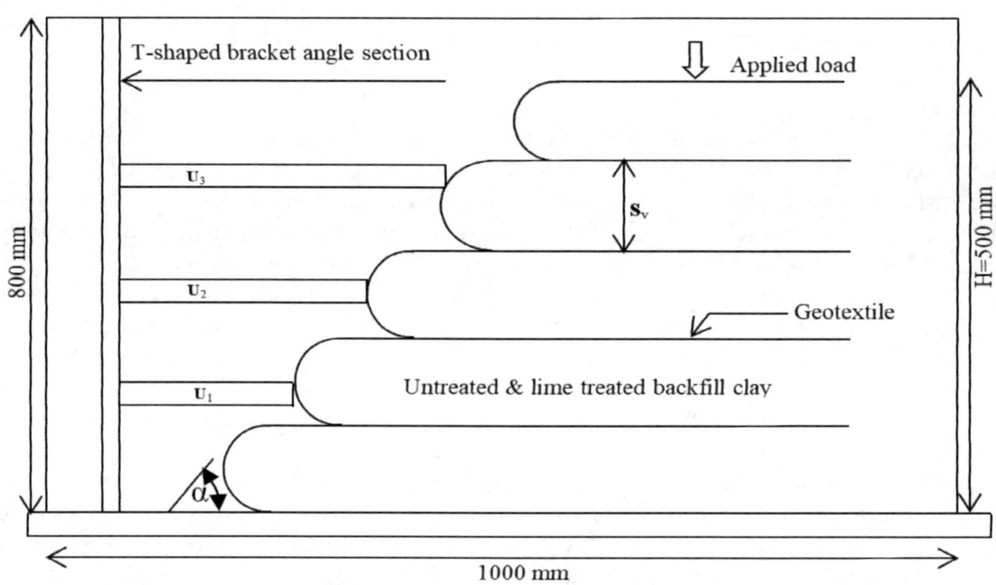

Figure 2. Schematic diagram of typical test rectangular wall configuration

2.2. Model Footing

The model footing was a steel square plate with plan dimensions of 120 mm x 120mm and with a thickness of 25 mm. A thin layer of geotextile was pasted on to the base of the model footing with Epoxy glue to create a roughness. The load was transmitted to the footing through a ball bearing that was placed between the footing and the proving ring.

IOP Conf. Series: Materials Science and Engineering **901** (2020) 012036 doi:10.1088/1757-899X/901/1/012036

2.3.Test Material

Locally available soft clay was used as the backfill material, and with its classification as CH according to the Unified Soil Classification System. The characteristics of untreated and lime treated clay used as a backfill are given in Tables 1 and 2. In the present study, 7% (by dry weight of clay) of lime was used to treat the clay that was used as a backfill material on wrap faced reinforced soil retaining wall (WRSW).

Table1. Properties of untreated clay.

Parameter	Value
Specific gravity	2.75
Liquid limit (%)	67.88
Plastic limit (%)	34.69
Shrinkage limit	19.50
Plasticity index (%)	33.19
Optimum moisture content (%)	26.00
Maximum dry density (kN/m^3)	15.33

Table2. Properties of lime treated backfill clay.

Parameter	Value
Specific gravity	2.62
Liquid limit (%)	60.59
Plastic limit (%)	43.86
Shrinkage limit	15.18
Plasticity index (%)	16.73
Optimum moisture content (%)	21.00
Maximum dry density (kN/m^3)	16.00

2.4.Reinforcement

The nonwoven polypropylene geotextile was used as the facia of wrap faced reinforced soil retaining wall in all the tests. Table 3 shows the properties of the geotextile used in the study. The wide-width strip method [13] was used to determine the tensile strength of the geotextile.

Table 3. Properties of geotextile.

Property	Value
Tensile strength(kN/m)	20.50
Warp/Machine direction	10.50
Weft/Cross direction	20.50
Elongation at break (%)	-
Warp/Machine direction	41.63
Weft/Cross direction	51.65
Thickness (mm)	1.50
Mass per unit area (GSM)	334.00

3. Model Construction and Testing Procedure

The model tests were conducted in the steel tank as discussed above. The model wall was built in lifts of equal height while reinforcing each lift with a layer of non-woven geotextile and keeping facia of the wall also of the same geotextile material (Fig.3). The overlapping length of geotextile was kept as 125 mm (for 4 layers wall), 100 mm (for 5 layers), and 83.33 mm (for 6 layers). The backfill soil was clay soil, which was compacted to its maximum dry density (MDD). The MDD was also verified in each layer by taking a representative sample in a core cutter of known volume. Wooden formwork was used [14] while constructing the model retaining wall so that all the lifts maintain the desired slope (α)with horizontal (Fig.2). The load on the model wall applied through a mechanical jack. The test tank was kept under a rigid loading frame of 10-ton capacity, to seek the reaction from it when the load was applied through the mechanical jack. A proving ring was also attached to record the actual value of the load. The settlement for each load interval was measured through two numbers dial gauges mounted through magnetic base fixed on one wall of steel tank (Fig. 3a and b).

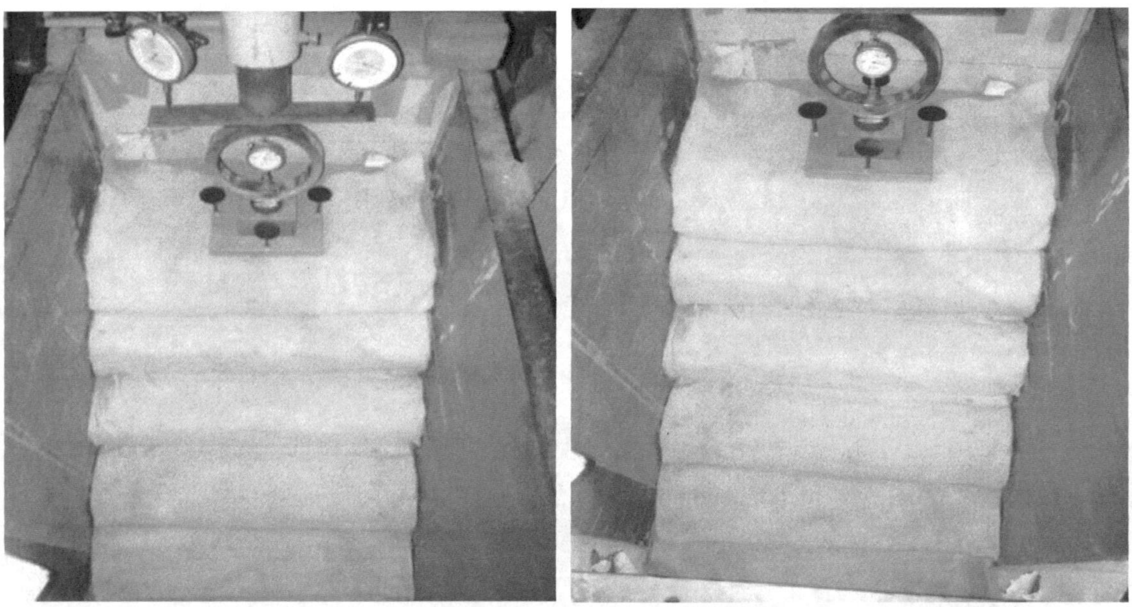

a) Wrap face wall b) Wrap face wall

Figure 3. Finished wrap-faced wall profiles: (a) five-layer configuration; (b) six-layer configuration.

4. Results and Discussion

In the model tests, the slope angle(α) was changed, and also the number of layers. Initially, all the tests were conducted on pure clay soil, and thereafter all the tests were also conducted on the backfill treated with lime. These tests are defined in Table 4. The vertical spacing between the two layers has been defined as Sv. The Sv is different for different numbers of layers. The value of Sv/H has also been defined in Table 4, where H is the total height of the wall, which was maintained as 500mm in all the tests. The horizontal displacement of the model wall was also measured (as a result of the application of load) through linear variable displacement transducers (LVDTs).

The results of horizontal displacement of the wall Vs. the height of the wall is shown in graphical form in Figure 4(for untreated backfill) and Fig.5(for lime treated backfill). It is seen that for a higher number of layers, the horizontal displacement is less for untreated and lime treated backfill. These figures give typical results for 60°inclination of the wall, through the same trend was seen at other inclination also.

Table 4. Test parameters.

Test no.	No. of layers	Slope angle	Sv/H	Test no.	No. of layers	Slope angle	Sv/H
Untreated backfill clay soil on WRSW				**Lime treated backfill clay soil on WRSW**			
N1	4	45	0.25	N7	4	45	0.25
N2	4	60	0.25	N8	4	60	0.25
N3	5	45	0.20	N9	5	45	0.20
N4	5	60	0.20	N10	5	60	0.20
N5	6	45	0.17	N11	6	45	0.17
N6	6	60	0.17	N12	6	60	0.17

It is worth noting here that horizontal displacements were recorded by using three linear variable displacement transducers (LVDTs) named U1, U2, and U3, positioned at 150, 300, and 450mm respectively, from the bottom of the tank (Fig. 2).

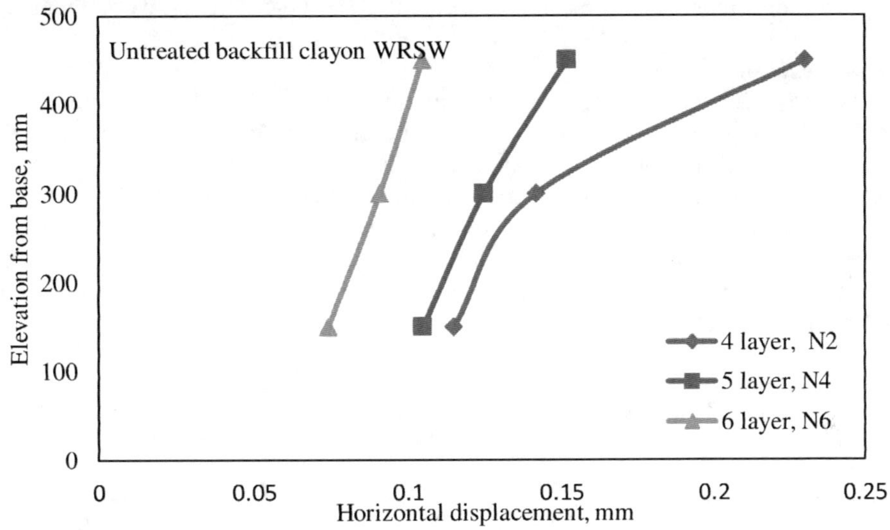

Figure 4. Effect of reinforcement layers on displacement profile at 60° (Untreated backfill).

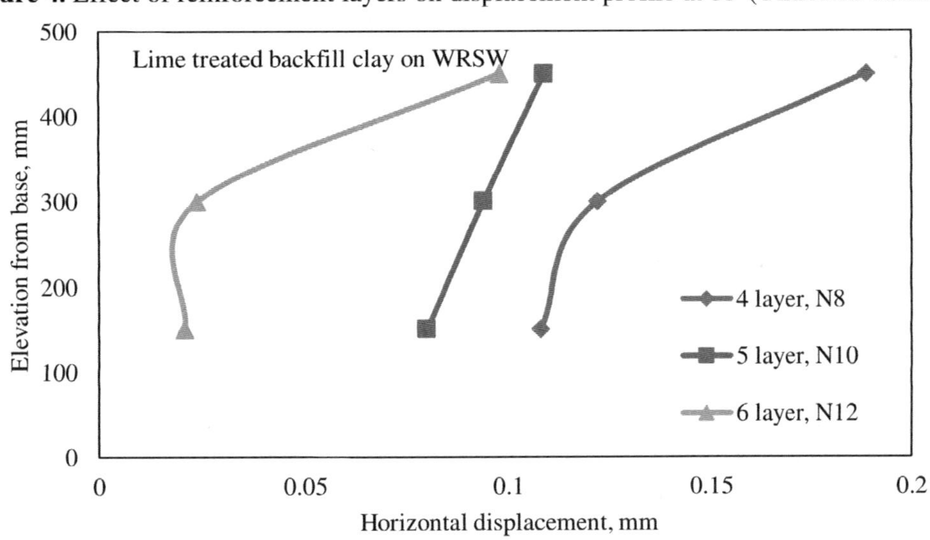

Figure 5. Effect of reinforcement layers on displacement profile at 60° (Lime treated backfill).

4.1. Plate loading tests

The test results indicate that there is an increment in bearing capacity when soil is treated by lime, and at the same time, there is a reduction in a settlement when soil is treated by lime. It can be understood by Table 5, which is for a typical case of 45^0 inclination. Horizontal face displacement was recorded at 450 mm height for comparison.

Table 5. Comparison of model tests results of the untreated and lime treated clay soil.

S. No	No. of layers	Reduction in horizontal Displacement (when treated with lime)	% Decrease in Settlement when treated with lime	Increase in bearing capacity when treated with lime
1	4	22.14	46.35	22.88
2	5	20.41	37.36	14.50
3	6	48.50	61.95	22.38

Remarks; The addition of 7% of lime significantly increases bearing capacity and reduces settlement

4.2. Numerical Modeling

The test results were validated by numerical modeling, using PLAXIS software. Due to the publication limit of pages, only one case of five layers is being presented in this paper. The model developed in PLAXIS is illustrated in Figure 6. Boundary conditions for the numerical were assigned to represent the real-life situation. The far end boundary of the backfill is fixed in the horizontal direction, displaying the rigid end of the tank. Construction sequence of the numerical model is adopted the same as that of physical model i.e, constructing wall sequentially with front lateral support, subjecting surcharge, and after that removal of supports from top to bottom after building a wall up to full height. The properties of untreated and treated backfill are shown in Table6.

Table 6. Material properties of untreated and lime treated clay soil.

Parameter	Name	Untreated clay	Lime treated clay	
		Value		Unit
Material model	Model	Mohr-Coulomb	Mohr-Coulomb	-
Type of material behavior	Type	Undrained	Undrained	-
Unit weight	γ	19.80	19.40	kN/m^3
Poisson's ratio	μ	0.35	0.30	-
Primary velocity	Vp	487.11	139.30	m/s
Secondary velocity	Vs	234.00	74.45	m/s

4.3. Horizontal face displacement and crest settlement on the numerical modeling

As discussed above, the horizontal face displacement and the crest settlement were observed for both untreated and treated backfill soils in the model tank. These results were validated through numerical modeling also. Face displacements along the height of the wall horizontally were observed by pinning the area in wrap-reinforced soil retaining wall depicted in Fig. 2 (represented and positioned by U1, U2, and U3) and vertical settlements also done to simulate wrap faced reinforced soil retaining wall on top of the wall. Test parameters adopted for numerical modeling are given in Table 5. The comparative result of physical model tests and numerical analysis are shown in Table 7. Horizontal face displacement was recorded at 450 mm height for comparison.

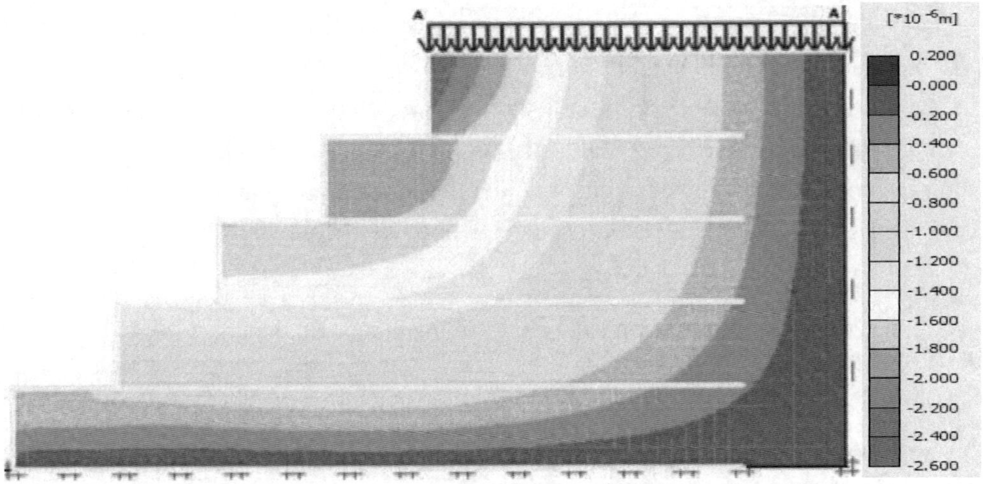

Figure 6. Distribution of horizontal displacement model wall with 5 layers reinforcement

Table 7. Comparison of test results.

Test no.	Crest settlement observed		Horizontal displacement observed	
	Physical model test(mm)	Numerical analysis(mm)	Physical model test(mm)	Numerical analysis(mm)
N1	0.23	0.32	0.13	0.35
N2	0.27	0.36	0.23	0.42
N3	0.18	0.21	0.10	0.22
N4	0.20	0.26	0.15	0.35
N5	0.09	0.12	0.07	0.16
N6	0.11	0.19	0.11	0.24
N7	0.13	0.26	0.10	0.29
N8	0.16	0.31	0.19	0.31
N9	0.12	0.18	0.08	0.17
N10	0.15	0.21	0.11	0.21
N11	0.04	0.09	0.03	0.14
N12	0.10	0.11	0.10	0.17

5. Conclusions

From the present experimental observations and numerical modeling, the following conclusions can be drawn:

- Crest settlement and horizontal face displacement decrease with an increase in the reinforcing layer on wrap-faced reinforced soil retaining wall (WRSW).
- Lime treated backfill clay material displays better performance than untreated backfill clay on WRSW for slopes $45^{\circ}$and 60°. The reinforced soil slopes of 45°performed better than the slope angle of 60°.
- The numerical modeling results indicate that vertical settlement is decreased by 18.75, 14.29, and 25% for 4, 5, and 6 reinforcement layers respectively for 45° slopes by using lime treated clay as a backfill material, while it is also reducedby13.89, 19.23, and 42.10% for 4, 5, and 6 layers respectively for the slope angle of 60°.

- The bearing capacity increases with the increase of the reinforcing layers for both lime treated and untreated backfill soil.
- Results from the physical model tests fairly match with the results from the numerical model for both lime treated and untreated soil backfill on WRSW.
- Wrap face reinforcing retaining wall can be a good alternative for problematic soil to be used as a backfill material, in places where such soils are abundant.

References

[1] Mittal, S. (2014) An Introduction to Ground Improvement Engineering, SIPL Publication, New Delhi, India.
[2] Koerner, R. M. and Soong, T. Y. (2001) Geosynthetic reinforced segmental retaining walls.Geotextextile. Geomembrane, 19(6), 359-386.
[3] El-Emam, M. and Bathurst, R. J. (2005) Facing contribution to seismic response of reduced-scale reinforced soil walls. Geosynthetics International Journal, 12(5), 215- 238.
[4] Juran, I. and Christopher, B. (1989) Laboratory model study on geosynthetic reinforced soil retaining walls. Journal of Geotechnical Engineering, ASCE, 115(7), 905–926.
[5] Benjamin, C. V. S., Bueno, B. S., and Zornberg, J. G. (2007) Field monitoring evaluation of geotextile-reinforced soil retaining walls. Geosynthetics International JOurnal, 14(2), 100-118.
[6] Tatsuoka, F., Hirakawa, D., Nojiri, M., Aizawa, H., and Nishikiori, H. (2009) A new type of integral bridge comprising geosynthetic-reinforced soil walls. Geosynthetics International Journal, ASTM, 16(4), 301-326.
[7] Yang, G., Ding, J., Zhou, Q., and Zhang, B. (2010) Field behavior of a geogrid reinforced soil retaining wall with a wrap-around facing. Geotech. Testing J., ASTM 33(1), 1-6.
[8] Suzuki, M., Shimura, N., Fukumura, T., Yoneda, O., and Tasaka, Y. (2015) Seismic performance of reinforced soil wall with untreated and cement-treated soils as backfill using a 1-g shaking table, Soils, and Foundations, 55, 626-636.
[9] Bathurst, R. J. and Hatami, K. (1998) Seismic response analysis of a geosynthetic-reinforced soil retaining wall. Geosynthetics International, 5(1–2), 127-166.
[10] El-Emam, M., Bathurst, R. J., and Hatami, K. (2004) Numerical modeling of reinforced soil retaining walls subjected to base acceleration. Proc. 13th World Conf. on Earthquake Engineering. Vancouver, Canada (CD-ROM).
[11] Fakharian K. and Attar, I. H. (2007) Static and seismic numerical modeling of geosynthetic-reinforced soil segmental bridge abutments, Geosynthetic International, 14(4), 228-243.
[12] Koerner, R. M. (1999) Designing with Geosynthetics, 4th edition, Prentice Hall, NJ, 761.
[13] ASTM D4595 (2003) Standard Test Method for Tensile Properties of Geotextiles by the Wide-Width Strip Method, ASTM International, West Conshohocken, PA, USA.
[14] Krishna, A. M. and Latha, G. M. (2007) Seismic response of wrap-faced reinforced-soil retaining wall models using shaking table tests, Geosynth. Int., 14(6), 355-364.

Sustainability Assessment System for Highrise Buildings in Iraq

Hussain M. Habib[1,a*], Kadhim R. Erzaij[1,b]

[1]Civil Engineering Department,University of Baghdad- Baghdad, Iraq.

[a]h.habib1001@coeng.uobaghdad.edu.iq, [b]Kadhim69@coeng.uobaghdad.edu.iq

Keywords: Sustainability, Highrise buildings, Iraq, Rating system

Abstract. This study aims to preparation a standards code for sustainability requirements to contribute in a better understanding to the concept of sustainability assessment systems in the dimensions of Iraqi projects in general and in the high-rise building. Iraq is one of the developing countries that faced significant challenges in sustainability aspects environmental, economic and social, it became necessary to develop an effective sustainability building assessment system in respect of the local context in Iraq. This study presented a proposal for a system of assessing the sustainability requirements of Iraqi high rise buildings (ISHTAR), which has been developed through several integrated steps, begin with investigates the well-known sustainability systems, experts consensus based on the Delphi technique involving 32 experts consider as sustainability world leading and local highly-informed to identify applicable assessment categories and factors for the Iraqi context and its weighting system, then analyze the results using the pairwise comparison and Analytic Hierarchy Process (AHP) technique.

Introduction

In the past two decades, it was increasing awareness of the importance of sustainable within the construction industry, it plays a major role in the use of natural resources and in the development of the quality of the natural environment. Thus, construction and civil engineering can contribute significantly to the sustainable development of our society [1]. This theme plays important roles in achieving quality, reliability and durability. The impact of the large construction industry on the sustainable actions of our society, evident in the vast amounts of resources consumed and the long-term impact on our environment [2]. These impacts were the motivation for the establishment of sustainable building standards, special certificates and classification methods and buildings sustainability assessment systems aimed at mitigating the impact of buildings on the natural environment, improving their performance through sustainable design [3].

Environmental assessment methods aim to create greener and sustainable buildings, creates economic opportunities and helps the surrounding socially and environmentally. The world's leading rating systems are such as BREEAM, LEED, CASBEE. etc. which provided various tools of assessment that can be used according to the conditions of the environmental and social construction area [2]. However, these well-established methods have not originally been designed to suit developing countries (including Iraq). Although Iraq has made efforts to observe the application of the principles of sustainability in the construction industry through legal legislation and the development of regulatory bodies and international partnerships. Recent years have seen an increase in high rise construction projects in the Iraqi construction industry market, providing economic opportunities and solutions to housing problems, but with serious sustainability challenges.

Traditionally, Highrise buildings requires a large number of resources, huge quantity of building materials, large amounts of energy and produce a large amount of waste when they are demolished in the end of their life cycle, therefore they are not considered as environmental buildings[3], current concerns and awareness of problems about rising energy costs, use of materials, limits on availability drinking water and waste disposal are affect the designers to combine more sustainable elements into high rise development, such as the correct location right mix between human design and green as possible[4] [5].

Methodology

The Methodology in this research is multi-dimensional involves quantitative and qualitative approaches. These approaches included structured and integrated stages as below

1) An extensive literature review as previous studies, scientific papers, books, journals, theses, conference documents.
2) Field approach: Included open questionnaire stage and closed questionnaire it been done by conducted structured and unstructured personal interviews with some professional and authorities in constructors' sector in Iraq which have an extensive experience in sustainability buildings and investigate the opinion of selected sample at different engineering professions and different academic levels.
3) Analysis the results: Using the statistical means to analysis the results of responds with representations the results in charts.
4) Developing a practicable model. After obtaining the experts consensus in results of the survey, a rating system model (ISHTAR) had been established as the most rating system applicable to the local range of the Iraqi construction industry
5) Prepare a list of recommendations to be submitted to the relevant authorities and institutions

Preparation of the Survey. The survey is an efficient and effective way to collect data and opinions, is widely used in various research studies and disciplines, this technique is appropriate in this research because it has enabled the inclusion of a wide range of sustainability relevant issues and factors. It was developed as a final result of iterative process to finalise the key sustainability category and factors in sustainability performance for Iraqi highrise buildings. The goal was to check and validate the selection and importance of each factor. Professional experts were involving certifying and assess the factors suitability and reliability and prove the process to achieve the objectives set in practice to achieve the objectives set.

Weighting Coefficient System. Developing weighting system of sustainability aspects, categories and factors is considered a crucial process for developing any assessment framework, it comes after establishing the systems compounds and levels. weighting system is the essence of any method of evaluating the building. The weighting coefficient system can identify the importance of each elements regarding the local context. Each valuation factor is weighted so that all weights are within the valuation category. Each factor scores are multiplied by the weighting coefficient and aggregated into summation.

AHP method is used to determine the weightings of items according to participant's interviews results. Expert Choice®2.10.0 Software used to establish square matrix structure to determine the relative importance (relative weight) of each category. Each element (category or factor) was compared with its level derived different values according to their importance in sustainability highrise building assessment tool in respect to Iraq local context. Finally, participants granted each item assessment and achieved the experts consensus by Delphi technique, then factor of each category were weighted in respect to their importance to that category based on total (1) full point.

Sustainability Categories

According to the sustainability aspects, the sustainability categories classified in 11 major categories as shown in Table 1, this had been made by experts consensus as achievement after four rounds of Delphi technique. The Delphi survey included a consensus on the factors that included in each category and their numbers shown in the Table 1. Super decision program used to find the final relative weights and the categories ranking as shown in Fig. 1, the matrix mode as shown in Table 2. Based on inconsistency value (which was 0.03567), we can argue that the results calculations reliable provisions. The reliability of judgments also tested by the outcome calculations conducted in SPSS program, it shows that the Reliability (Alpha) Value was 0.677 which is accepted. The experts consensus on final categories listed below with the weight and descending ranked as shown in Figure (2) and these essential categories ranked based on their contribution in overall assessment of the sustainability in the highrise building. This should be a clear image for all stakeholder and professionals in the sustainability with Iraqi context.

Table 1. Sustainability categories criteria.

No	Categories	No. of Factors	Cronbach's Alpha	Std. Deviation	Z calculated
1	Site and Location	21	.909	0.772	20.06981
2	Energy efficiency	7	.802	0.834	18.57176
3	Water efficiency	12	.868	1.342	11.54701
4	IEQ	17	.906	0.799	19.3938
5	Management	22	.943	1.401	11.05862
6	Waste Management	9	.911	1.263	12.26343
7	Pollution	16	.922	1.025	15.11858
8	Materials	10	.869	1.100	14.08635
9	Economic Aspects	10	.888	1.389	12.54796
10	Socials Aspects	8	.895	1.223	12.66924
11	Innovation and regional contribution	7	.811	0.915	16.92228

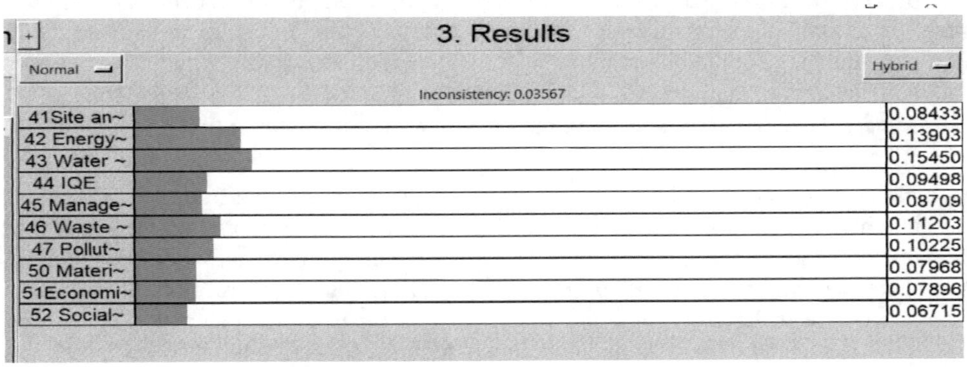

Figure 1. Main Category respect to the goal.

NO.	Category	Weight%
1	Water Efficiency	15.4
2	Energy Efficiency	13.9
3	Waste Management	11.2
4	Pollution	10.2
5	IQE	9.5
6	Management	8.8
7	Site and Location	8.4
8	Materials	8.0
9	Economic Aspects	7.9
10	Social Aspects	6.7
	Total	**1.00**
11	Innovation and Regional contribution	5 Extra Bonus

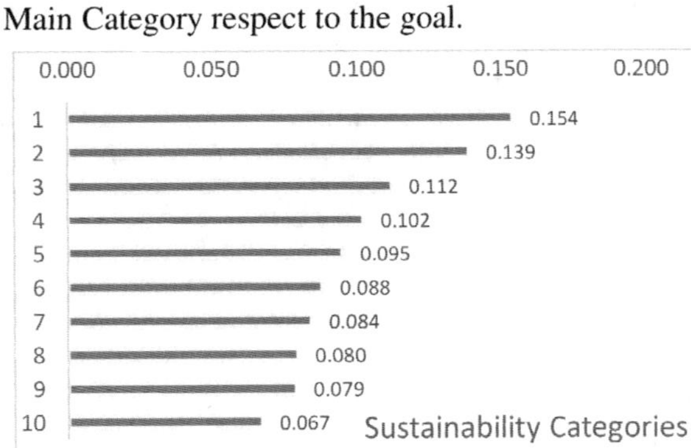

Figure 2. The main Sustainability categories criteria.

Table 2. Main categories pairwise matri.

	41Site and	42 Energy	43 Water E	44 IQE	45 Manage	46 Waste I	47 Pollutio	50 Materia	51Econom	52 Social A
41Site and	1	0.744	0.38	1	0.865	0.477	0.694915	1.028398	2.030928	1.489758
42 Energy	1.344086	1	1.3125	1.867	1.844	1.78125	0.968504	1.625	1.49925	1.746649
43 Water E	2.631579	0.761905	1	2.25	2.09375	1.84375	1.90625	1.34375	1.490313	1.702128
44 IQE	1	0.535619	0.444444	1	2.15625	1.0625	0.53125	1.15625	1.333	1.702128
45 Manage	1.156069	0.542299	0.477612	0.463768	1	0.625	0.571	1.684211	2.171646	1.49925
46 Waste I	2.096436	0.561404	0.542373	0.941176	1.6	1	1.53125	2.0315	0.841257	1.632653
47 Pollutio	1.439025	1.03252	0.52459	1.882353	1.751313	0.653061	1	1	0.84125	1.081481
50 Materia	0.972386	0.615385	0.744186	0.864865	0.59375	0.492247	1	1	1.030928	1.489203
51Econom	0.492386	0.667	0.671	0.750188	0.46048	1.188697	1.188707	0.97	1	1
52 Social A	0.67125	0.572525	0.5875	0.5875	0.667	0.6125	0.924658	0.6715	1	1

Factors of Sustainability Categories

Water Efficiency. Through the pairwise comparison results analysis, the Water Efficiency category shows a highest weight equal to 15.4%, which is consistent with current concerns about water scarcity in the Middle East region, especially neighboring countries. Increasing the concerns of specialists is the quality of water and the strategic policy to manage it both locally and internationally. This category embraces 12 factors. By using Likert scale the participants in the general survey asked to assessment the importance of each factors to the category. The External water use had the heighst importance and the rates for all factors in this category was between (4.02) which is very importance for (External water use) and (3.46) which is moderate importance for (Planters and vertical landscaping) resp. The weights and ranking this category factors shown in Fig. 3.

Energy Efficiency. Came in second in participant concern with contribution equal to 13.9% , this category contents of 7 factors are closely related to energy consumption in highrise buildings, the participants assess their importance from 4.16 which it very important for (Indoor and outdoor lighting (Make daylight calculation) to 3.57 for (Deterioration of disinformation) which means moderately important on Likert scale as shown in Fig. 4. Due to Iraqi fluctuated weather, the (Heating and cooling system) factor has a high concern of the stakeholder in term of its installation and efficiency in operation and energy consumption the factor was in second highest importance in this category with rate value of (4.09). This category embraces a unique factor which is the vertical transport, its related to this type of facilities, the integration of elevators and escalators in the high rise buildings causes increases the initial cost and energy consumption to make this facility smoothly operate .where it contributes to a considerable percentage of energy consumption (3-9%) of total energy use [6]. This factors importance had been assessed by the respondents in the survey in value equal to (3.73), moderate important.

WasteManagement. Experts agreed that one of the most important obstacles facing the implementation of sustainability projects in Iraq is waste management. This difficulty lies on the scarcity of waste separation and recycling facilities and lack of availability of landfills conforming to standard specification, this category took almost weight 11.2 %. The importance for its factors was between 4.56 very important to (Waste and rubble collection and transfer) factor and 3.05 moderately important to (Construction waste materials reuse) factor. Figure (5) illustrates the importance for the Waste management category.

Pollution. It's clear from Figure (6) that the most important factor in this category is the(Air conditioning system) that coming from the participants serious concern to this factor effect in the operating of the highrise building, the important rate for this factor was (4.56) very important according to participants records in Likert scale , the lowest important rate was (2.79) gave to (Drying space) factors.

NO.	Factor	Mean	Credit
1	Reduce landscaping water use	3.94	2
2	Gray water recycling	3.85	2
3	External water use	4.02	2
4	Water meters	3.79	2
5	Leak detection system	3.9	2
6	Surface water runoff	3.96	2
7	Surface water runoff	3.69	2
8	Surface and ground water contamination during construction	3.71	2
9	Planters and vertical landscaping	3.46	1
10	Landscaping compatibility and integration with surroundings	3.52	2
11	Equipment for water conservation	3.83	2
12	Water system	3.79	2
TOTAL			23

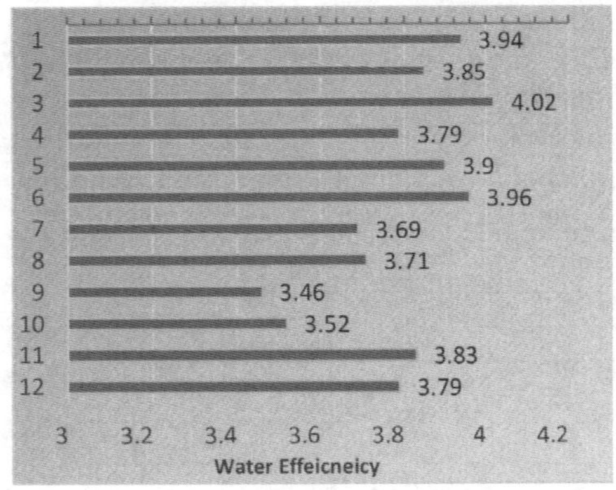

* Credits allocation will explain in section (5-2)

Figure 3. Water Efficiency factors criteria.

NO.	Factor	Mean	Credit
1	General building energy performance	3.96	2
2	Heating and cooling system	4.09	2
3	Indoor and outdoor lighting (Make daylight calculation)	3.95	2
4	Hot water system	3.65	2
5	Renewable energy technology	4.16	2
6	Vertical transportations	3.73	2
7	Deterioration of disinformation	3.57	2
TOTAL			14

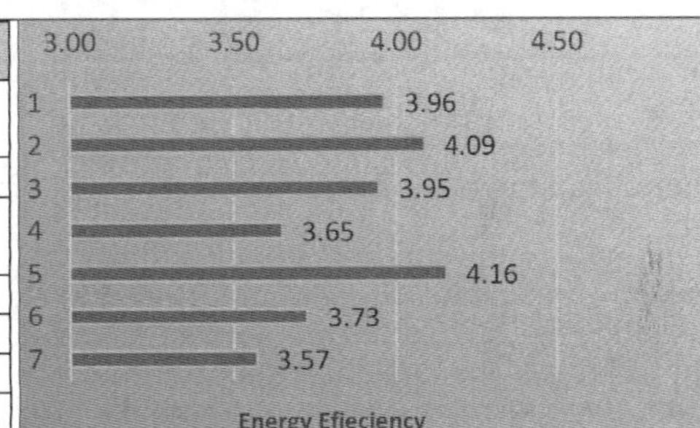

Figure 4. Energy Efficiency criteria.

NO.	Factor	Mean	Credit
1	Building waste management plan	3.96	2
2	Waste and rubble collection and transfer	4.56	3
3	Construction waste management plan	3.91	2
4	Construction waste materials reuse	3.05	1
5	Construction waste disposal	3.82	2
6	Solid waste treatment	3.75	2
7	Waste separation equipment	3.73	2
8	Recycling collection and storage locations	3.77	2
9	Recycling Facilities	3.73	2
TOTAL			18

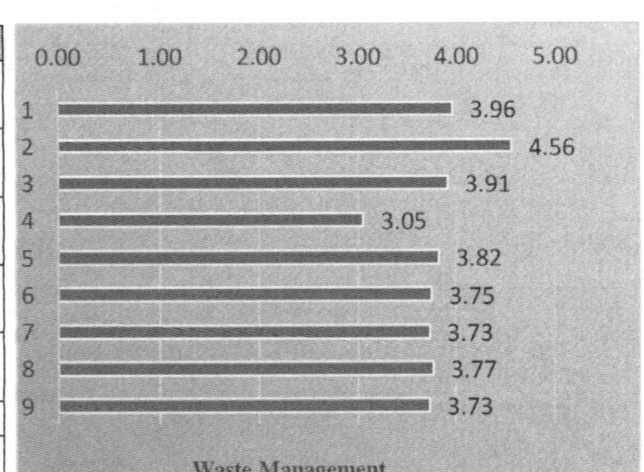

Figure 5. Waste Management criteria.

Indoor Environmental Quality: Although this category contains important factors, but the results of the pairwise comparison was surprising in terms of ranking with only (9.5) importance value, Delphi panel experts attributed that to the Iraqi conditions imposed the importance of other factors affected in the assessment of sustainability. This category contains 18 factors are concerned with wellbeing and living healthy inside the highrise building. The (quality of drinking water) factor was in the most important with rate 4.65 and lowest important rate was 2.95 for (Textile/ Envelope/ Thermal construction). factor as shown in Fig. 7).

NO	Factor	Mean	Credit
1	Renewable technologies	3.83	2
2	CO2 Emissions	3.96	2
3	Light pollution reduction	3.93	2
4	Global warming potential	3.83	2
5	Nitrogen Oxide (NOX) emission	3.65	2
6	Natural ventilation	3.94	2
7	Air conditioning system	4.53	3
8	Prevent cooling leakage	3.76	2
9	MVHR-mechanical ventilation and heat recovery system	3.70	2
10	Building fabric/ envelope/ thermal comfort	3.70	2
11	Environmental impact of materials	3.81	2
12	Fire hazard	3.98	2
13	Risks resulting from disasters such as flooding	3.80	2
14	Protection from storms	3.60	2
15	Heat island effect	3.69	2
16	Drying space	2.79	1
	TOTAL		32

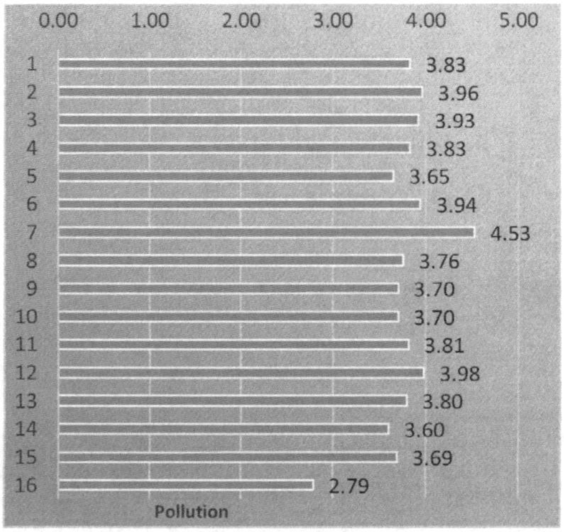

Figure 6. Pollution criteria.

No.	Factor	Mean	Credit
1	Sound and noise control	3.67	2
2	The surrounding view	3.70	2
3	Control and measurement of lighting	3.64	2
4	Control and measurement of glow	3.46	2
5	Take advantage of daylight	4.07	2
6	Natural ventilation	3.86	2
7	Mechanical ventilation	3.45	1
8	Indoor air quality sensors - CO2 ratio control	3.77	2
9	Organic volatile compounds	3.67	2
10	Microplastic contamination	3.93	2
11	Indoor chemical and radiology pollutant detector	3.88	2
12	Ventilation of the building -dust prevention	4.09	2
13	Heat control systems	3.79	2
14	Temperature control systems	3.93	2
15	Quality of drinking water	4.65	3
16	Plantations and internal reforestation	3.64	2
17	Textile/ Envelope/ Thermal construction	2.95	1
	TOTAL		33

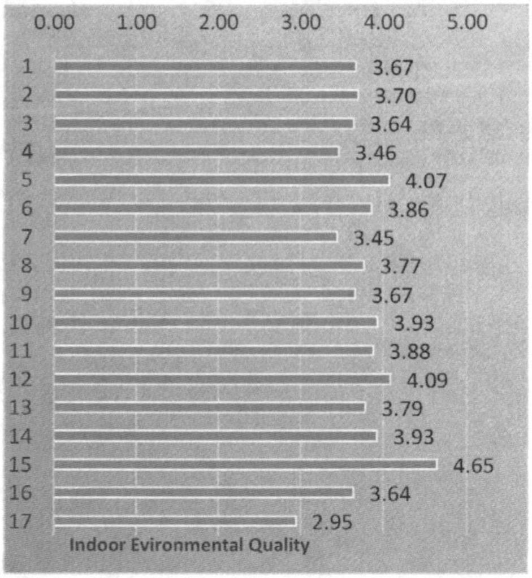

Figure 7. Indoor Environmental quality.

Mangement. This category comprises 21 factors, Fig. 8. In pairwise comparison, the Management came sixth in sequence of importance in the process of assessing sustainability in highrise buildings with 8.8 weight value. Safety and infrastructure efficiency factors were in the top priority in this category with importance 4.66 for both, then Security factor in evaluation importance value 4.54, the (Supply chain management) was in the bottom of priority with 3.59. It is clear that these results are affected by the local circumstances surrounding the Iraq construction market.

Site and Location. The Delphi panel give site and location category value of importance equal to 8.4%, is included 21 factors their importance ranging from 4.13 for (Pollution prevention) to 3.32 for (Improve site aesthetics) in the bottom of the grading list (Figure 9).

This category embraces one unique factor (Rezoning the property) added by the experts with to respect to Iraqi condition in construction industry and its important was assessed as (3.38). This reflects the experts growing concern of from the phenomenon of investing agricultural land in residential and commercial projects, which negatively affects sustainability in general the Iraqi environment.

Materials. category included some important factors. The material choosing has a key role in all sustainability aspects in high rise building, the participants rated the importance for this category factors form 4.54 (very important) for (low environmental impact) to 3.14 (Material designed to deal with future climate changes). Delphi panel has been mentioned that the use of produced local material should have positive impact on the sustainability performance for the high rise building regarding the local context, this factor came in second priority with important value equal to (3.53). Figure (10) illustrates its criteria.

Economic Aspects. The results of this questionnaire were identical to the context of global sustainability assessment systems and the same problem with this category, which despite its importance but neglected or came in low stages in the sequence of importance as its share 7.9% in the assessment of general sustainability. Figure 11 shows the rates of the importance for this category factors was between very important (4.53) for the (Cost of life cycle) and the lowest rate was (3.16) (mordantly important) for the (Building adaptive capacity) Factor.

Social Aspects. The social life in Iraq has an impact on sustainability projects and higher construction more clearly so the experts agree not to remove any of the factors of these eight category factors and be subject to evaluation as mandatory requirements. The importance evaluation was between 4.14 very important for (Providing employment opportunities for local workers) and 3.28 mordantly important for (Community facilities) as shown in Fig. 12.

Innovetation And Regional Contributation: This category, after discussion and deliberation among the experts, was obtained by consensus as an advocacy tool for the sustainability application and is calculated as additional points above the general assessment. This category includes seven factors that encourage innovation in all three aspects of sustainability and (Site orientation) was the highest rated by 3.93 and (Area of planting) factor Lowest rated by 2.91, as illustrated in Fig. 13.

No.	Factor	Mean	Credit
1	Safety	4.66	3
2	Security	4.54	3
3	Arrangements for disabled people	3.80	2
4	Construction management plan or scheme	3.80	2
5	Supply chain Management	3.62	2
6	Sound insulation	3.70	2
7	Lighting control (Un-occupied area)	3.78	2
8	Heat control	3.63	2
9	Energy labelled white goods	3.59	2
10	Orientation and passive solar gains	3.69	2
11	Daylighting	3.67	2
12	Extent of communication	3.74	2
13	Pedestrian facilities	3.67	2
14	Contracting and procurement methods	3.69	2
15	The extent to which stakeholders focus on the project	3.65	2
16	Equipment management	3.70	2
17	Integration of services	3.87	2
18	Function and ease of use	3.70	2
19	Flexibility and the possibility of air conditioning with the changes	3.71	2
20	Durability and reliability	3.78	2
21	Maintenance and good implementation	3.84	2
22	Infrastructure efficiency	4.66	3
TOTAL			47

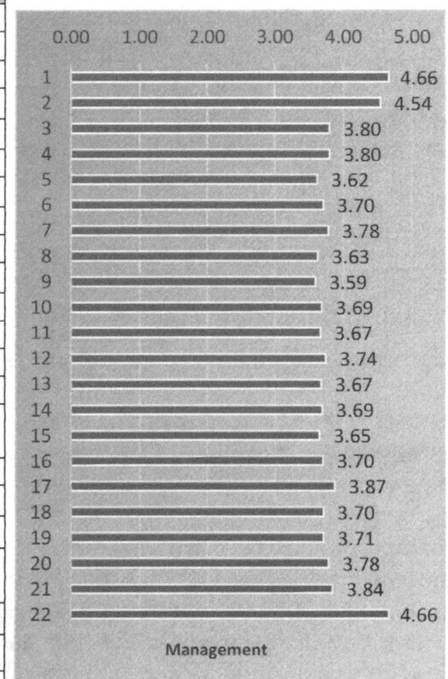

Figure 8. Management criteria.

NO.	Factors	Mean	Credit
1	Urban redelvopment	3.91	2
2	Pollution prevention	4.13	2
3	Mitigation of environmental impact	4.04	2
4	protection of biodiversity and Ecological features	3.83	2
5	Ecological value of site	3.65	2
6	Ecological enhancement	3.60	2
7	Shading and Afforestation	3.85	2
8	Redelovpment of brown land	3.53	2
9	Easy access to the site	3.76	2
10	Density in development area	3.57	2
11	Community communication	3.52	2
12	Car parking areas	3.33	1
13	Transportation alternatives	3.61	2
14	Impact on ecosystems and water ways	3.55	2
15	improve site aesthetics	3.32	1
16	Promote corrosion reduction	3.37	1
17	community consultation	3.37	1
18	Weather hazards	3.42	1
19	Improve positive effects	4.08	2
20	Building footprint	3.60	2
21	Rezoning of property	3.38	1
TOTAL			36

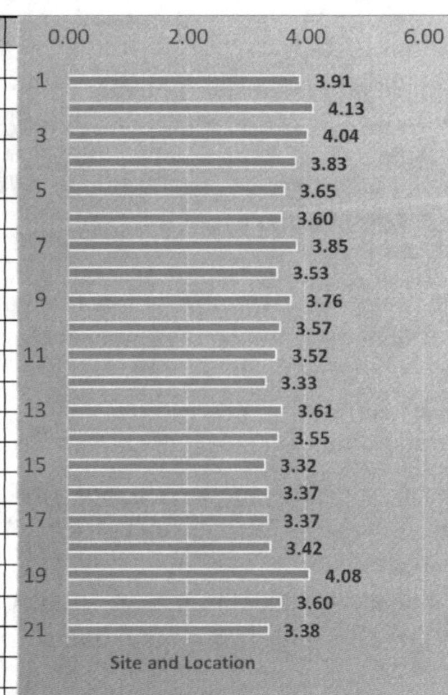

Figure 9. Site and Location criteria.

IOP Conf. Series: Materials Science and Engineering **901** (2020) 012037 doi:10.1088/1757-899X/901/1/012037

NO.	Factor	Mean	Credit
1	Local production	3.53	3
2	The materials are recyclable	3.68	2
3	Reuse of building structure	3.70	2
4	Material efficiency within the life cycle (LCA)	3.75	2
5	Material designed to deal with future climate changes	3.14	2
6	Reliable sources of materials	3.79	2
7	Materials with a high thermal mass	3.28	2
8	Low environmental impact	4.54	2
9	Quality finishing materials	3.81	2
10	Components of insulation materials	3.64	1
TOTAL			20

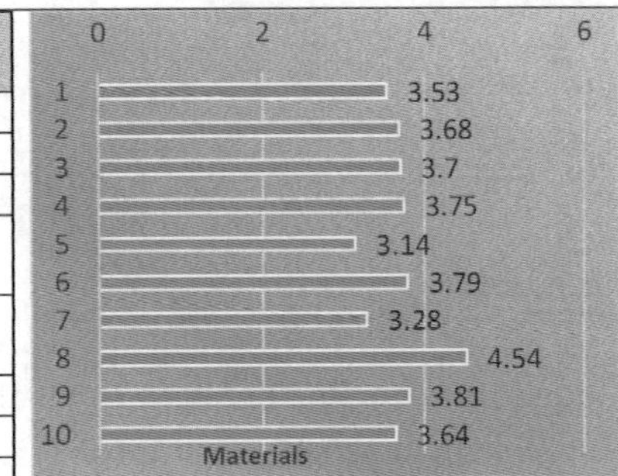

Figure 10. Materials criteria.

NO.	Factor	Mean	Credit
1	The cost of construction	3.93	2
2	Cost of life cycle	4.53	3
3	Cost of operation maintenance	3.86	2
4	Levels rent offers	3.38	1
5	Using the total cost of the stages of the life cycle of the building	3.48	1
6	Identify the lowest operating cost in the level of rent	3.54	2
7	Economic activities of the local population	3.46	1
8	Building adaptive capacity	3.16	1
9	Carbon credits/ Carbon reduction commitment	3.63	2
10	Commissions	3.29	1
TOTAL			16

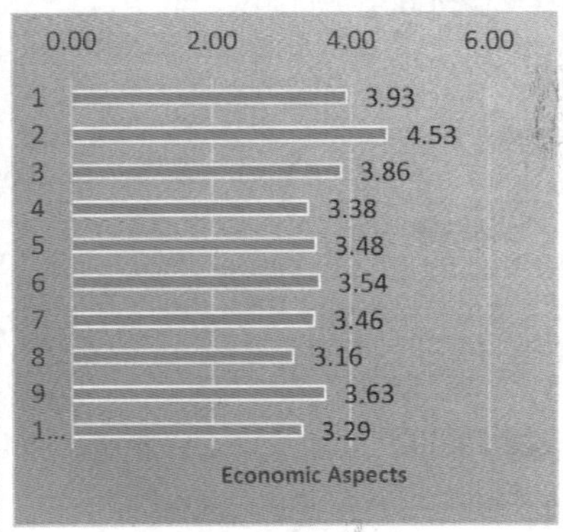

Figure 11. Economic Aspects criteria.

NO.	Factor	Mean	Credit
1	Knowledge of the historical and cultural community	4.05	2
2	Providing employment opportunities for local workers	4.14	2
3	Social equity and contributing to the fight against poverty	3.42	1
4	The influence of hobbies and habits on the community building	3.44	1
5	The personal behavior of the building occupants and its beneficiaries	3.33	1
6	Recognize and encourage the use of the building and the site as an educational resource to establish environmental awareness	3.42	1
7	Community facilities	3.28	1
8	Areas of community privacy	3.33	1
TOTAL			10

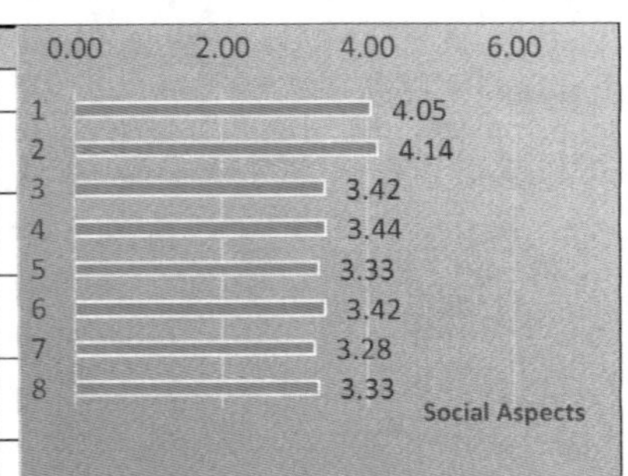

Figure 12. Social Aspects criteria.

NO.	Factor	Mean	Credit
1	Site orientation (Innovative design)	3.93	2
2	Internal space management	3.26	1
3	Area of planting	2.91	1
4	Regional or local material	3.46	1
5	Recycled content	3.75	2
6	The team members professional sustainability on the project	3.71	2
7	school or university as an educational tool	3.79	2
TOTAL			11

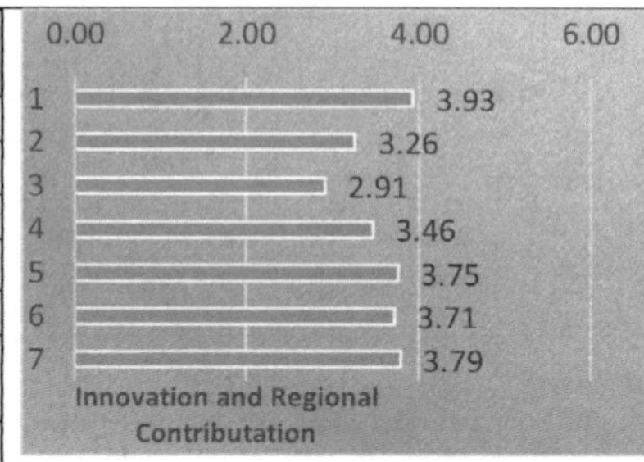

Figure 13. Innovation and regional contribution criteria.

Developing (ISHTAR Model) Sustainability Assessment for Iraqi Highrise Building

Based on the previous findings that included the assessment categories, assessment factors, and their weighting coefficient, a basic framework had been proposed for the coherent and comprehensive system of the sustainability assessment in Iraqi highrise buildings, and have been developed and establish its criteria by the informed consensus of international and local experts. It suggest to be called (ISHTAR), inspired on the gate of Ishtar, the eternal symbol of Babylonian civilization, which built the oldest sustainability buildings that known to man.

By systematic consultation with informed experts in reviewing the results, the framework of the assessment system composed of (11) categories, ten of them are mandatory they are (Water Efficiency, Energy Efficiency, facility Management, Site and Location, Materials, Pollution, Waste Management, Indoor environmental Quality, Economic Aspects and Social Aspects) and one encouraging category which is (Innovation and regional contribution),these categories embraces different numbers of factors their total is (139) factors. The expert's consensuses that the suggested green building assessment framework is comprehensive, efficient and most appropriate for Iraqi local context. They agree on that all the items (categories and Factors) are crucial and should be adopted in the sustainability assessment processes in the highrise buildings.

Practical ingredients for ISHTAR. ISHTAR system had been developed by a methodology process included a set of sustainability elements (categories and factors) derived from literature and further complemented by piloting the practical application through questionnaire, Delphi panel judgements and a case study, to optimise items weightings and subsequent ranking. Different statistical tools were used to finalise the eleven categories and 139 factors of ISHTAR system which including social, economic and environmental parameters

1) 1.This framework been developed based on scientific research and technical knowledge.
2) 2. Local and international Experts, world well-known rating systems agency actors, elected official's decision makers, designers, and stakeholder with different roles others participated in the development as key participants in the process of this method.
3) 3.The sustainability strategies and objectives were defined and addressed as a key objective.
4) 4.The rating system must design to fit the country's local context, depending on its culture, issues, actors, practices and institutions, this should not prevent from that countries can learn from each other's in investing the experts work and ideas as inputs in the system developing process.
5) 5. ISHTAR as a rating system not only allow for the assessment of the buildings sustainability, but also it could be used as a decision-making tool in identification the most suitable development alternative to achieve high rate of sustainability as well as the improving of building operation and efficiency.
6) 6. The main approaches for assessment the sustainability building performance in ISHTAR was the Total Quality Assessment (TQA) systems, which evaluate environmental (ecological), economic and social aspects, same means in well-known Sustainability rating systems such as

LEED and BREEAM

Credits Allocation. Credits in the ISHTAR system will allocate based on the Likert scale Rate assessment obtained for each factor and reached by consensus of the Delphi panel on the relative importance of the factor, the ISHTAR credits will present ways of distinguishing between these factors. Figure 14 as an example, illustrates the (Economic Aspects) credits allocation. In order to distinguish these factors, therefore, a three-tiered credit allocation was proposed A factor that exceeds 50% of the scale range or 2.5/5 is applicable to highrise buildings in Iraq. In other words, criteria rated more than 2.5/5 can grant one credit; more than 3.5/5 classification criteria can be granted two credits; classification criteria over 4.5/5 can be awarded three credit. Then the total available credit in this category will be equal to 1*3+4*2+5*1=16 credits. This method is promoting the increasing compliance with higher requirements According to the weighting system derived from AHP, the ISHTAR scores, which reflect the rate or the sustainability level in the Iraqi Highrise building, can be achieved by means of the following procedure:

1) Determine the number of available credits for each category,
2) Assess the number of reached credits for each factor of the category
3) Calculate each category score by equation (1), ISHATAR has 10 categories: that's mean 10 different rating scores will be calculated
4) Adding any innovation points that may have been achieved.

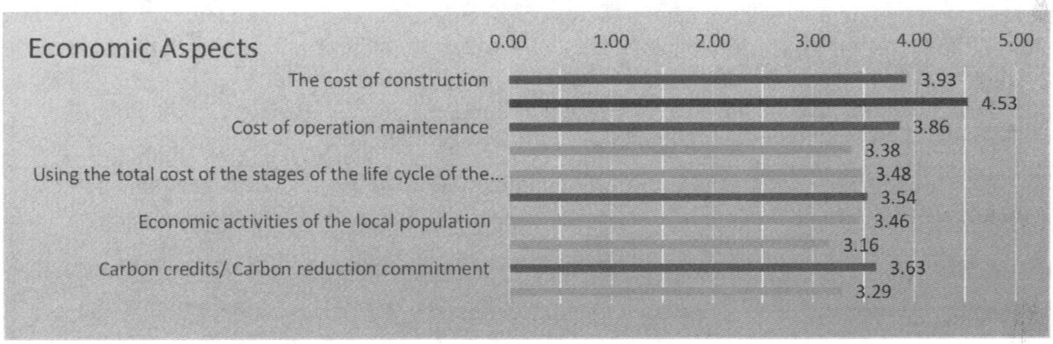

Figure 14. Credit allocation for Economic Aspects factors.

The calculation is as follows: the overall rating score will provide by the summation of these 11 rating scores (Eq. (2))

$$CS = \frac{Cr}{Ca} * W * 100 \quad (1)$$

Where:
CS: Category Sustainability Scores
Cr: Credits Reached
Ca: Available Credits
W: Weighting Coefficient

$$SBR = \sum_{n=1}^{11} CSn \quad (2)$$

Where:
SBR: Building Sustainability Rate
n: The category numbers included the Innovation category

ISHTAR Rating Benchmarks. ISHTAR Followed (BREEAM and LEED) in the environmental assessment method where the result of sustainability building assessment process is converted into a single ranking expression. The assessment score in BREEAM and LEED, have been calculated out of 100 percent and then converted into a single rating expression. The Delphi panel in this search had reached consensus to use of a percentage-based classification scale as illustrated in Figure (18) divided in five levels of certification as follow

ISHTAR classification starts from 40 as base line, it's the level of meeting the basic criteria, therefore any building rated below 40 will be considered "UNCLASSIFIED";
Level 1: Buildings rated 40 ≤ 50 will considered "PASS and grant "BLUE Ishtar"
Level 2: Buildings rated 51≤ 60 will grant "GREEN ISHTAR"
Level 3: Buildings rated 61 ≤ 70 will grant "BRONZE ISHTAR"
Level 4: Buildings rated 71 ≤ 80 will grant "SILVER ISHTAR"
Level 5: Buildings rated 81≤ 100 will grant "GOLD ISHTAR"

Conclusions

The main conclusions of this study can be drawn as follows:

- The largest concerning in most assessment systems was to energy conservation, but in ISHTAR, the largest weight was for both water efficiency (15.4%) and energy efficiency (13.9%) among all the performance categories. It demonstrates the Iraq crucial circumstances of water and energy conservation, due to its mostly semi-dry climate.
- The proposed rating system comprised some vital categories and factors were not considered in
- several assessment systems such as "Economics" and the "Waste Management". Furthermore, some extra factors were eliminated or merged together in the proposed assessment framework of Iraq
- As a result of the research findings it was observable there is awareness lack about the importance of applying the sustainability approach in Iraqi construction.
- Adopting sustainability practice in highrise buildings well provide extra work opportunities for local community and reduce the unemployment such as the process of waste management and materials recycle.

References

[1] M.A. Rosen and H.A. Kishawy, Sustainable Manufacturing and Design: Concepts, Practices and Needs, 2012 154–174.

[2] S. Maltese, L.C. Tagliabue, F.R. Cecconi, D. Pasini, M. Manfren, and A.L.C. Ciribini, Sustainability Assessment through Green BIM for Environmental, Social and Economic Efficiency, Procedia Eng., 180 2017 520–530.

[3] R. Wener and H. Carmalt, Environmental psychology and sustainability in high-rise structures, Technol. Soc., 28(1–2) 2006 157–167.

[4] K. Yeang, L.D. Yeang, S.G. Energy, K. Yeang, A.A. Dip, and D. Lit, Ecoskyscrapers and Ecomimesis: New tall building typologies Architectural, 2008.

[5] S. Yaser Mousavi, Sustainable high-rise building (Case study: three example of sustainable high-rise building in Iran), Int. Res. J. Appl. Basic Sci., 9(11) 2015 2027–2033.

[6] M.R. Akbar, Evaluating Sustainability In High-Rise Developments, 2012.

AUTHOR INDEX